CAMBRIDGE LIBRARY COLLECTION

Books of enduring scholarly value

Botany and Horticulture

Until the nineteenth century, the investigation of natural phenomena, plants and animals was considered either the preserve of elite scholars or a pastime for the leisured upper classes. As increasing academic rigour and systematisation was brought to the study of 'natural history', its subdisciplines were adopted into university curricula, and learned societies (such as the Royal Horticultural Society, founded in 1804) were established to support research in these areas. A related development was strong enthusiasm for exotic garden plants, which resulted in plant collecting expeditions to every corner of the globe, sometimes with tragic consequences. This series includes accounts of some of those expeditions, detailed reference works on the flora of different regions, and practical advice for amateur and professional gardeners.

Notes of a Botanist on the Amazon and Andes

Having previously embarked on a collecting expedition to the Pyrenees, backed by Sir William Hooker and George Bentham, the botanist Richard Spruce (1817–93) travelled in 1849 to South America, where he carried out unprecedented exploration among the diverse flora across the northern part of the continent. After his death, Spruce's writings on fifteen fruitful years of discovery were edited as a labour of love by fellow naturalist Alfred Russel Wallace (1823–1913), whom Spruce had met in Santarém. This two-volume work, first published in 1908, includes many of the author's own illustrations. Showing the determination to reach plants in almost inaccessible areas, Spruce collected hundreds of species, many with medicinal properties, notably the quinine-yielding cinchona tree, as well as the datura and coca plants. Featuring four maps, Volume 2 includes discussion of the Peruvian and Ecuadorian Andes and the cinchona forests of western Chimborazo.

Cambridge University Press has long been a pioneer in the reissuing of out-of-print titles from its own backlist, producing digital reprints of books that are still sought after by scholars and students but could not be reprinted economically using traditional technology. The Cambridge Library Collection extends this activity to a wider range of books which are still of importance to researchers and professionals, either for the source material they contain, or as landmarks in the history of their academic discipline.

Drawing from the world-renowned collections in the Cambridge University Library and other partner libraries, and guided by the advice of experts in each subject area, Cambridge University Press is using state-of-the-art scanning machines in its own Printing House to capture the content of each book selected for inclusion. The files are processed to give a consistently clear, crisp image, and the books finished to the high quality standard for which the Press is recognised around the world. The latest print-on-demand technology ensures that the books will remain available indefinitely, and that orders for single or multiple copies can quickly be supplied.

The Cambridge Library Collection brings back to life books of enduring scholarly value (including out-of-copyright works originally issued by other publishers) across a wide range of disciplines in the humanities and social sciences and in science and technology.

Notes of a Botanist on the Amazon and Andes

Being Records of Travel on the Amazon and Its Tributaries, the Trombetas, Rio Negro, Uaupés, Casiquiari, Pacimoni, Huallaga and Pastasa

VOLUME 2

RICHARD SPRUCE
EDITED BY
ALFRED RUSSEL WALLACE

CAMBRIDGE
UNIVERSITY PRESS

University Printing House, Cambridge, CB2 8BS, United Kingdom

Published in the United States of America by Cambridge University Press, New York

Cambridge University Press is part of the University of Cambridge.
It furthers the University's mission by disseminating knowledge in the pursuit of
education, learning and research at the highest international levels of excellence.

www.cambridge.org
Information on this title: www.cambridge.org/9781108069212

This edition first published 1908
This digitally printed version 2014

ISBN 978-1-108-06921-2 Paperback

The original edition of this book contains a number of colour plates,
which have been reproduced in black and white. Colour versions of these
images can be found online at www.cambridge.org/9781108069212

NOTES OF A BOTANIST

ON THE

AMAZON AND ANDES

MACMILLAN AND CO., LIMITED
LONDON · BOMBAY · CALCUTTA
MELBOURNE

THE MACMILLAN COMPANY
NEW YORK · BOSTON · CHICAGO
ATLANTA · SAN FRANCISCO

THE MACMILLAN CO. OF CANADA, LTD.
TORONTO

NOTES OF A BOTANIST

ON THE

AMAZON & ANDES

BEING RECORDS OF TRAVEL ON THE AMAZON AND ITS TRIBUTARIES, THE TROMBETAS, RIO NEGRO, UAUPÉS, CASIQUIARI, PACIMONI, HUALLAGA, AND PASTASA; AS ALSO TO THE CATARACTS OF THE ORINOCO, ALONG THE EASTERN SIDE OF THE ANDES OF PERU AND ECUADOR, AND THE SHORES OF THE PACIFIC, DURING THE YEARS 1849-1864

BY RICHARD SPRUCE, PH.D.

EDITED AND CONDENSED BY

ALFRED RUSSEL WALLACE, O.M., F.R.S.

WITH A

BIOGRAPHICAL INTRODUCTION

PORTRAIT, SEVENTY-ONE ILLUSTRATIONS

AND

SEVEN MAPS

IN TWO VOLUMES—VOL. II

MACMILLAN AND CO., LIMITED

ST. MARTIN'S STREET, LONDON

1908

Lo to the wintry winds the pilot yields
His bark careering o'er untrodden fields;
Now on Atlantic waves he rides afar
Where Andes, giant of the western star,
With meteor standard to the winds unfurl'd,
Looks from his throne of clouds o'er half the world.

CAMPBELL.

The sounding Cataract
Haunted me like a passion; the tall rock,
The mountain, and the deep and gloomy wood,
Their colours and their forms, were then to me
An appetite, a feeling, and a love.

WORDSWORTH.

CONTENTS

CHAPTER XV

FROM BARRA DO RIO NEGRO TO TARAPOTO, PERU

CHAPTER XVI

RESIDENCE AT TARAPOTO: EXPLORATION OF THE EASTERN ANDES OF PERU

CHAPTER XIX

BOTANICAL EXCURSIONS IN THE ANDES OF ECUADOR: AT BAÑOS AND AMBATO

CHAPTER XX

AMBATO AND THE CINCHONA FORESTS OF ALAUSÍ

CHAPTER XXI

THE CINCHONA FORESTS OF WESTERN CHIMBORAZO

CHAPTER XXII

SPRUCE'S LAST THREE YEARS IN SOUTH AMERICA: ON THE SHORES OF THE PACIFIC

CHAPTER XXIII

LETTERS AND ARTICLES RELATING TO HIS AMAZONIAN TRAVELS

CHAPTER XXIV

ON ANT-AGENCY IN PLANT-STRUCTURE

CHAPTER XXV

ON INDIGENOUS NARCOTICS AND STIMULANTS, WITH THEIR USES BY THE INDIANS

CHAPTER XXVI

THE WOMEN-WARRIORS OF THE AMAZON: A HISTORICAL STUDY

CHAPTER XXVII

THE ENGRAVED ROCKS OF THE RIO NEGRO AND CASIQUIARI

CHAPTER XXVIII

A HIDDEN TREASURE OF THE INCAS

ILLUSTRATIONS

MAPS

CHAPTER XV

FIFTEEN HUNDRED MILES UP THE SOLIMOẼS: FROM MANÁOS TO TARAPOTO, PERU

(*March* 14 *to June* 22, 1855)

[This chapter consists largely of a full and very descriptive Journal, which required comparatively little pruning; and this is supplemented by letters to Messrs. Teasdale and Bentham, giving to the former vivid sketches of scenery and of the passengers on the steamer, and to the latter an account of one of the numerous personal dangers of which Spruce had his full share, though from all of them he escaped with his life.]

Voyage up the Solimoẽs

(*Journal*)

March 14, 1855.—Embarked on the *Monarca*, an iron steamer of 35 horse-power, built at Rio de Janeiro. We left the port of Barra at six the next morning, and I enjoyed much the rapid run up the Solimoẽs, contrasting strongly with the painful way in which we crept up in a canoe in 1851, when it took a week to reach Manaquirý́. In the steamer we spent but ten hours. The river appears more than half full, and the current is strong. There are numerous floating trunks and small grass-islands. At night it was very dark, and we frequently struck against these trunks, sometimes with a considerable shock which made us all run on deck, but no damage was done. On the afternoon of the 17th we passed the mouth of the large river Purús, which enters from the south. It is not wide but brings down a large volume of white water.

Between Coary and Ega there is a long range of cliffs, which are much bored by kingfishers and by a small white-bellied sand-martin, scarcely larger than a humming-bird. . . .

On the 25th we reached Saõ Paulo d' Olivença about noon. It stands on very high land, rising abruptly from the river about a hundred feet, but the site is flat and the village contains several regular streets, though the houses are mostly miserable. The great concourse of people here is owing to its being the residence of a padre who suits them excellently and conforms in everything to their way of life, *i.e.* he is a gambler and indulges in every other vice of the country.

I took a turn in the forest. The soil is a deep clay, in hollows scarcely passable in rainy weather. The valleys are all traversed by streams of clear water, and abound in tree-ferns, but apparently all of one common species. The caapoera vegetation is very luxuriant and comprised much that was new to me, especially a shrubby papilionaceous climber with delicate pinnate leaves (resembling *Abrus tenuifolius*) and largish scarlet flowers, which hung in large masses from the lower trees and bushes. Also a low Nonatelia (Cinchonaceæ) with large corymbs of pretty purple flowers. On one clayey slope was a large bed of Umirí (Humirium sp.) with ripe fruit, which the numerous cattle (belonging chiefly to the Padre) pick up as they fall. Two Monimiaceæ, one with very large Melastoma-like leaves and large fruits, I have not seen before. The other is very near a Uaupés species.[1]

Notes on the Vegetation of the Solimoẽs

The sloping banks clad with long grass form a strong contrast to those of the Rio Negro. On the islands the chief vegetation is *Salix Humboldtiana* and a Cecropia, with a rather inelegant bamboo supporting itself on them. The white trunks of the trees are very remarkable—actually white with a crust of rudimentary lichens, especially those of Cecropia. The foliage at this season is rather ragged and scanty, but when the rising or setting sun illuminates the white skeleton, the dots of green on the extremities of the branchlets have a pretty effect. This is particularly noticeable in places where the winds have broken off the tops of the trees.

Of palms the Murumurú is abundant. An elegant Bactris (probably *B. concinna*, Mart.) about 18 feet high grows in broad patches. It is abundant at Yurimaguas on the Huallaga.

A Loranthus with large red flowers tipped with yellow grows on many different trees—very often on Imba-úba and a species of Maclura. Several Ingas are in flower, and *Triplaris surinamensis* (Polyonaceæ) is frequent. The Arrow-reed abounds on low coasts and islands, and in similar places there are often low trees whose trunks are draped with a species of Batatas. Here and there in the gapó is to be seen a Nutmeg tree 50 feet high or more, its branches nearly horizontal, but often bent up abruptly into a vertical position about midway.

From the Mouth of the Purús to that of the Coary

Very frequent in clumps is the fine Pao Mulatto, 50 to 70 feet high, with lead-coloured bark and large umbels of white flowers. A

[1] [Readers of Bates's *Naturalist on the Amazons* will remember that this was his farthest station on the river, that he stayed here five months (a year later than Spruce's visit), and that he speaks of its luxuriance in every department of natural history with the greatest enthusiasm, adding, that five years would not be sufficient to exhaust its treasures in zoology and botany. In particular, the numerous pebbly streams, and the magnificent vegetation on their banks, surpassed anything he had seen during his ten years of forest ramblings.—Ed.]

thickish Imba-úba (Cecropia) has the bark mottled with red and white as in the Bread-fruit tree. In some places is an Anonaceous tree, about 30 feet high, with a profusion of flowers in small axillary clusters on the upper side of the long branchlets. The solitary tall Assaí palm is very scarce, occurring only towards the mouth of the Coary.

A remarkable tree occurs below Coary, 50 feet high, the top spreading, the lanceolate pale green leaves clustered on the ends of slender twigs, the flower-stalks long, descending then ascending, growing on the main branches and trunk nearly to the base, fruits pendent, globular, size of an orange, but said when ripe to be much larger, having a hard shell with four seeds. It is probably a species of Couroupita (Lecythideæ).

Much wild Cacao is seen on the margins and as far within as the inundation extends—conspicuous from its young red leaves. There is generally much Castanha (Brazil-nut) in the forests.

At Tabatinga I gathered flowers of a small Composite tree growing 6 to 15 feet high and looking very like a willow. It is the *Tessaria legitima*, DC., and had been noticed from the mouth of the Japurá upwards.

A Serjania (Sapindaceæ) with large masses of red capsules is now very frequent, and a low Copaifera in flower grows here and there by the water's edge. The Pao Mulatto continues very abundant and our firewood consists wholly of this species. There is no handsomer tree in the gapó. It sometimes reaches near 100 feet high. It is branched from about the middle, and the top forms a narrow inverted cone. The surface of the trunk and branches is somewhat wavy or corrugated, but the bark is quite smooth and shining. When I went to Manaquirý in June 1851 the trees were shedding their bark, the process being a longitudinal splitting up in one or more places, and a rolling back from both edges of the rupture. The young bark thus exposed is green, but it speedily assumes a deep bronze or leaden hue, and finally a chestnut colour—hence its name.[1] Some small Rubiaceous trees have the same property; for instance, *Eurosmia corymbosa* and a tree in the forest at Yurimaguas, with leaves resembling those of a Nonatelia, but the bark is greener than that of the Pao Mulatto. With this latter tree, on the Solimoẽs, frequently grows the Castanheiro do Macaco, with globular brown fruits, probably a species of Couroupita.

JOURNAL (*continued*)

March 27.—At 4 P.M. we reached Tabatinga, the frontier town of Brazil, situated on the north

[1] [This tree was, later, collected by Spruce, and being new was named by Mr. Bentham *Enkylista Spruceana*. It belongs to the Cinchonaceæ.]

bank, a miserable place containing scarcely any houses but those of the garrison, though a little to the eastward, across a small valley, is a village of the Tucáno Indians. The barracks consist of two small, low ranchos, and there is no fort, though I saw two or three pieces of cannon laid on the ground. The soil is clayey and the vegetation luxuriant.

Early on the 29th we reached Loreto, the first town in Peru and decidedly better than Tabatinga, having some good houses. The white inhabitants, however (even the Governor), are Portuguese.

March 30.—Coasting the south bank of the river, the land being somewhat high and settlements more frequent. The vegetation here was more new and striking than any I had seen during the voyage. A little inland grew a very handsome palm (Attalea), resembling the Palma Yagua of the Orinoco, but rather smaller and with pendulous bunches of small hard red fruits.

Here I first saw the Bombonaji, a palmate-leaved Carludovica. It grows on steep red banks, and is submersed when the river is at its height. Several other trees in flower and fruit were quite new to me.

In the afternoon we reached Cochiquina on the south bank, inhabited by Mayironas, that is, Indians from the Rio Mayo. At this season there is a small lagoon between it and the river which makes it difficult of access. The Indians are numerous, and apparently very submissive to the Gobernador (the only white inhabitant) and to their Curácas or chiefs, who go about with polished walking-sticks headed with silver. There are plenty of pigs and fowls. The houses are kept in better repair and the weeds kept down more than in Brazilian villages. About 1000 sticks of firewood were embarked here in two hours.

On April 1 we reached Iquitos, a considerable village on the north bank at the mouth of a small stream of black water. It contains many people of mixed race, besides a great many Iquitos Indians who inhabit the western portion of the village.[1] Here I first saw the fruit of a remarkable palm-like Pandanaceæ (Phytelephas) allied to the plant that produces the vegetable ivory.

On April 2, reached Nauta, on the north bank, a few miles above the mouth of the Ucayáli, which enters from the south—a river equal in size to the Marañon itself. Nauta stands on rising ground from 30 to 60 feet above the river. The soil is sandy with some mixture of clay near the river. At the back the ground goes on gently rising for a considerable distance, only interrupted by rivulets. In the second growth on old clearings, the most curious feature is the absence of Selaginella, so constant in such places on the Amazon and Rio Negro. There is, however, a common Adiantum and a low tree-fern.

[As the steamer went no farther, Spruce had to wait a fortnight at Nauta before he could hire two canoes with the necessary Indians to take him and his goods up to Yurimaguas on the river Huallaga. In the intervals of this work he collected such

[1] Iquitos is now a town of about 10,000 inhabitants. It is the capital of the Peruvian province of Loreto, and the centre of the rubber trade of the Ucayáli, the Napo, and all the higher tributaries of the Amazon. There is a monthly communication with Pará by river steamers, while at longer intervals steamers make the through journey from Liverpool to this inland port within sight of the lower ranges of the Andes.

plants in flower as were new to him, and noted several others, but as he does not seem to have reached the virgin forest these were not very numerous. He notes generally that the river-bank vegetation was here identical in its main features with that of the river below. In a small side-channel near the village he noted a twining Bignoniacea with long white flowers in axillary clusters resembling those of a Posoqueria; a sweet-smelling Calyptrion (Violaceæ); a Maclura laden with pendent catkins, like those of a hazel; a spreading tree with clusters of winged fruits, apparently one of the Ulmaceæ, and several others not in flower which were quite new to him.

The Journal of his voyage (now in canoes) continues:—]

April 16.—Left Nauta at noon. Passed along low shores. Besides the *Salix Humboldtianæ*, two other willow-like trees were noticed for the first time. At 8 P.M. reached four low huts or tambos, where we stopped for supper and for the night. I went back to the canoes, but the zancudos were terrible and I got no sleep. Next day the river continued rising, but last year's flood-mark is still 6 feet higher.

April 18.—At 8 P.M. reached San Regis, one of the most ancient pueblos (villages) on the river. I slept in the convent, which dates from the old missionaries. The roof was of very neatly woven Irapai (a species of Pandanaceæ).

April 19.—Just before 6 P.M. we reached some dry ground, where among lofty trees a space had been cleared sufficient to accommodate a few palm-leaf shelters. Under one of these I slung up my mosquito-net, large enough for a whole family. Charlie and I stowed ourselves beneath it, having first spread on the wet ground three layers of palm-leaf mats, and over these our blankets. The heat was almost insupportable early in the night, but afterwards the temperature was agreeable, and the shelter from dew and gnats was a luxury, and I enjoyed a fair night's sleep.[1]

April 21.—Reached Parinari, a rather populous pueblo on a low site scarcely raised above the river at flood. The inhabitants are all Indians except the Governor, who is a Zambo named Don Domingo Mayo. We found the people beginning their Easter feast, the Cura of Nauta being expected on the following day. Both men and women had their faces painted red or white in lines and dots, while many were already half intoxicated.

The Governor was not an amiable character. He was very distrustful, and was especially afraid that on account of his colour due respect should not be paid to him as governor, and was also jealous of his wife and of her daughter (a girl of fourteen). He was also in constant fear of his life (though, I believe, resolute to defend it), and not without reason, for his rule over the Indians was a most severe one. I could not help admiring the facility with which he, alone and without assistance, kept some hundreds of Indians in order. He told me, however, that he had several times had to defend his life against them, and not long ago a number of them came on him with pikes; but the

[1] "Charlie" was an English sailor Spruce had found at Barra do Rio Negro and had engaged as an assistant. His story and fate are described later in letters to Mr. Bentham.

mere pointing his gun at them generally sufficed to put them to flight. Once, however, at San Regis, they closed upon him and he had to stab one of them with a sword and then stamp upon his body, at which sign of determination the rest fled.

He told me that the inhabitants of Nauta, San Regis, Parinari, and Urarinas are Cucãma Indians from La Laguna and Santa Cruz on the Huallaga river. It is remarkable that the language of the Cucãmas is so like Tupí (or Lingoa Geral of Brazil), that when I made use of the little I had learnt of the latter on the Rio Negro, the Cucãmas understood me perfectly, and I in like manner understood most of what they said. The nouns are often absolutely identical, the verbs mostly differing only by a few letters, and the grammatical construction similar. The only other remnant of the Tupís I have heard of is the small tribe of Tupinambaras at the back of Villa Nova on the Amazon, but they seem to have become so mixed with black and white, that in 1850 I sought in vain for any pure Indians of the tribe there. These Cucãmas have no record of their origin, as have those of Yurimaguas.

Left Parinari late on the 22nd, and the next day passed along a coast rich in palms, such as the Paxiúba and Urucurí (Attalea sp.), and on the very margin clusters of the elegant *Bactris concinna*, its slender stems of some 6 feet high crowned by pale green regularly and closely pinnate fronds, beneath which hung on a short stalk dense clusters of black fruits. Very rarely I saw another Attalea more resembling the Jagua of Venezuela.

April 25.—Stopped to cook our breakfast this morning on a bit of dry land (inundated only in the highest floods) where the forest was lofty and not much obstructed by twiners. One very fine Pao Mulatto, perhaps not less than 100 feet high, had a mass of broad strips of shed bark at the base. I picked up a piece of this, and while examining it heard a rattling in the place whence I had taken it. Stooping down, I saw that I had uncovered a large rattlesnake, who was raising himself up and poising his head for a spring at my leg, which was not more than two feet off. I retreated with all speed and fetched my gun from the canoe, but on returning the snake had disappeared.

On the 26th we reached Urarinas, a small pueblo about the size of San Regis, and already referred to as having a common origin.

April 28.—About noon to-day we spied a band of peccaries crossing the river towards our side, and already beyond the middle. With considerable difficulty we secured nine of them by the use of our guns and cutlasses. One of the largest boars, when wounded, was very fierce and tried to climb into the canoe, and had he not been speedily killed might have wounded some of the men seriously with his large keen tusks, of which, as is well known, even the jaguar is afraid. As we did not reach a place where we could prepare and cook them till early the following afternoon, the meat had already become too tainted for salting, but we had a meal of it, and the remainder was all cooked and eaten during the succeeding night by my Indians and the villagers.

We had entered the Huallaga river during the

night and the village was La Laguna, so called from a large lake a little behind it, but not visible from the village, which is reached by a very narrow side-channel. There are perhaps a hundred families in fifty houses built irregularly around a square open space. There is a very large church dating from the time of the Jesuits. The walls are of adobe and the roof is supported on pillars formed from large trees. The Cura was absent at Moyobamba.

May 4.—This day (about 4 P.M.) we passed some rather high land about 12 feet above the highest floods, and the first uninundated land I had seen on the banks of the Huallaga. It had been very wet, but after 5 P.M. it cleared up and I enjoyed my first view of the Andes. The part seen is called the Serra de Curiayacu, and in form and extent reminded me much of Duida as seen from the Casiquiari, showing a table-like summit with several outlying peaks on the right. Yurimaguas was reached the next day at 10 A.M.

We were very kindly received by the priest (Dr. Don Silverio Mori), and as I had decided to wait here until I could get Indians from Chasuta to take us up the pongo, he installed us in the cuartel, a commodious building of three rooms, but much infested by rats.

Yurimaguas is a small place (about equal to San Regis), but is pleasantly situated on ground rising abruptly but to no great height. It is one of the most ancient missions in Maynas, and according to information derived from the priest, it was founded in 1709 by Spanish Jesuits, who, accompanied by a few armed whites, descended the

Amazon as far as Parinari, a little above Ega. Thence they ascended the Yapura river, where they found a tribe of Indians called Yurimaguas, and after a time persuaded these to return with them up the great river and the Huallaga to the present site, where they have remained. They were induced to do so the more readily on account of the constant enmity of a neighbouring more powerful tribe. At present these Indians all use the Inca language, and only a few of the older ones have an imperfect knowledge of their original language.

The church here perhaps is the most ancient, and is certainly the best built of any I have seen in Maynas. It is built of adobes in a style very similar to that of churches in Lima, having a very high-pitched roof. The floor is of tiles. The priest's house seems to be of the same date, and has been much ornamented within by cornices, etc., painted in various colours—the work of the last priest. Over one of the doors is inscribed in Latin the verse of Proverbs: "Give me neither poverty nor riches."

[During Spruce's stay here he made a very careful pencil-drawing of the church, with its well-designed entrance of the simplest native materials. The figures on each side of the door are those of St. Peter and St. Paul, executed in coloured earths, while on the left is the belfry with its ladder—the campanile of South Europe reduced to its simplest elements. The figures of an Indian man, woman, and boy, with the priest going to the church, are characteristic; while the background of forest, with its various forms of trees, completes the picture. I

FIG. 1.—YURIMAGUAS, ON THE RIVER HUALLAGA. (R. Spruce, May 19, 1855.)

am indebted to my friend Mr. Young, a good artist, for strengthening the shading, defining the outlines, and putting in the foreground, so as to render the drawing suitable for reproduction to half the original size.]

Don Silverio makes an admirable priest for the Indians, as indeed he would for people of any colour. Low in stature and not stout, but firmly knit, with a rather dark but ruddy complexion; a small well-formed mouth, which even in its most severe expression speedily relaxes into a benevolent smile; a sonorous and untiring voice; added to this an irreproachable conduct very unusual in South America, and an untiring vigilance over the moral and physical condition of his parishioners. Every day, both morning and afternoon, he has in his house all the boys, both Indians and Mestizos, whose parents will allow them to be taught, and takes all possible pains to teach them to read and write, with such success that nearly all can do both intelligibly. Their writing-books are mostly nothing but slips of plantain-leaves, on which when fresh the ink-strokes are very distinct. He is much put about to find them reading-books, in lieu of which old newspapers, letters from his friends, and, in fact, any scraps of MSS. or print are made to serve. He finds it, however, very difficult to get them to speak Spanish, as out of school they speak only Lingua Inca with their families and playmates. Every evening, except Sundays, all the young girls present themselves in the corridor of his house, where they repeat to him the "Doctrina" at length.

At feast times there is mass every morning, and

at other times every Wednesday and Saturday morning. On Saturday evening nearly the whole population assists at vespers—the Litany to the Virgin—when the altars are decked with small vases filled with flowers of *Poinciana pulcherrima* (called Uaita-sissa, *i.e.* swimming flower), and at the conclusion the patron saint, mounted on a stage, is carried in procession round the streets, the Padre and his people chanting as they march. After each mass in the morning, and after the Ave Maria in the evening, the chief officials of the town present themselves to the Padre to receive his orders, and he is fortunately not trammelled by the presence of any interested white man under the name of Gubernador, this office being filled by an intelligent old Indian.[1] His rule is strict without being severe, and I have nowhere seen the Indians so docile. True, they are a rather sluggish race—poor oarsmen—and many of them have the skin disfigured by black and red blotches from the leprous disease called purupurú in Brazil.

Outside the pueblo is the cemetery, surrounded by an adobe wall, with gates under a porch. It is usual to bury a man in his old canoe, cut up into something like a coffin. The houses at Yurimaguas, as in most other places on the rivers of Maynas, are built of Caña brava—a stout reed—stuck close together in the ground and crossed by others near the top and bottom. The doors are made of the same material.

Stages (called barbacoas), on which the inmates

[1] The officials of Yurimaguas, in the order of their rank, are Curaca, Capitan, Alferes, Alcaide, Procurador, Regidor, Alguazil major, and two Alguaziles minor—in all nine.

sleep, are made of the Tarapoto palm (*Iriartea ventricosa*) split and flattened out into slabs. These beds are raised about 3 feet from the ground. A mat of one or more layers of Tururí (bark cloth) is laid on the barbacoa, and the whole is enclosed in a quadrilateral bag of Tocuyo (a coarse native cotton cloth), supported on a framework of reeds, to serve as a mosquito curtain. It effectually keeps out insects but is very hot. Benches, both inside and outside the houses, are made in the same way, but the latter sometimes of an old canoe, the bottom forming the seat and one side the back, like a settle.

The industry of Yurimaguas, besides the salting of fish, which is done during summer, is chiefly the fabrication of painted ollas and cuyas (pots and calabashes), and numerous old calabash trees scattered about the pueblo form one of its most picturesque features.

The Padre's house is much better than the rest —built as in Brazil on a framework of rods filled in with clay, and painted white, outside and in, with gypsum. It contains several tables, the tops of which are single slabs, one 4 feet across. The rooms are ceiled with Caña brava, closely laid across the beams and covered above with a thin layer of clay.

A peculiar utensil seen here and elsewhere in Maynas is a large flat shallow dish, of the form of the tin vessels used by gold-washers; it is made of the sapopema of some light-wooded tree, and I have seen one above 5 feet in diameter. It is used chiefly for crushing maize with a stone for the fabrication of chicha (native beer), but is also used for grinding coffee, etc.

The animal food at Yurimaguas, besides pigs and fowls raised on the spot, is chiefly fish, game being very scarce. In the summer many large fish are obtained, but when the river is full only small ones can be had.

About a quarter of a mile below Yurimaguas a deep valley enters on the west side called Paranapura, which is the route to Balsapuerto and Moyobamba, and thence by Chuchapoyas and Truxillo to the coast. The navigation of the river is uncertain and perilous, not on account of rapids, of which there are hardly any, but because of its often rising a great height in a few hours (or even minutes) from the sudden swelling of mountain streams consequent on heavy rains. When in its best state the voyage from Yurimaguas to Balsapuerto takes six days, but when full the current is very strong, and when low channels have to be dug through sandbanks, so that several weeks are sometimes required.

A little way within the Paranapura there is a village a little larger than Yurimaguas called Muniches, which may be reached by a good track through the forest in four hours. This track crosses several elevations and valleys, each of the latter with a stream running in a sandy or pebbly bed. Along this track the land has been almost all formerly cultivated and there are still several fields of Yucas and Plantains.

About the same distance above Yurimaguas there is a very similar but smaller stream called Chamusi, which affords a route to Tarapoto and Lamas, occupying usually six days, of which three are by water. But the Chamusi has the same impediments to navigation as the Paranapura, and the road overland is more elevated and very rough.

Voyage up the Huallaga to Chasuta, and thence to Juan Guerra and Tarapoto

On Tuesday, June 12, at 7.30 A.M., we left Yurimaguas for Chasuta, myself and goods occupying two ubadas (large dug-out canoes), one with nine, the other with eight Indians. The river had been sinking for some time, but for four days much rain had fallen and the river had risen again. When we started it was 8 feet below high-water mark.

On the next day at 4.30 P.M. we reached the mouth of the Cainaiuche, up which there is a way to Tarapoto when the Huallaga is so full as to render the pongos of Chasuta impassable. As rain seemed coming on, we remained for the night on a sand-bank, where it took us near an hour to erect some twenty tambos (shelters) of palm-leaves, under which we hung our mosquito-nets, and so many green tents scattered over the sand had a pretty appearance, the picture being completed by two fires blazing in the midst, around which crowded the Indians until rain compelled them to turn in. After the rain a very strong and cold south wind sprang up—more searching than any I have felt since I left England. A good many waterfowl begin to appear on the beaches as the receding waters gradually expose them. Among them we had numbers of Jabirus and Garcias (cranes and herons), and one day two majestic Tayuyus (the giant stork *Mycteria Americana*) were seen, but were too wary to be shot.

June 15.—The river now reminds me of the Upper Rio Negro—similar banks sloping steeply to the water's edge, inundated in winter and clad with black rootlets. In many places the perpendicular cliffs of earth are speedily covered with rudimentary mosses. The little Oxalis also reappears accompanied by patches of a grass and a small Composite herb. The wind has been very cool these two days, and in the morning actually cold.

June 16.—This morning we passed, on the north bank, a line of cliffs about a quarter of a mile long, the upper 12 feet being red earth in scarcely distinguishable horizontal layers, while the remaining 20 feet were in distinct layers inclined about 30° to the horizon. These were also of red earth, but in two places a few beds of greyish sandstone occurred. A little below the entrance to the pongo we came to a large clearing on the north bank, partially planted with Yucas and Plantains.

June 17.—Soon after starting this morning we reached the pongo, where the river is much narrowed and confined in one channel by steep hills on each side. The margins were at first rocky, with large blocks irregularly scattered, soon changing to low walls of thick rock-strata.

An hour and a half within this channel we came to streams of hot water, pouring in four or five slender rivulets from a black cliff perhaps 20 feet high and 20 or 30 yards from the river's margin. Each flowed in a slight hollow marked by vapour that constantly rose from it. The cliff itself was draped with a curtain of twiners which I had not time to penetrate. The water was quite clear and destitute of taste or smell, but so hot at 20 feet from the source that I could not bear my finger in it.

About noon we entered a long narrow channel between loftier rocks and steeper hills above them, where the currents and whirlpools gave us some trouble. At its upper entrance stands a steep cerro where the rock is only partially clad with vegetation, and is stained in bare places with blotches of red or dull purple. It is called Uámar-uássi or Eagle's house, from having been once the habitation of an immense eagle which guarded the pass, and the purple patches are blood-stains—the blood being of those who were so rash as to attempt the pass in its guardian's despite. The scenery throughout this pongo is beautiful, though the enclosing mountains do not exceed 500 to 800 feet in height. The strata are sometimes almost vertical, and are then partially naked, the scanty vegetation being upheld (as I have noticed in other places) by masses of Bromeliaceous plants.

The next mal paso is called "Arpa," because just above it there is a rock supposed to resemble a harp. The current round the rocky point was so strong that the canoes had to be dragged along by stout creepers. Afterwards we came to grey friable rock in very thin layers, and this was succeeded by a slaty-looking dark rock, and then the friable grey rock again appeared. These shales are all Triassic, and produce salt. Two other rapids of less importance were passed before dark.

June 18.—We slept in a chacra (shed) just below the worst fall, called Yurac-yacu (white water), because the water here bursts into foam over rocks strewed in the river at a narrow curve. An hour farther there is another similar mal paso (called Curi-yacu), where a stream comes in on the left bank, said to contain gold. Some way below Chasuta we passed, on the left bank, a considerable ravine with still black water called Yanacána-yácu (Ladder River), from its running over steps in its upper part as it comes down from Curi-yacu. This mountain, whenever we came in sight of it, had its summit wrapped in mists and showers, from which it is said to be never free.

After passing the rapid of Curi-yacu the river gradually opens out wider, but still in many places runs rapidly over sharp gravel. Mountains appear on every side—Curi-yacu on the right, the low, rounded, acute-edged cerros of Chasuta nearly in front, and the lofty Morillo (yielding only to Curi-yacu in height) in front and rather to the left. On our left, directly across the river, are only lower hills.

Alligators, turtles, and pirarucú exist in the Huallaga as far as the rapids of Yurac-yacu. The small alligator is found all the way up to Huanúco, as is also the fresh-water dolphin of the Amazon. Electric eels are frequent in the Huallaga and Ucayáli, and still more in the lakes connected with these rivers.

June 23.—We reached Chasuta on the evening of the 18th. It is a considerable village on the left bank of the Huallaga, at the mouth of a rather large ravine, and from being situated at the very foot of abrupt rocky hills, while loftier ones appear on every side, it is one of the most picturesque places I have seen. Its population is entirely of Indians, though many show evident traces of white blood, and they are among the tallest and handsomest I have met with—especially the women. Even the Governor is an Indian—an old man, formerly a soldier, in which profession he learnt his Castilian. The pueblo numbers less than 300 married men, and about 1500 souls. All speak the Inca language, and very few have a smattering of Spanish.

Our Indians from Tarapoto were paid to take us up as far as Juan Guerra—a small pueblo at the junction of the Combasa and Mayo rivers above the pongos of the Huallaga. We found it, however, impossible to persuade them to proceed beyond Chasuta, the reason given for deserting us being that the Indians of Chapaja, a pueblo in the pongo, were awaiting their arrival to fall on them unawares and kill them, as there had been a quarrel between them a short time before and serious wounds had been given on both sides. It was plain, however, that they also wanted to escape the labour, as there are three of the worst passes on the Huallaga a little way above Chasuta, where the whole cargo has to be carried overland among large blocks of rock for some hundred yards or more, and we had found the Tarapotinos much disinclined to work hard. There being no authority at Chasuta able to make them fulfil their contract, we had no alternative but to engage other Indians at Chasuta for the rest of the voyage. We had already paid a dollar apiece to our men, and we now had to give a cutlass to each man of our new crews.

Most of them were half tipsy, as they had been preparing rum for the feast of their patron saint on June 29, and it was with some difficulty we got them embarked on the afternoon of the 19th. The actual distance from Chasuta to the mouth of the Mayo river could be passed in three or four hours were it not for the rapids, which are at about equal distances apart. The second of these is difficult to pass all the year round, the first is worst when the river is rather full, and the last when it is nearly dry. We found the first the most difficult of approach and ascent, and the last the easiest, but in all of them it is difficult and dangerous work for the Indians who carry the cargo across the rocks. The empty canoes are dragged up with stout creepers, and though they fill with water they suffer no injury.

The falls resemble in some respects the first fall of the Uaupés, but with less water and on a rather smaller scale, while the whirlpools below are much less dangerous. The scenery of the falls of the Huallaga is, however, far more picturesque, from the steep and lofty mountains which rise on each side of the river, and the dense tapestry of mosses on the moist rocks and inundated branches at the very edge of the water. There is much similarity in the shrubs and trees growing about both, though the species are, I believe, entirely different, and the palm of botanical novelty must perhaps be given to the Uaupés. The most striking difference is perhaps the vast abundance of *Neckera disticha* (or an allied species), forming a dense beard to branches of trees hanging into the water, as Hydropogon does on the Upper Rio

Negro and Casiquiari, while I only saw in one or two places scraps of a Selaginella—a genus which is represented by several beautiful species growing in great quantity about the falls of the Uaupés.

Night came on immediately after we had passed the first fall. We slept on a sandbank shaded by overhanging trees, which did not prevent our feeling the strong and cool south wind which blew all night. Our men worked well in the morning, and by 10 o'clock we had got the cargo carried safely up above the last fall, and we then set on to cook our breakfasts with light hearts. Into all the falls there enters a stream of clear cool water tumbling down among mossy rocks, in the first and last fall from the left, and in the second from the right. In all these falls stones which have 12 feet or more of water over them in flood are often coated by a black varnish, as in the cataracts of the Orinoco, but those higher up the slope, and therefore under water for a shorter period, rarely show this peculiarity.

Above Estero-yacu (the highest fall), the Huallaga is again broader and stiller, though running rapidly at points; the mountains recede from the river-margin, and the vegetation puts on the same aspect as below the pongo. About an hour more brought us opposite Chapaja, an Indian village of a few scattered huts, whence there is a track leading to Tarapoto, occupying about three hours with mules. Another hour and we had entered the mouth of the Mayo, a somewhat smaller stream than the Huallaga, which it quite resembled. Here were banks of mud and sand, sometimes covered with pebbles, as on the Huallaga from Yurimaguas to the pongo, and on banks grow abundantly the Gynerium, Enkylista, Lythracea, and other species frequent also on the Marañon and Huallaga. It was a tedious navigation up the winding Mayo to the mouth of the Cumbasa. There were, the Indians said, twelve turns, and we had expected but two or three, and it was accordingly near sunset when we got to that river. To our great annoyance we found that it had fallen so much that there was no possibility of our getting our laden canoes up to the pueblo of Juan Guerra, which is nearly a mile within. We slept therefore at the mouth, and the next morning had the cargoes carried overland to the village.

[A letter to his friend Teasdale describes the more personal and social aspects of the voyage up the Solimoẽs, and will supplement the purely geographical and botanical notes in the Journal.]

To Mr. John Teasdale

TARAPOTO, *July* 1855.

. . . .

I had a long and wearisome voyage from the Barra to this place, lasting from March 15 to June 21. I was eighteen days in getting up to Nauta—a distance of some 1500 miles — in the steamer *Monarca*; a wonderful difference this from the sixty-three days spent in getting from Santarem to the Barra, a distance scarcely one-fourth so great. When we were going smartly along by day it was really delightful, though the coasts are exceedingly flat—much more so than those of the Amazon below the Barra. I was, however, never tired of admiring the ever-varying forest-panorama—the broad beaches densely clad with Arrow-reeds growing 20 or 30 feet high, behind which extended beds of slender and graceful willows (*Salix Humboldtiana*), their yellow-green foliage relieved by the occasional admixture of the broad white leaves of *Cecropia peltata* (a tree of the Mulberry tribe), while beyond rose abruptly the lofty virgin forest, composed of trees of the most different types growing side by side. Add to this the noble river, the innumerable islands (fixed and floating), the cranes and herons, the never-failing alligators, the fresh-water dolphins chasing one another and turning "summersets," besides numerous other sights and sounds which I cannot here enumerate—the whole viewed leisurely and ociosamente ("at one's ease"), free from any tormenting recurrence of mosquitoes, and you will understand that a voyage up the Amazon in a steamer has enjoyments peculiar to itself, although one's nerves may be occasionally shaken by the vessel scraping on a snag, or by the sudden assault of a violent thunderstorm. Oh that these had been the only troubles! But as we were only about half the time under way—the other half being spent in embarking firewood, a cargo of mosquitoes always coming on board, uninvited, along with the latter (and I think the higher you ascend the Amazon the more numerous and voracious they become)—you may say that we were half the voyage in paradise and the other half in purgatory.

The *Monarca* is a small but strongly-built iron steamer, with low-pressure engines of 35 horse-power which occupy so much space as to leave very little for cargo. The firewood also took up a deal of room, and, besides what could be stowed down below, had generally to be piled to an inconvenient height on deck. We used to embark as much as would last us from thirty to thirty-six hours, and we consumed on an average seventy sticks an hour, the sticks being a Portuguese vara (five spans) long and three or four inches thick. Piles of firewood are established at convenient distances all along the banks. The wood which is most largely consumed is that of the Mulatto tree, so called from its shining bark, which is sometimes of a leaden-coloured hue, at others verging on red. It is one of the most abundant and at the same time handsomest trees all along the Amazon, growing often to 100 feet high, and in the spring-time bearing a profusion of white flowers which may be compared to those of the hawthorn for size and odour. The tree, however, belongs to a very different tribe, and is closely allied to the Cascarilla or Peruvian Bark tree. It was unknown to botanists until I sent specimens from Santarem, and Mr. Bentham has called it *Enkylista Spruceana*. The wood causes a good deal of flame, and burns nearly as well when green as dry. . . .

Imagine the cabin passengers of the *Monarca* stretched in their hammocks under an awning in the poop eagerly listening to one of their number reading from an old black-letter copy of the fabulous history of "Carlos Magno," and amongst those listeners were a Juiz de Direito, a Procurador Publico, two military Commandants going to take charge of garrisons at Ega and at the mouth of the river Içá, and an English botanist whom, at least, one would have supposed far in advance of such old-world fooleries. When I reached San Carlos in

Venezuela the only books in the Spanish language existing there were "*El Sepulcro*, por Anna Radcliffe," and a translation of one of the Duchesse d' Abrantes' novels. They are scarcely more numerous at Tarapoto, where one of the most famous books is "*Waverley ó ahora sesenta años*, por Sir Gualterio Scott." In short, so far as I can judge of South America from having seen only the most thinly-inhabited portions of it, I can truly say that Mrs. Radcliffe, Walter Scott, and Alexandre Dumas are far more popular there than Cervantes and Camoens. To the credit of the Brazilians, they are far more familiar with the *Lusiads* than the Spanish Americans are with *Don Quixote*. . . .

Well, we reached Nauta, beyond which the Brazilian steamers do not proceed. Nauta is an Indian village established about twenty years ago just above the mouth of the Ucayáli. It is a good way within the frontier of Peru, but is at present the seat of the frontier garrison (of twenty-five men) and also of the government of a department with provisional limits and a provisional name (Dept. del Litoral do Loreto), nearly conterminous with the ancient province of Maynas. Two steamers were got out here two years ago from the United States where they had been purchased for two or three times their value. They were intended to navigate the Huallaga and Ucayáli; but proved such trashy things—slightly built of pine wood, and containing large, coarsely-made, high-pressure engines that were continually shaking the boats leaky—that the Peruvians could make nothing of them, and they are at this moment lying rotting in the port of Nauta, manned by a crew of rats and mosquitoes. The state of these steamers was a great disappointment to me, as I had calculated on getting up as far as Yurimaguas on the Huallaga in one of them, and I had now no alternative but to continue my voyage in canoes, in the rainy season and with the river full. I got a couple of canoes, and after a fortnight's delay in putting them in order and getting crews of Indians to navigate them, I took my weary way up the Marañon. . . .

[Part of a letter to Mr. Bentham carries on the narrative by describing an incident at Nauta that might have had very serious consequences, or even caused the death of the traveller.]

To Mr. George Bentham

YURIMAGUAS, PERU, *May* 27, 1855.

. . . . . .

I left the Barra on March 15 in the steamer, and reached Nauta on April 2. Had it not been for the delays in taking in firewood every day or nearly so, the voyage might have been made in half the time. At Nauta I was detained a fortnight getting together Indians and a couple of canoes to continue my voyage. From Nauta to Yurimaguas took me till May 5—a voyage made sufficiently uncomfortable from abundance of mosquitoes by day and night, and rendered perilous by frequent falling in of the banks of Marañon and Huallaga, and by the risk of upsetting when the deeply-laden canoe struck on some hidden stump, which happened every day.

My repose in the Barra had been of great service to my health, but I reached Yurimaguas pretty nearly done up, and on the very day I arrived I was seized with diarrhœa—caused probably by drinking the saline waters of the Huallaga. I had scarcely shaken off this when I was taken with influenza, which still holds me. To these inopportune bodily ailments have been added no small mental trouble. You will perhaps have heard in England of the number of adventurers of all nations, but principally English and Americans, who, misled by a false report of gold on the Upper Marañon, went thither seeking it. Many of these had passed the Barra before I arrived there, but I still met several, and amongst them an English sailor who seemed a very quiet fellow, and whom I engaged to accompany me to Peru, thinking that a stout companion like him would be invaluable to me in a country where, as report truly said, there was no law but that of the strongest, and acts of atrocity were of frequent occurrence. I might, with a little more forethought, have considered that a man who had once become imbued with the idea of acquiring riches by some sudden fortune (for I knew he had been a "digger" in Australia) was never likely to take steadily to any work which brought him in but small, though certain, gains; but I could not tell beforehand what I know now, that my companion had marked by violence his course through Peru, and had been in prison at Lima for murder. When we reached Peru, and had consequently passed the limit of any efficient police, his nature began to show itself, and I had proof that he sought occasion to murder me and decamp with the money I carried with me. On the way here from Nauta he ill-treated the Indians, and being rather deaf and understanding scarcely anything of Spanish, he fancied that every one whom he saw laugh was ridiculing him. A few days after we got here an old Indian, who officiates as sacristan to the Padre, was conversing with other Indians in the square, when my man went up to him, seized him by the neck, and with his right fist broke his mouth in. On the following day, when we were at dinner with the Padre, where was also a Portuguese who had travelled along with us nearly all the way from Nauta, the latter was telling some tale about the students at Coimbra which set us a-laughing; my man thought the laugh was directed against him, got up from table and challenged the Portuguese to fight him with his fists. Attempts at explanation only infuriated him more, and seizing a pickaxe he aimed a blow with it at the Portuguese, which I happily averted by lifting up the handle. The Portuguese then, at the Padre's request, entered an inner room and fastened himself in, the other still attempting to burst open the door in order to wreak his vengeance. It was, of course, quite impossible for me to excuse or palliate such conduct as this to the good Padre, who had treated us most kindly, and as it is equally impossible for me to follow my pursuits without keeping on good terms with all, my separation from such a companion became imperative. I do not trouble you with a detail of the reason I had for concluding that he contemplated violence towards myself, and which for several days had induced me to sleep always with a revolver under my pillow. Suffice to say that with much trouble and no small sacrifice on my part we succeeded in getting him sent off. I paid him three months' wages and the passage from the Barra and back—in all 140 milreis—and on the whole I am some £20 out of pocket by the speculation.[1]

Many of the gold-seekers marked their way through Peru by violence, and some of them came to violent ends: an Englishman was killed in Chasuta by the Indians, an American was drowned in a stream which enters the Huallaga within sight of Yurimaguas, and many others perished miserably in one way or another. All were known to the natives under the generic name of "Ingleses," who are consequently by no means in good odour.

You will perhaps not be surprised to hear, after what I have above stated, that I am inclined to repent having come on this expedition, which is proper only for a person enjoying the best bodily health and strength. I have still considerable expense and risk before me—to get to Tarapoto will cost me fifty dollars, though it is so near in a straight line that I can nearly see it from a little way down the river. But the delays always annoy me more than the expense, especially when I cannot work. The great bulk of my baggage is paper, which it is of the first necessity to bring, as I understand I could not procure any from nearer than Lima, where I have no funds and no corre-

[1] In a letter written shortly before he quitted Tarapoto, Spruce gives the termination of this man's career as follows:—

"In my letter from Yurimaguas I spoke of an English sailor who came up with me from the Barra, and whom I was obliged to dismiss for his violent conduct. He has lately been cruelly murdered by two Indians who navigated his canoe a little below the mouth of the Ucayáli, much in the same way that Count D' Osery was, some distance higher up the river. Though, from his own confession to me, I have no doubt that the same measure has been meted to him as he had meted to others, I am not at all satisfied that his murderers have been set at liberty without punishment."

spondents. At Nauta I collected scarcely anything, for fear of adding to my already unwieldy baggage, and I could not leave any dried plants there, where they would be wasted. The same reasons, added to illness, have limited my gatherings at Yurimaguas, for I cannot hope to gather sufficient to make it worth while sending a collection from here direct to England. Towards the sources of these rivers it would be easier to collect in descending than in ascending were it practicable to remain a few days in the promising localities; for in coming down the size of one's canoe may be as large as one chooses, but in going up one must necessarily use the smallest canoes, and even then be content to get on at a very slow rate.

. . . .

[The letter to Mr. Teasdale now takes up the narrative again :—]

The banks of both Marañon and Huallaga continue flat all the way to Yurimaguas, but at about two days below this place *I enjoyed my first view of the Andes*! It was on the 2nd of May —we had had terrible rain from midnight to noon, and it still kept dropping until 5 P.M. About half-past five the sky cleared to N.W., distinctly revealing a line of blue mountains which might be some 4000 feet higher than the river. They are called the Curi-yacu (Mountains of the River of Gold), and extend along the western side of the pongos of the Huallaga.

You are, I daresay, aware that the Marañon, the Huallaga, and their tributaries have the peculiarity of issuing from the mountains into the plains through deep narrow rifts called pongos. From the steep perpendicular walls which confine these narrows, the Peruvians say very expressively that the rivers in such places are *boxed in* ("encajonado"). The pongo of the Huallaga commences a little above Yurimaguas, and it takes two days to ascend it when the river is pretty low—when it is high the pongo is impassable. Above the pongo are three of the worst malos pasos (rapids and falls) in the whole river. . . .

The principal inhabitant of Tarapoto is a Spaniard (a native of Mallorca) named Don Ignacio Morey. We had known each other by name some years, and he had signified to me that if I would visit Tarapoto he would assist me as far as lay in his power. From Yurimaguas I had advised him of my approach, and he was kind enough to send a couple of mules to meet me at Juan Guerra. When you consider the amphibious life I had led for six years, during a great part of which period I had not so much as set sight on a horse, and that for several years before leaving England I had discontinued equestrian exercises, you will understand that I found the transition from a canoe to a horse rather abrupt. I am, however, too old a traveller to be taken aback by anything, and I immediately made choice of one of the two animals sent me—a large white macho, whose stride was as long as that of a racehorse, and whose caparisons were altogether strange to me, especially the large wooden stirrups, in form of a square pyramid, with a hole on one side for inserting the foot; the whole curiously sculptured. An English horse would have felt weary with such trappings, but he would have stared in dismay at the road, though one of the best in the country. At the commencement it was pretty level, though very muddy in places, and much obstructed by roots of trees and even by fallen trunks stretching across the path; while overhead the branches and twiners hung so low that I was compelled every now and then to duck my head to avoid a fate similar to that of Absalom. They who opened the road had never calculated that a long fellow like myself would have to traverse it. Farther on were ups and downs strewed with stones and often skirting declivities. We traversed three considerable streams, tributaries of the Cumbasa. The track invariably led straight down to the water without any winding, and the mules partly slid, partly walked down.

CHAPTER XVI

EXPLORATION OF THE EASTERN ANDES OF PERU: RESIDENCE AT TARAPOTO

(*June* 22, 1855, *to March* 22, 1857)

[DURING the period comprised in this chapter Spruce appears to have kept no regular Journal, and though there are many scattered notes referring to his various excursions, they are so imperfect, and sometimes so condensed and enigmatical, that I was at first in despair as to how I should find materials for an account of what was, to himself, one of the most enjoyable and interesting portions of his travels, as well as one of the best districts for a botanical collector which he met with during his fourteen years' residence in the equatorial regions of South America.

Fortunately, his letters to Mr. Bentham and to his friend Mr. Teasdale were so full as, to some extent, to supply the place of Journals, while one of his most interesting botanical excursions was described in some detail in an article he contributed to the short-lived periodical the *Geographical Magazine* of July 1873. With these materials, and by making use of some of the descriptive notes mentioned above, and greatly assisted by a very rough sketch-map of the district which I found among his papers, I have, I hope, succeeded in giving a tolerable idea of this interesting locality, which forms the most important centre of population in North-Eastern Peru, and which seems to be still very little known to European, and certainly to British scientific travellers.]

To Mr. John Teasdale (continued)

March 23, 1856.

On reaching Tarapoto about sunset, Don Ignacio placed his well-furnished table at my disposal, and he had already secured me an unoccupied house in a situation exactly corresponding to my wishes. It is away from any street, in the midst of a garden, and only a dozen yards from the edge of a declivity which barely allows the canes and plantains to take root on it; at its base the turbulent Shillicaio seeks its course among rude masses of rock, its sparkling waters appearing only here and there, because hemmed in by a dense hedge of low trees and twiners. It much reminds me of the Pyrenean "gaves." There is no other house nearer than fifty paces, and this, though conducing to my more perfect quiet, may be a disadvantage if it should happen that I have come among ill-disposed folk. The garden is planted with sugar-cane, yuca-dulce, cotton, sweet potatoes, frijoles (beans), and calabash trees. There are also several clumps of herbs (including at least three distinct species of Capsicum), and two or three young trees of *Yangua tinctoria*.

Across the stream is the pueblo of Cumbasa—a sort of suburb to Tarapoto, inhabited chiefly by descendants of two powerful tribes of Indians who occupied the same site when the first whites came from Lamas, about seventy years ago, to found Tarapoto. Looking over the pueblo from my house, I am reminded by the general aspect of an English village in some agricultural district, though the accessories are different. Here and there in the forest (which is mostly low, though there are a few lofty relics of the old primeval woods) are verdant spots whereon pasture various domestic animals: horses, mules, cows, pigs, turkeys and other fowls. On their margins, or from amid the forest, peep out the straw roofs of cottages, often accompanied by plantain gardens and by orange and other fruit trees. Beyond the pueblo stretches a plain towards the S.E. and S., while towards the E. and N.E. the ground gently rises, to fall again into the deep valley of the Aguashiyacu. The plain is bounded by a low ridge of Lamas shales, whereon a red loam predominates and gives to the ridge the appellation of Púca-lama. A broad red road is seen winding over it which leads to the fields and gardens of Aguashiyacu.

The track leading to Chasuta passes through the village of Cumbasa in an easterly direction. After crossing the Aguashiyacu it emerges on a very wide plain of loose sand, covered chiefly with coarse grasses and low scattered trees.[1] This pajonal (open campo) is not visible from Tarapoto, but it extends nearly to Puca-yacu. Immediately across this stream and a little more than two

[1] Among these are *Curatella americana*, a Tecoma with yellow flowers, a strange-looking Tiliacea, and a prickly Xanthoxylum which gives out an abominable odour of bugs when bruised.

leagues from Tarapoto begin to rise the abrupt ridges of Guayrapurima ("where the wind blows"), which are crossed to reach Chasuta.

More to the north is a rather lower ridge whose top, bare of trees, gives to it the name of Cerro-pelado (the bald hill). Over this passes the track leading to a noted fishing stream called Tiracu, whose sources are near those of the Aguashiyacu in the high mountains N.E. of Tarapoto. From this mountain come more storms than from any other quarter. A long day of painful ascents and descents brings fishermen to Tiracu, where they sometimes remain a week, exposed to almost daily rain and barely sheltered at night in a rude rancho of palm-leaves. Some way lower down the Tiracu are cliffs of white salt. The inhabitants of Lamas make frequent visits there, and when I visit the Guayrapurima mountain I never fail to encounter one or more troops of them.

[The accompanying view of Tarapoto from the southern entrance shows the straggling suburbs backed on the north-east by the grand mountain of Guayrapurima, to which Spruce made many excursions. The conical peak on the left is probably the same as that shown in another drawing (at p. 94) as the singular Cerro Pelado when seen from a different point of view, perhaps from the village of Morales.]

The sound of the waters of the Shillicaio generally reaches my ears in a soft murmur, often mingled with the less musical sounds of a cane-mill on its opposite margin; the squeaking of the cane-crushers; the shouts of the men who goad along the poor oxen or mules in their painful round;

Fig. 2.—Tarapoto, on entering from Juan Gerra. (R. Spruce, 1856.)

the grunting of pigs, which chew the crushed canes as they are thrown out; and very often the laughter and playful screams of boys and girls bathing in the stream. But when heavy rain falls on the hills to the northward, the swollen stream comes rushing down with a roar which drowns every other sound, bearing along with it logs and trunks of trees, and sometimes tearing loose from its banks a large mass of rock which falls with a thundering crash. At such times all communication is suspended between the town and the village. The poor people who are returning from their farms on the opposite side, with their load of plantains or other vegetables, have then to wait perhaps a couple of hours shivering on the bank ere they can cross. Their natural apathy prevents the people from obviating this inconvenience by throwing a bridge across the narrow stream, which would be easily done, as the channel is in many places scarcely ten yards across, and the banks are so high that the adjacent ground is never inundated by the highest floods, which always subside a few hours after the rain ceases. A bridge was indeed commenced in 1856, but the foundations were so ill-laid that the first flood swept them away.

At some seasons, especially during the rains, scarcely any colour but green, of various shades, can be discerned in the landscape, save that in the morning the lower part of the course of the Mayo and Cumbasa are marked by a line of hovering mist, and that a tall column of grey smoke may be seen rising in the forest from some newly-made clearing; but a few sunny days after rain variegate the forest here and there with the flowers of several trees and twiners, and the colours are gayest and most varied in the months of July and August. Then are scattered over the plain, especially where the soil is sandy, dense posies of the "Purple flower," a species of Physocalymma (Lythraceæ), and the less conspicuous ones of the "Yellow flower" (Vochysia sp.); more sparingly is seen mixed with these a larger mass of the orange flowers of *Vochysia ferruginea*, and these are everywhere set off by white bunches of Myrtles and Melastomas. Near the Shillicaio rise here and there magnificent trees of Ama-sisa (*Erythrina amasisa*, Sp.), which have been spared by the axe of the first settlers—some of them as much as 80 or 100 feet high, and twice in the year, at intervals of six months, clad with large flame-coloured or vermilion flowers, sometimes with no accompaniment of leaves and sometimes with young leaves of a most delicate green just appearing. I have been delighted to walk by the Shillicaio at sunset and observe the tracery of the crown of the Ama-sisa, with its copious red tassels, projected on the pale blue eastern sky, when the flowers of almost every tree showed a different shade of yellow-red, not, however, paling to yellow on the one hand or deepening to scarlet on the other. It continues in flower nearly two months, and before it has well done flowering the ripened follicular pods splitting up on one side only, and with the two or three seeds still adhering, begin to strew the ground. The trunk is more or less closely beset with shortly conical, sharply cusped prickles.[1]

[1] On this account it is constantly selected by the sagacious troopial (*Cassicus icteronotus*) for its long pensile nests; though, as if doubting that this were

On the declivities sloping to the Shillicaio and too steep for cultivation there are other trees of the primeval forest which flower along with the Amasisa, especially the Lupuna (*Chorisia ventricosa*), a Bombaceous tree with prickly trunk swollen above the base, producing abundance of large rose-coloured flowers, and a tree of moderate growth bearing large panicles of rather small white odoriferous flowers (allied to Loganiaceæ or Gentianeæ). Some two months later a low spreading Bauhinia, abundant on the rocky margin of the stream, appears every morning sprinkled with large white flowers resembling a Prince's feather in form. I know not at what hour they open, but it is certainly before daylight, as I always see them fully expanded at earliest dawn. A Capparis which often grows near it has large white inodorous flowers which begin to open at sunset, and at daybreak the stamens and petals are falling away. It flowers more or less all the year round, and the Bauhinia does not go out of flower for full eight months.[1]

Tarapoto is situated in a large pampa or plain

sufficient to render them inaccessible, it hangs them on the very points of the outermost twigs. All the species of troopial I have seen on the Amazon and Rio Negro show similar foresight in selecting a place where to rear their infant colonies; and the robber who, observing no impediment from below, ventures to climb to their eyrie finds to his cost that it is defended by some large wasps' nest, or by hordes of stinging ants.

[1] [It is interesting to note how often Spruce mentions white flowers as night-blooming, but these two cases are especially interesting because one opens in the evening, the other apparently during the night or long before dawn. This accords with the fact, communicated to me by Mr. K. Jordan of the Tring Museum, that their moth-collector in South America has found that besides the species of moths that come to light or to flowers in the evening and principally up to about midnight, there are other species which only appear probably an hour or so before dawn till near sunrise. Presumably these latter moths are those which fertilise these early flowering white-blossomed trees and shrubs observed by Spruce.—A. R. W.]

of such dimensions that London might be set down in it entire, and so completely encircled by mountains as to form a vast natural amphitheatre. It is about 1500 feet above the sea, while the encircling ridges are 2000 or 3000 feet above the plain, and some of the peaks one or two thousand feet more.

The town dates only from some seventy years back, yet according to a census made since my arrival it numbers nearly 12,000 souls, including two small hamlets which form a kind of suburb to it. The dominion of the Incas does not seem to have extended much to the eastward of the central ridge of the Andes, and the Spaniards found this part of the montaña occupied by independent Indian tribes, of which considerable remnants still exist, both pure and mixed. The first town established by the conquerors was Lamas, which stands on the top of a curious conical hill five leagues (seventeen miles) westward of Tarapoto and 1500 feet above the pampa. From my house I can, with the telescope, distinctly see the white houses glistening in the morning sun. I have also visited it, and may have something to tell you of it in a future letter. It numbers now only from 6000 to 7000 inhabitants, but Tarapoto and several villages on the Mayo and Cumbasa rivers are all colonies of Lamas. Moyobamba, more to the westward, among the mountains, has about 20,000 inhabitants; it is the great centre of the manufacture of those beautiful straw hats sold extensively in Brazil under the name of "Chapeos de Chile," and of which the finest sell for an ounce of gold, or even more. They are made from the same plant as the Panama hats. All these places are inhabited by the same mixed (and I must say very degenerate) race, who have nothing about them of the European but a whitish skin; their ideas, modes of life, and language being still entirely Indian.

Tarapoto is regularly built, and covers a good deal of ground, as the houses mostly stand in gardens. . . . They are all of a single story with thick walls of adobes and palm-thatch roofs.

The climate is much drier than that of the Amazon, but this depends entirely on the peculiar position of the town, for while heavy rains are frequent on the hills, they are very rare at Tarapoto, and we see and hear almost every day violent thunderstorms skirting the pampa, but only occasionally giving us a slight taste. Fogs, however, are frequent in the mornings, and no doubt make up for the deficiency of the rains.

As to temperature, I have once had the pleasure of seeing the thermometer at Tarapoto down to 61° at daybreak. The sensation of cold was so great that had I been in England I should have looked to see the mist deposited on the trees in the shape of hoar-frost. More commonly at that time of day the thermometer marks from 72° to 75°. At two in the afternoon it gets up to 84° to 87°, and in my house to 95° to 98°—on one occasion to 100°. On the hills it is much cooler, and even here we have generally a strong northerly breeze from 10 A.M. to sunset, which tempers the heat. In the months of November and December, I spent three weeks on the Cerro de Campana, at three days' journey to the west of this, and two days from Moyobamba. Here I got nearly 4000 feet higher than the Pampa of Tarapoto, and the cold was sometimes sensible enough, but I could not take my thermometer in my excursions to the highest points. . . .

I have been much interested to meet here several tribes of plants which I had not seen since leaving England. I have got, for instance, a Poppy, a Horsetail, a Bramble, a Sanicle (exceedingly like your common wood Sanicle), some shrubs of the Bilberry tribe with edible fruit similar to that of the English species, a Buttercup (very like the minute *Ranunculus hederaceus* which grows by Ganthorpe Spring), a Hydrocotyle rather smaller than the *Hydrocotyle vulgaris* whose round shining leaves you must have noticed in boggy parts of Welburn Moor, a Chaffweed like the minute *Centunculus minimus* which grows rarely on Stockton Common, and some others. In a deep dell on the way to Moyobamba I was delighted to find a few specimens of that rare plant the Chickweed: its seeds had most likely been brought in the dung of mules which travel that way. . . .

[The following letter now takes up the narrative from the point of view of the botanical collector:—]

To Mr. George Bentham

TARAPOTO, PERU, *Dec.* 25, 1855.

. . . I did not get away from Yurimaguas till the 12th of June, and on the 21st reached the end of my long voyage. Yurimaguas has the most equable temperature I have anywhere experienced, the thermometer sometimes not varying more than 8° in twenty-four hours, but I have found no place so relaxing, and the addition of a severe attack of diarrhœa and catarrh had reduced me pretty low when I left. Periodic returns of this diarrhœa, and ulcerated feet caused by walking in the cold waters of mountain streams, are the chief inconveniences I have experienced at Tarapoto. In other respects I am more agreeably placed than anywhere previously in my South American wanderings. I am among magnificent scenery and an interesting vegetation, and there are a few pleasant people with whom to converse. The pampa or plain of Tarapoto is a sort of amphitheatre entirely surrounded by hills; its position is in the lower angle of the confluence of the Mayo and Huallaga, and the town itself is about three leagues (ten miles) from the latter river. The hills are an offshoot from the main ridge of the Andes, and from being watered by the Mayo and its tributaries I must call them, for want of a better name, the Mayensian Andes. The ridges rise to some 3000 feet above the pampa, and some points are probably much higher.

Good botanising ground is unfortunately rather distant. The pampa either is or has been wholly under cultivation, with the exception of the precipitous banks of the rivulets, and it is a long way across it to the foot of the hills. The summits of the hills have most of them never been reached, and they are clad with the same dense forest as the Amazon, showing rarely scattered bald grassy places (called pajonales or pastos). Where there are no tracks one must ascend by the beds of the streams, all of which, including the Huallaga, have the peculiarity of being, as the Peruvians say, boxed in ("encajonado") between steep walls of rock, where they issue from the hills. These steep narrows are called pongos, and often include falls and rapids. They are rich places for ferns, but it is both difficult and dangerous getting along them, now and then scrambling over large slippery rocks which block up the passage, or wading up to the middle through dark holes with the water below 70°. An exploration of one of these places generally costs me a week's suffering in the feet. I have at last got into a fern country, and I have already gathered more species than in all my Brazilian and Venezuelan travels. Mosses also are more abundant, and there is a greater proportion of large species.

Among the flowers I believe you will find a good share of novelty. I expect I have two new genera of Rubiaceæ, both very fine things, one of them allied to Calycophyllum but with large flowers almost like those of Henriquezia. There are new things also in several other tribes. The general character of the vegetation is, as might be expected, intermediate between that of the valley of the Amazon and of its alpine sources. As evidences of an approach to cooler regions, and to a flora more European in its affinities, I may mention having met here, for the first time in my American travels, a Horsetail, a Poppy, a Bramble, a Crosswort, and a Ranunculus (a minute species, trailing over moss by mountain streams, and looking quite like a Hydrocotyle). The ferns may possibly include some new species, especially among the larger ones, which are likely enough to have been passed over on account of their bulkiness. The fronds of one of these are 22 feet in length, though it never shows more than a rudimentary caudex: its affinity seems to be with Cyathea. In my collection are a good many species of Grammitis, Meniscium, Davallia, Diplazium, Litobrochia, Aneimia, etc., together with several pretty Selaginellas and an Adder's-tongue. A small species of Grammitis growing on trees in the mountains is very odoriferous when dry, and the Indian women put it in their hair, calling it Asiníma.

These things have not been got together without greater trouble than I had calculated on. I expected to find roads on which I could take long journeys with mules, but though there are a few mules there are no roads on which they can be taken with cargoes. Between Moyobamba and the Huallaga all cargoes must be carried on Indians' backs, and indeed throughout the eastern slope of the Cordillera the roads rarely admit of any other mode. The number of Indians is constantly diminishing, and barely suffices for the ordinary traffic of the district. I have ridden a few times across the pampa to the hills, but for longer excursions this mode does not suit. The journey alluded to at the opening of my letter was to visit a mountain lying beyond the Mayo, at two days from Moyobamba and three from Tarapoto. It is called the Campana, from some fancied resemblance to a bell, and the road crosses it at about 3500 feet[1] (by barometer) above the plain of Tarapoto; but there is a peak to northward of the pass rising a thousand feet higher. It differs notably from the adjacent mountains by being nearly all pasto, only the valleys and ravines towards its base being filled with forest, in which abundance of palms are conspicuous. The only habitation there is a chacra on the side next Moyobamba, at 1500 feet below the pass, and with no other dwelling nearer than a day's journey. Here I established myself with a stock of paper, and with provisions for three weeks, which I had taken the necessary precaution of carrying with me from Tarapoto. My cargoes loaded five men on the way thither and six on the return. I have reason to be satisfied with my success at the Campana, and I should probably

[1] Perhaps 5000 feet above the sea, but I have no barometric readings below the mouth of the Rio Negro.

have brought away more specimens had not my host, a few days after my arrival, been severely bitten by a snake, the cure of whom prevented my leaving the house for several days.

[An exceedingly interesting account of this whole excursion, and of the special incident above referred to, forms part of a lengthy article in the short-lived and long-extinct periodical, the *Geographical Magazine.* It is unfortunately almost the only portion of his Tarapoto journal that he wrote out in full, and I therefore insert it here.]

After exploring the most accessible hills and gorges within a day's journey of Tarapoto, I decided to devote a month to a mountain called La Campana or the Bell, three days' journey away to westward. It was just visible from Tarapoto, and was described to me as abounding in ferns and flowers, and having on its flanks large pajonales or natural pastures, embosomed in virgin forest. As all loads must be carried on men's backs in that region, I had first to get together a sufficient number of cargueros, as they are called, for the transport of my baggage, which included salt beef and fish, as I did not calculate on finding much beside vegetable food on the mountain, and I intended to give up my whole time to plants, and not to waste any of it in hunting game for my dinner, as I had often had to do on the Rio Negro. I started therefore on the 20th November (1855), accompanied by my assistant—a young Englishman named Charles Nelson[1]—and by six Indian

[1] ["Nelson" is here mentioned for the first time, and I can find nothing more about him except that he was English, and stayed with Spruce till he left Tarapoto.—ED.]

cargueros. Our first day's journey, of about 15 miles, brought us to Lamas[1]—a town of 6000 inhabitants, near the top of a conical hill, that reminded me of similarly situated towns and villages in Valencia, as they are depicted in Cavanilles' History of that province of Old Spain.

The Hill of Lamas is plainly volcanic, although there is no evidence of eruptions in the shape of lava, or any obvious crater, unless certain small lakes without inlet or outlet a little below the summit may be considered such. The fertile soil which covers its flanks, and yields abundant crops of every esculent that will bear the climate, especially of the indispensable poroto (a kind of kidney-bean), consists almost entirely of decomposed shales of divers colours—sulphur-yellow, vermilion, purple, slate-blue, and black. These shales belong to the Triassic series—near Tarapoto I found ammonites of immense size in them—and have apparently been broken up by the protrusion of a columnar jointed trap-rock, which is here and there exposed in the shape of a sloping floor, divided with much regularity into squares, rather less than a foot on the side, and called by the natives ladrillos or bricks. The slope of the floors is always towards the apex of the mountain, and is inclined to the horizon at from 10° to 30°. Overlying the shales there has been a soft white sandstone, in thick strata, great part of which has been decomposed and carried into the hollows, and even into the plain below, by the torrential rains, leaving only a few scattered blocks of more tenacious material than the rest.

[1] Lamas: lat. 6° 5′ S.; alt. (convent) 2594 E. ft., (hill-top) 2849 ft.

The town occupies a series of terraces, from 200 to 300 feet below the hill-top; but except in what is called the plaza, where the church, convent, and government house—the last appropriated to the lodging of strangers—occupy three sides of a square, scarcely anywhere is there the semblance of a street or square. The nature of the ground is partly the cause of this, for the rains have worn narrow zigzag ravines, called zanjas, 40 feet or more deep, and with perpendicular sides, that radiate from the convex summit in all directions; so that two houses only a few paces apart may be separated by an impassable gulf, and even in the daytime it is necessary to take heed to one's steps, while by night the town is actually impassable for a stranger. It should be added that a bridge, even in the shape of a simple plank, is a luxury unknown in the land of the Motilones. The scanty clothing worn for decency's sake in that warm region is soon dried up by the sun and wind after wading through one of the streams, even up to the neck. The zanjas widen downwards, and from their sandy bed distils a deliciously cool and clear water, which is made to collect here and there in little wells, covered in with a flat stone, and is used by the inhabitants for all domestic purposes.

[The drawing here reproduced was made by Spruce during his two days' stay here (as stated on p. 60). It shows the plaza from a slight elevation, the irregular houses around it, the two-towered church and convent, with a detached bell-tower at some distance, as at Yurimaguas; the whole backed by the forest-clad Tarapoto mountains. This was

FIG. 3.—LAMAS, LOOKING NORTH-EASTWARD. (R. Spruce, 1855.)

delicately outlined by Spruce and the shading added by Mr. Young under my directions.—A. R. W.]

The river Mayo—a broad, shallow stream, whose sources are in the summits of the Eastern Cordillera—runs half round the base of the hill of Lamas, first from north to south, then eastward to unite with the Huallaga.

The inhabitants of Lamas are a mixed race, descended partly from Spanish colonists, partly from the ancient Indian inhabitants, of the tribe of Motilones or Shaven Crowns; so called by the first Europeans who visited them from their custom of cutting off the hair close to the head, with the exception of a fringe left hanging in front to the level of the eyebrows. The custom is still common among the Indians and half-Indians throughout that region; but nowadays the barber's tools are scissors—anciently they were sharp-edged mussel-shells. In 1541, only a few years after the conquest of Peru, Felipe de Utre (or Von Huten) set out from Coro in Venezuela, in quest of El Dorado and the Omaguas, and after travelling southwards ten years, reached the province of the Motilones in Peru, by way of a large river that flows thence to the Amazon. That large river we now know to be the Huallaga. Some years later (in 1560) the famous expedition headed by Pedro de Ursua, and numbering many hundred men, reached Lamas, described as a small village of Motilones, on the banks of the river Moyobamba, where he delayed to build vessels for navigating the Amazon. In his train was the infamous Lope de Aguirre, whose name—synonymous with "traitor" throughout that region—is still given to one of the malos pasos in the rapids of the Huallaga, two days' journey below Lamas. It was not there, however, that he assassinated his patron, Ursua, but on the Amazon itself, at some place not well made out, on New Year's Day, 1561.

Ursua has not been the only adventurer whose miscarriage dated from Lamas. When I embarked at Liverpool, in June 1849, for the mouth of the Amazon, I was shown by the Messrs. Singlehurst great piles of a spurious Peruvian Bark, which had been found to contain no particle of quinine or of any cognate alkaloid, and was therefore quite unsaleable. Its history, as I made it out many years afterwards, was as follows:—A certain Don Luis ——, a young Peruvian, of good address and figure, energetic but restless, and sadly deficient in knowledge and prudence, whilst occupied as intendant of a mine near Cajamarca, had heard reports of the abundance of bark-trees in the lower part of the valley of the Huallaga, and having obtained specimens of the leaves and bark, he rashly pronounced them identical with true Cascarilla, such as he had seen at Huanuco. Forthwith he persuaded several other young men—some of them of good family—to join him in an expedition in quest of it. They found it in greatest abundance on the hill of Lamas, where they collected what they considered would make a shipload of it, embarked it on the Huallaga in rafts, and thus conveyed it all the way down the Amazon—some 2000 miles—to the port of Pará. In all the towns on their route their bold venture created a great sensation. At the city of Barra (now Manáos), at the mouth of the Rio Negro, they delayed long enough for Don Luis to win the heart of and actually marry the daughter of the oldest Portuguese colonist, Senhor Brandão, who (as he himself has told me) considered himself of the same race as our ancient Dukes of Suffolk. Arrived at Pará, the resident merchants and druggists, deceived by the appearance of the bark, and probably at that epoch unable to test it chemically, offered to buy the whole cargo at a price that would have amply remunerated the adventurers, who, however, now thoroughly persuaded of the genuineness of their bark, and believing they could obtain a far higher price for it in England, determined to proceed with it to Liverpool. They accordingly freighted a vessel of Singlehurst's, partly on borrowed money and partly on credit of the proceeds of the sale they hoped to effect. It must have been a sorrowful moment for them when their bark, having been analysed at Liverpool by competent judges, was pronounced to be utterly worthless, and not Peruvian Bark at all. When ulterior analysis only confirmed the sentence, nothing was left for them but to abandon their hoped-for source of wealth and return to their own country, which they were only enabled to do by the beneficence of the merchants of Liverpool. Mr. Singlehurst had the unsaleable bark left on his hands, in lieu of £400 due to him on freight from Pará, and for expenses incurred in England.

At Lamas I was shown the spurious bark-tree, still growing in tolerable abundance, and recognised it as one I had gathered in flower and fruit on hill-sides at Tarapoto. It is the *Condaminea corymbosa* of Decandolle, and belongs to the same family as the Cinchonas, some of which it sufficiently resembles in both leaves and flowers, but differs generically in the seeds being wingless; and the bark, although slightly bitter, has none of the febrifugal and antiperiodic properties of the Cinchonas. There had not been wanting people on the spot who warned Don Luis of his mistake; but he was too opinionated to listen to them, and persevered to his disastrous overthrow.

My host at Lamas was the venerable vicar, Padre Antonio Reategui, and he must needs have me stay all the following day with him; but the time was not lost to me, for I botanised the whole hill-top, made a sketch of the curious town, and on the two evenings of my stay profited by the intelligent conversation of the Padre. It was from him I got the first trustworthy account of the mountain I was bent on visiting. A small colony had recently been established on the flanks of the Campana, consisting of an Indian named Chumbi and his family, and of his two sisters, their husbands and young children. To Chumbi the Padre gave me a letter of recommendation, and assured me I should find in his hut at least good shelter, and store of plantains to eat along with my charqui.

Having lingered so long at Lamas, I must hasten over the remainder of the journey. On the 22nd we reached Tabalosos, a small Indian village on the opposite side of the deep valley of the Mayo, and at about the same distance from Lamas as Tarapoto. At Tabalosos I passed the night in the house of some relations of Chumbi, my bed being merely a hide spread on the earthen floor, like those of the other inmates. The next day's journey was a very long one, and when I returned, with heavier loads, I found it expedient to divide it into two. It would take several pages to describe the savage, rocky and wooded gorges, with rugged ascents and descents; and the torrents that traversed them, and must be crossed and recrossed, as the cliffs rose from the water's edge, first on one side, then on the other. A turbid saline stream of considerable volume, called Cachi-yacu (Salt River), had to be waded through eleven times in the space of half a mile. When we reached the grassy rounded summit of the pass of the Campana, at about 5000 feet, the sun was fast declining, and we had still a long and devious descent on the other side of the mountain to Lirio-pampa[1] (as Chumbi had called his chacra), which we reached about nightfall. On receiving the Padre's missive, Chumbi, with a profound bow, begged permission to open it, and when he had read it and applied his lips to the signature, he placed himself, his house, his wife, and his little ones at my entire disposal.

Lirio-pampa was a nearly level strip of fertile land adjacent to a considerable stream (the Aláu) that ran not into the Mayo, but into the Sisa, the next river entering the Huallaga to southward. It was all forest, save where Chumbi's colony had made their little plantations of plantains and other esculents, including a plot of thriving sugar-cane, of which the first crop was expected to be ripe by the time the mill they were putting up with wooden machinery should be ready to grind it. At a short distance a spur of the Campana ran down into the

[1] Lirio-pampa: lat. 6° 25′ S., long. 76° 50′ W., alt. 3335 E. ft. Campana: alt. (pass) 5144 E. ft., (mountain-top) 6000 ft.

valley, partly bare of wood but clad with natural meadow, where Chumbi had placed a few young cattle. The dwelling-house, being at a little more than a thousand metres above the sea, was in a very pleasant climate. The temperature at sunrise was usually from 64° to 68°—once down to 61¼°—and the maximum rarely exceeded 81°, though it once rose to 87°. The weather was fine and dry during the three weeks of our stay, except one day of heavy rain with thunder. When we had been there a few days, incessantly occupied from earliest dawn till nightfall in collecting and preserving specimens of the beautiful plants that everywhere abounded, I began to grow tired of the salt beef and fish which, with plantains and yucas, were our only fare; and as Chumbi told me there was plenty of game in the woods, I sent him out one morning before daybreak to shoot paujíles (curassows or wood-turkeys). At 5.30 A.M. Nelson and I had our coffee, and then set off to herborize. Fortunately I indicated to Chumbi's wife the direction we should take, and we had been gone but a little while when her son came running after us to beg that we would return instantly, as his father had been stung by something in the wood and had reached home in a dying state. We hurried back, and on arriving at the house found Chumbi sitting on a log, looking deadly pale, and moaning from the pain of a snake-bite in the wrist of the right arm. He told us in a few broken words that he was creeping silently through the bush to get within shot of a turkey, when, on pushing gently aside an overhanging branch, he felt himself seized by the wrist, and was immediately attacked with so terrible a pain that he ran off in the direction of his house as fast as he could. He judged an hour might have elapsed since he was bitten, and the hand and arm as far as the elbow were already dreadfully swollen and livid, while the pulse even in the left arm was scarcely sensible. We bandaged the arm above the elbow, and as Mr. Nelson averred that his mouth was perfectly sound I allowed him to suck the wound, which was merely two fine punctures in the wrist on a line with the little finger; but the time was evidently past for either suction or bandaging, for Chumbi declared he felt excruciating pain in every part of his body. I also made him swallow three wine-glasses of camphorated rum, and we bathed the arm with the same spirit. Then we got him on his feet, and, one of us holding him on each side, we walked him up and down by the house. After a few turns he declared he could walk no more, and begged us to let him sit down; but after sitting a few minutes the pain returned with redoubled violence, and the pulse, which had beat a little stronger with the stimulant and the exercise, again became imperceptible. So we forced him up again, and made him walk as long as we could; then wrapped up the wrist in cotton soaked with spirit, and every now and then gave him a glass of the same, into which I threw a quantity of quinine. At short intervals we also gave him strong coffee, which evidently enlivened him. Still, with all we could do, and although we contrived to keep up the circulation, the swelling gained on us, and by night the whole arm up to the shoulder was so much swollen and discoloured as more to resemble the branch of a tree than anything human, and the hand was most like a turtle's fin.

Whilst this was going on, the relatives of the poor man kept up a continual wailing, as though he had been already dead; and he himself, although he submitted patiently to our efforts to procure him relief, had lost all hope of living. He indicated the spot where he wished to be buried, and gave what he considered his last directions to his wife about his children and property. He also sent off a messenger to his mother and brothers at Tabalosos, telling them that he was dying, and offering them his last adieux.

Towards evening, although the pain was still intense, the beating of the heart had become fuller and more regular, so that I felt sure the progress of the poison had been arrested, and I was now only afraid of mortification supervening in the arm. I therefore set Chumbi's wife and daughter to grind a quantity of rice, and enveloped the hand and wrist in a thick poultice, and had the rest of the arm fomented with an infusion of aromatic herbs at short intervals throughout the night. When the poultice was taken off in the morning, it was saturated with blood and putrid matter from the wounds, which had become much enlarged. The swelling was sensibly diminished, and the arm had become covered with pustules containing bloody serum, which we evacuated by puncturing them. A ready-made rice-poultice replaced the one taken off, and we kept up the fomentation and the poulticing until, at the end of forty-eight hours, the swelling had entirely subsided. The blood, besides breaking out at the skin, had also got mixed with the excretions. To remedy this, I prepared a decoction of an aromatic pepper (a species of Artanthe) that I had seen growing close by, and knew to be a powerful diuretic, and made him drink largely of it. In twelve hours the skin and the excretions were restored to their normal state.

On the second day he could take a little broth, and on the third he again ate heartily. For a month afterwards he had occasional acute pains in the arm and about the region of the heart, but at the end of two months he was quite restored, and avowed that his arm was as strong as it had ever been.

Chumbi had caught a glimpse of the snake, and recognised it as the Urrito-machácui or Parrot snake, so called from being coloured like the common green parrot, and thus rendered scarcely distinguishable from the foliage among which it lurks. It grows to a yard or more long, and its bite is considered incurable. Several fatal cases had occurred in the country adjacent to the river Mayo.

It may well be imagined that until Chumbi was fairly out of danger I felt no small anxiety, and it was not lessened by gathering from the whispers of his relatives that they considered me responsible for the accident that had befallen him. *I* had sent him into the forest, and had *wished* that the snake might bite him. If he had died, my life would have been in imminent danger. Nelson and I could probably have defended ourselves against any open attack of the few inhabitants of Lirio-pampa, but we could hardly have made head against Chumbi's numerous relatives at Tabalosos.

When I reached Lamas on my way back, I was

warmly received by the worthy Padre, who had heard of what he considered the wonderful cure of the snake-bite; but when I told him all the circumstances, and especially that Chumbi had been bitten *when on my errand*, he looked very grave. "If Chumbi had died," said he, "I should never have seen you more. Chumbi's relatives would have poisoned you. I in vain preach to them," he continued, "of what the Bible tells us about the entrance of sin and death into the world, and appeal to their reason to note how the body wears out with age, and how it is constantly exposed to accidents which may suddenly bring its machinery to a dead stop; they still in their inmost hearts believe—as their pagan ancestors believed—that death is in every case the work of an enemy."

Chumbi himself was very grateful to me, and during the remainder of my stay at Tarapoto often sent me little presents, especially of cakes of chancáca or uncrystallised sugar, the produce of his chacra; and he told to all the passers-by the story of his narrow escape from death by a snake-bite, through the skill (as he was pleased to say) of an Englishman.

Venomous snakes become rarer in the Equatorial Andes when we ascend beyond 3000 feet, and at about 6000 feet disappear altogether—at least I never saw or heard of one above that height. The natives believe the snakes of the sierra to be just as venomous as those of the plains, and that it is the cold that renders them bobas (stupid)—of course a mistaken notion, like most other popular beliefs.

The superstition that *it is unlucky for a woman*

to kill a snake I have found among the native races all the way across South America, but nowhere so strong as in the roots of the Andes. A woman must never kill a snake when she can get a man or boy to do it for her. In some places it must not only be killed but buried. When among the wild Jibaro and Zaparo Indians in the Forest of Canelos, I have sometimes had to kill two or three snakes a day for the women. How is it that the woman and the serpent are in mysterious relation in the early traditions of many civilised nations, and in the actual customs of savage nations even at the present day?

[It may be as well to continue here Spruce's experience of the results of the bites and stings of venomous insects, especially as they include one during his residence at Tarapoto which had results as bad as those of his Indian host above described.]

After snakes, the venomous animals most to be dreaded are the large hairy spiders, especially the species of Mygale, of whose bird-hunting propensities Mr. Bates and others have told us. I never saw a case of their sting, and all I ever heard of proved fatal except one, and that was of a woman at San Carlos, who was bitten in the heel and immediately dropped, with a shriek, as if shot. She lay at the point of death for ten days, but finally recovered. I have been bitten by spiders, but never seriously. At Tarapoto a smallish green spider abounded in the bushes, and would sometimes be lurking among my fresh specimens. It bit furiously when molested, with an effect about equal to the sting of a bee. At the same place, cockroaches were a great pest in the houses, and bored holes in the wall, which looked as if some one had amused himself by thrusting his finger into the adobes while still fresh and soft. They had a great enemy in the large house-spider, which springs on its prey from concealment, but spins no web. I had a tame spider for above a year, which used to come every evening for its supper of cockroaches. When I lighted my lamp, it would be waiting behind and upon the open door for the cockroach, which—dazzled by the glare—I had no difficulty in catching with my forceps. Sometimes, after an hour or two, it would come back for a second cockroach. Once, as I offered it the cockroach, I suddenly substituted my finger, which it seized, but immediately released without wounding, although this spider can bite severely when irritated.

Next to snakes and spiders come the ants, which are so numerous, and so ubiquitous, that no one escapes them. Their stings are of all degrees of virulence, but rarely prove fatal. Many ants bite fiercely, but not venomously. I could fill many pages with my experiences of these pugnacious and patriotic marauders, and of the nearly-related wasps. I once sent an Indian up a tall laurel, a hundred feet high, to gather the flowers. At half-way up was the first branch, and a large paper wasps' nest in the fork, hidden from view by the ample leaves of an Arad. As he passed it, the angry insects swarmed out, but he gained the top of the tree without a sting, broke off some flowering branches and threw them down. Unfortunately, there was no friendly liana by which he could slide down or pass to a neighbouring tree, and he must needs descend the way he ascended. He did so, through a perfect cloud of wasps, and got horribly stung. When we had got away from the foot of the tree, and had beaten off the wasps that followed him, I saw that his face and his naked back and shoulders were covered with knobs from the stings. He staggered and looked wild, and was evidently in great pain. I took out my flask, and was about to pour some spirit into my hand to bathe the stings, when he said, "If it's all the same to you, patron, I'd rather have it inside." I gave him the flask, and he took a good pull. No doubt he preferred the remedy that way because he liked the taste of it. Anyhow, it was the right way, and he went on through a long hard day without a word or gesture of disquiet, and when we reached home declared himself quite well again.

I have been stung by wasps I suppose hundreds of times—once very badly, having above twenty stings in my head and face alone. Yet I have always admired their beauty, ingenuity, and heroic ferocity; and I have twice in my life lived on good terms with them for months together. At San Carlos I had several little colonies of the large brown house-wasps, which hung their nests—like inverted goblet-glasses—from the rafters, and outside the house under the eaves. They never once stung me, not even when they had so multiplied as to become troublesome, and I poked down and swept out several of their nests. They seemed to recognise me as the real owner of the house, where they existed only on sufferance. But a stranger who should imprudently linger in the doorway would be sure to be attacked by them. Stedman, in his *Expedition to Surinam*, gives an amusing account of an impertinent intruder on his dormitory who was ignominiously tumbled down the ladder by his house-wasps. They serve to keep down the pest of large flies and cockroaches, and it is amusing to watch them at work, both as butchers and as builders.

On the Casiquiari, when we were one day hooking along my piragoa against the rapid current, one of the hooks caught a branch on which was a large wasps' nest. The wasps sallied out in thousands, and the men threw down their hooks and leaped into the river. I was at work in the cabin, and had just time to throw myself flat on my face, when the fierce little animals came buzzing in, and settled on me in numbers, but not one of them stung me. The boat drifted down the stream, and in a few minutes all the wasps had left it, when the men clambered on board and pulled across to the opposite bank. Another day I had got on the top of the cabin to gather the flowers of a tree overhead, and the first thing I hooked down was a wasps' nest, which I kicked into the river, and then went on gathering my specimens—battling all the while with the wasps and getting severely stung—for I saw the tree was new (it is *Hirtella Casiquiarensis*, n. sp. hb. 3196), and was determined not to leave it ungathered.

Scorpions and centipedes are formidable and repulsive enough to look at—I have seen the latter 11 inches long—but their sting or bite is rarely fatal. When it is so, the last stage of suffering is always lockjaw; and it is the same in death from ant and wasp stings. I have been a few times stung by scorpions, but only once badly, in a finger which was benumbed for a week afterwards. That was at Guayaquil, where the scorpions are of different species from those of the Amazon, and more virulent. It is a common thing there for a person stung by a scorpion to have the tongue paralysed for some hours. This property suggests a new version of *The Taming of the Shrew*, much to be commended to Guayaquilian Petruchios.

The stinging properties of the large hairy tropical caterpillars are well known. The venom resides in the long fascicled hairs, and the pain of the sting is so like that of a nettle—although often far more acute, and extending far beyond the surface stung—that it is presumable the hairs are hollow, with a poison-bag at the base, like the stinging hairs of nettles. But an hour's careful examination of the hairs in the live animal would settle this question, so that it is useless to theorise about it. I have had rather too much experience of mere mechanical stinging by vegetable hairs, which are usually minute or scabrous bristles, closely set on the leaves, pods, or other parts of a plant, and so deciduous that a touch brings them off. The pods of Mucunas (*i.e.* Cowitches), the spathes of some palms, the spathe-like bracts and stipules of Cecropias and some other Artocarps, are beset with this sort of pubescence, and I have often got considerably punished in collecting and preparing the specimens. In all these the bristles, or at least their points, remain sticking in the skin, and it is this that causes the irritation; but after the sting of a caterpillar nothing is visible in the skin, beyond the inflamed surface.

Leguminous trees are peculiarly liable to become

infested with stinging caterpillars. Children who play under the Tamarind trees at Guayaquil often get badly stung by hairy caterpillars that drop on them. I had always made light of caterpillars' stings until one evening at Tarapoto, in gathering specimens of an Inga tree, I got badly stung on the right wrist, at the base of the thumb; and when the pain and irritation at the end of half an hour went on increasing, I applied solution of ammonia pretty freely, and it proved so strong as to produce excoriation. The next morning the wound (for such it had become) was inflamed and very painful, but I tied a rag over it and started for the forest, accompanied by three men. We were out twelve hours, and had cold rain from the sierras all day; and when I reached home again my right hand was swollen to twice its normal size, and the swelling extended far up the arm. That was the beginning of a time of the most intense suffering I ever endured. After three days of fever and sleepless nights, ulcers broke out all over the back of the hand and the wrist—they were thirty-five in all, and I shall carry the scars to my grave. For five weeks I was condemned to lie most of the time on a long settle, with my arm (in a sling) resting on the back, that being the easiest position I could find. From the first I applied poultices of rice and linseed, but for all that the ulceration ran its course. At one time the case looked so bad that mortification seemed imminent, and I speculated on the possibility of instructing my rude neighbours how to cut off my hand, as the only means of saving my life. I attributed my sufferings almost entirely to the ammonia—or rather to my abuse of it—and to the subsequent chill from exposure to wet; for had I not been impatient of the pain of the sting, I have little doubt it might soon have subsided of itself.

[A few more passages from the letter to Mr. Bentham, illustrating the difficulties a collector has to encounter, are now given. It is probable that the same condition of things still exists there.]

TARAPOTO, *Dec.* 1855.

I have been most put about here for materials of which to make boxes, as such things as boards are not to be had. The only use the inhabitants have for a board is to make a door, and this is either cut out of some old canoe or they cut down a tree in the forest, roughly carve out a door from it on the spot, and bring it home on their backs. For other purposes, such as benches, shelves, bedsteads, etc., the never-failing Caña brava (*Gynerium saccharoides*) is all they require. After trying in vain to buy boards, I went to two ports on the Huallaga and in each of them bought an old canoe. I had then to go again with a carpenter to cut them up into pieces of convenient size, which had to be conveyed to Tarapoto on Indians' backs, and afterwards laboriously adzed down into something like boards. All this, with the trouble of looking up Indians, the making of two boxes and preparing boards for other two, left me little leisure for anything else for the space of near a month.

. . . .

I propose extending my stay at Tarapoto to a little over the twelvemonth—say to somewhere in

August. I shall thus be able to gather a few things which illness and fatigue obliged me to leave at the time of my arrival. I have been on the top of three mountains, and their vegetation is so nearly identical, that I should hardly find work at Tarapoto for a second year. . . .

[The next letter from Tarapoto to Mr. Bentham, dated April 7, 1856, is chiefly personal and botanical gossip relating to his work and future travels. After describing how a box from England was damaged and nearly lost by the boat being wrecked in the rapids of the Huallaga, he adds: "The difficulty, risk, and expense of getting plants from here all the way down to the mouth of the Amazon are so great, that I see my Tarapoto collections are not likely to repay more than the expense of collecting."

The letter concludes with a reference to the news he had just received of the ravages of yellow fever at the Barra, and then gives a short biographical note about a bird-collector, whose name and specimens must be well known to most English ornithologists. I therefore give it.]

"I am sorry to say that Hauxwell is about perdido (lost) as far as natural history is concerned, which is a pity, as no one has come here who puts up birds so beautifully as he does. He has got an Indian squaw and a child, and is turned 'merchant.' I am surprised he writes English (with a small taste of 'Yorkshire') so well as he does. His parents removed from Hull (where he was born) to Oporto when he was a little boy; thence he came out to the coast of Brazil as merchant's clerk, and anon turned naturalist."

[The next letters to Mr. Bentham are nearly a year later, and from these I give a few more extracts of general or botanical interest.]

To Mr. George Bentham

TARAPOTO, PERU, *March* 10, 1857.

. . . . . .

I am still a prisoner here, what with revolutions on the one hand which render the Sierra very unsafe to pass, and with the swollen rivers on the other; as soon as the latter abate we hope to be off.

. . . I cannot collect more, because excursions to be profitable would be long and expensive, and I want to save my money for my Ecuadorean expedition; so I am ruminating on dried herbs, and working off arrears in my Journal.

. . . . . .

To Mr. George Bentham

TARAPOTO, PERU, *March* 14, 1857.

I believe I told you some time ago of my intention of proceeding to Guayaquil in company with two Spaniards (Don Ignacio Morey and Don Victoriano Marrieta), who are going thither to purchase hats. . . . We had made our arrangements for going overland, but the revolution which has become almost general throughout Peru, and which nobody thinks can be closed in less than six months, renders the roads impassable. We have therefore reverted to our original project of proceeding up the Pastasa. . . . The advantage of this route is that one thus avoids the yellow fever of the coast of Peru and Ecuador, and its disadvantages are the chance of being killed and eaten by the "Infieles" on the Pastasa, or of being prostrated by ague.

. . . . . .

I think that on the whole my Maynensian collection may contain as many new genera as that of the Uaupés, but proportionately fewer new species. I have been much interested in it, because to many plants of Amazonian type it unites a good many characteristic Peruvian. Such are Weinmannia, the ivy-like Cornidia (three species), an arborescent Boccinia, the curious Proteaceous genus Embothrium (one or two species), and several others.

[The "revolution" just mentioned in the letter to Mr. Bentham is more fully described in the following letter to Mr. Teasdale written a few days later. This letter also contains an account of some of the industries of Tarapoto, and serves to complete the rather meagre narrative of Spruce's residence at this place. There are, however, a considerable number of "notes" on various aspects of the town and its inhabitants, and there is even a list of headings for chapters, showing that he had the idea of some day writing a very complete account of the district which was at that time the most easterly outlier of civilisation in Northern Peru, and one of the places least known—as it still seems to be—to European, or at all events to English travellers.]

To Mr. John Teasdale

TARAPOTO, PERU, *March* 16, 1857.

I have been waiting here to proceed to Quito since November last. Money which I had been expecting for months from Pará did not come till the end of the year, and by that time nearly all Peru was in a state of revolution. The first wave of insurgency rose in this very province, but was soon stilled. The Governor (Colonel Ortiz) was on his way from Tarapoto, where he had been sojourning a while, to Nauta, his usual place of abode. He went by way of the river Ucayáli, and ere he could reach Nauta, the garrison of that place had deserted, and set off for Tarapoto by way of the Huallaga. From Nauta he pursued them, but they reached Tarapoto before him and took it without resistance. They got here by night, made the Commandant prisoner in his bed, and the small garrison left here by Colonel Ortiz deserted to the insurgents. It was festival time at Tarapoto, and the town was full of people. As day broke they were preparing to resume the festivities—for the insurrection had been accomplished so quietly that few but the actors knew of it—when all at once the cry arose "Viene el *reclutamiento*!" The horror of that word to a Peruvian may be comprehended when I add that "recruiting" in Peru is something like what the pressgang used to be in England, only much more barbarous. Somebody had caught sight of the soldiers' uniforms and at once concluded it to be a recruiting party. Immediately all was panic and confusion, and in less than an hour nearly the whole population was in full flight. As I sat with my door open, quietly working at my plants, I could see a continuous stream across the pampa of people laden with their household gods, as if emigrating; and the drums, fiddles, and guitars which had been so noisy the three previous days were all silent. Two men ran by my house to hide in the sugar-canes on the hill-side, but so terrified were they that they could not reply a word to my inquiries. I got my breakfast, and about noon walked into the town to see if I could make out what had happened. The hot sun beat down into the streets, in which no living thing was to be seen save a few lazy dogs and pigs lying under the projecting eaves, and the houses were all closed as if some inmate had died. I walked on and on till I came to the house of Don Ignacio Morey, who I knew had gone down to the Amazon some weeks previously; but I found his wife and trembling children, naturally full of anxiety. From them, however, I learnt that it was probably no recruiting force, but a revolutionary one, that had arrived. I returned to my house, and shortly afterwards news was brought me that the insurgents had sacked the Commandant's house, not leaving therein so much as a cup, and that they were preparing to sack other houses. I loaded my six-shooter and my double-barrelled "Nock," and prepared to defend my house; but at this juncture a report reached the insurgents that a messenger had arrived from Colonel Ortiz, to warn the local authorities of what had occurred; and, armed with bayonets, they proceeded to search the houses where they supposed he might be hidden, but without finding him. Then, fearing on the one hand the arrival of Colonel Ortiz in their pursuit, and on the other that news of their uprising should reach Moyobamba before them, they began to prepare for departing, and at nightfall started for Moyobamba—five days' journey away at the least—where they calculated their numbers would be swelled by all who were disaffected towards the Government. But as soon as they were gone the loyal people of Tarapoto sent off a courier who passed them on the road, unseen, that very night, and reached Moyobamba long before them; so that the sub-prefect of that city, warned of their approach, placed an ambush in the way, which poured in a deadly fire on the insurgents, killing or wounding all the leaders. The rest fled into the forest, but after several days' chase were all captured. Among the slain was a young lieutenant, Don Domaso Castañon, who had been my particular friend at Tarapoto—a man of some talent, but of an ardent, impatient spirit. I had lent him two numbers of the *Illustrated London News*, and when he left Tarapoto in hot haste, he still found time to make a roll of them and write on it, "Esto es de Don Ricardo" —the last words he wrote in this world, poor fellow!

Thus ended this ill-concerted attempt at revolution. Its originators proposed to place General Vivanco in the presidency, in the room of the actual president General Castilla; and they expected that in all the towns on their route to the capital they would be joined by numbers who desired a change of government, so that by the time they reached Lima their forces would exceed anything Castilla could bring against them.

After this, there was an uprising throughout the south of Peru with the same object in view, and at this moment it has become nearly general in the country. Those who adhere to Vivanco are so numerous that it is thought Castilla must ultimately fall, although no one expects the struggle will be over in less than six months. [No! Castilla proved too strong for them.] Meantime, an innocent traveller, who may be supposed to possess anything worth robbing, runs the risk of being accused as a partisan, either of Vivanco or of Castilla, according to the colour of the revolutionary band he falls in with; so that even Peruvians, who have anything to lose, put off their journeys to an indefinite date. I had lately a dispute with the present Commandant of Tarapoto—a presumptuous, ignorant young fellow—wherein he propounded the doctrine "En tiempo de revolucion *todos los bienes son comunes*!" I told him the intent of such revolutions was simply indiscriminate plunder.

On the last day of the carnival (Shrove Tuesday) we had an uprising of the Indians, and there was a struggle between them and the soldiery in the square. Several Indians received bayonet-wounds, and one died of his wounds the second day.

A few days ago a tiger[1] was killed within forty paces of my house. I was sitting in the doorway at daybreak, sipping my chocolate, when I heard a multitude of people running down the valley and uttering the most infernal cries, among which I at length distinguished the word "puma" many times repeated. I seized my pistol and ran to the edge of the barranco, where I saw the puma coming straight for my door; but he missed the narrow track among the canes—the only practicable ascent—and got to the foot of the barranco, where it rose in a perpendicular wall 30 feet high. There he was

[1] [This term seems to be applied to both the puma and the jaguars—very distinct animals.—ED.]

speedily dispatched with bullets and lances. He made indeed no sign of resistance, and seemed stupefied by the savage shrieks and cries of his pursuers, who must have been near upon a thousand. They then carried him out to a piece of open ground, skinned, roasted, and ate him. This unfortunate tiger had been surprised while quietly breakfasting on a fat turkey. Tiger-skins—both of the red puma and the spotted jaguar—may be bought here for the merest trifle—a knife or a handkerchief. They serve me for cushions and mats, and my dog's bed is usually a tiger's skin—stretched across the doorway by night, for I generally sleep with the door wide open on account of the heat. The dog amuses himself by gnawing at them, and in this way has eaten me up three tigers' skins.

In a box of plants I am dispatching to Mr. Bentham I have enclosed a small parcel for you containing two "monteras," which are broad-brimmed cloth hats of many colours, worn by all the women of Tarapoto in out-of-door work. If they reach you safely, will you keep one of them for Mrs. Teasdale and keep the other for my sister Lizzie. Although they may never be worn, they will serve as memorials of the usages of a strange land, and of a friend whom you may never see again. They will probably seem to you outrageously gaudy and harlequin-like, but somehow they harmonise excellently here with everything around them. They are worn by the women chiefly when spinning cotton yarn in the streets or in the open grounds near the town. The mode of spinning is this. A little child sits under the projecting roof of the house, or anywhere in the shade, turning a wheel with one hand; and as he turns he gaily sings, or now and then munches at a truncheon of inguíre (boiled green plantain) he holds in his other hand. An upright piece attached to the frame of the wheel carries one or several spindles, and from each spindle a woman spins away in a right line, all she has to do being to draw out the cotton (which she carries in little rolls in her girdle) to a uniform thickness. Here and there forked sticks, 6 or 7 feet long, are stuck up, over which the lengthening thread is passed, so that pigs and other animals running about may not get entangled in it. The work of spinning begins at daybreak, and as the morning mist rolls away hundreds of spinners are to be seen on the pampa —each crowned with her gay montera—drawing out their long gossamer lines. As the sun rises higher, and even the broad montera cannot wholly shade the spinner's face from the intense heat of his rays, the task is laid aside, to be resumed towards evening, and sometimes, when there is a bright moon, continued till a late hour.

Cotton-spinning is the principal industry of the women of Tarapoto. The thread is remarkably strong, and is woven by the men into a coarse cloth called "tocuyo," which used formerly to be much exported to Brazil; but latterly English and American unbleached cottons (called "tocuyo Inglez") have come hither so cheap that the native manufacture has greatly fallen off.

A Sketch of Spruce's Botanical Excursions while residing at Tarapoto

(By the Editor)

[Among Spruce's MSS. are a number of loose sheets (about sixty or seventy) in a stiff paper cover, inscribed "Notes for Description of Tarapoto, in the Andes of Maynas or Eastern Peru." These are grouped under twenty-five headings, including topography and geology; the inhabitants, their industries, customs, amusements, etc.; the climate and natural history of the surrounding country; languages, government, etc. etc., evidently showing that he intended writing a full account of the interesting and little known district. But the "Notes" themselves are very fragmentary, and quite unfitted for any one but a person with full local knowledge to make any use of. Some are mere headings of subjects to be treated, others are very brief memoranda of facts or figures, while wherever there is any consecutive description this has been often utilised for some of the letters or extracts already given.

Besides these loose memoranda, there is the small "Note-book" already referred to, which gives a list of all his more important botanical excursions, generally a mere bald statement that such a valley or mountain was visited on such a day, week, or month, with, very rarely, a note added of some special feature of the excursion. As a supplement to this, we have a few scattered sheets giving "notes" of some of the more interesting excursions; but these, too, are quite fragmentary and very often break off in the middle, and appear never to have been finished.

I have also a rude sketch-map of the plain of Tarapoto, and of the chief villages, streams, and mountains around it, drawn from his own compass-bearings taken from various elevated spots and mountains, and by a few latitudes and longitudes from his own astronomical observations. This I have endeavoured to fill up from the notes and descriptions so as to include all the chief places he visited during his explorations. This will, I hope, enable the reader to follow more easily the references to places in his letters, and the short sketch I may be able to give of his botanical work in this very rich and then almost unknown district.

Among the notes for his account of Tarapoto there is a rather full description of the roads, where there were any, along which he had to pass to and fro in various directions. This is not only instructive and interesting in itself, but is essential to a proper comprehension of the difficulties under which his collections were made, even in this outlying portion of the Andes, where the mountains were very little higher than those in our own country. I will therefore give it in full.]

THE ROADS COMMUNICATING WITH TARAPOTO

The roads between the towns mostly occupy ancient Indian tracks, and it is easy to see how they were originally made out. Some bare grassy summit which will admit of a view being taken ahead, and which is nearly in the direction of the place to be reached, has been sought to be obtained by following a ridge separating two streams. The summit attained, another similar one has been picked out and reached in a similar manner, often no doubt with much trouble, and after considerable entanglement in the valleys. Thus the roads here, like the first-made roads in all parts of the world, go straight over the tops of hills, instead of winding around their base. The dense forest makes the finding of a way among hills infinitely more difficult when no compass is used, and though it would seem more feasible to have sought out a passage along the watercourses, a very little practice shows the impossibility of this. Besides that the vegetation is much ranker near water, the course of the streams —not merely their bed, but the whole of the narrow valley in which they run—is so obstructed by large masses of rock and stones as to be all but impassable, and completely so when the valley narrows to a gorge with perpendicular sides which merely admits the passage of the stream in an alternation of cascades and deep still pools. To avoid a pongo—as these gorges are called—one must climb a mountain-side and then go down again, and perhaps steep cliffs render descent impossible for a long distance. Hence it may be seen how, by seeking out the sharp ridge of a mountain, when not too steep, we really avoid invincible obstructions, although we have to ascend and descend great heights. It is true that a little previous surveying and a little good engineering would smooth down most of the difficulties that offer themselves, and I have no doubt that good winding mule-roads, at a slight inclination, might be made in any part of the mountains I have yet seen; but here, where not even spade or pickaxe are used, much less has it ever been attempted to move a rock by gunpowder, what can be expected? All that is generally done is to clear away the forest with axe and cutlass, and that often imperfectly, stumps of trees being often left some inches above the ground, while the branches and twiners overhead are cut away only to such a height as may be reached by an Indian, so that a tall horseman has to look out continually to save his head from entanglement. Rarely is any attempt made to level the road with a rude hoe, and the tropical rains are left to smooth or furrow it according to the locality. In steep hollow ascents logs are sometimes laid across, against which sand accumulates with the rains, and thus a sort of stair is formed. The idea of a cutting along the face of a declivity, or even the rudest bridge over the streams, never occurs to any one. No one is charged with the repair of the highways, and it is only once a year that the inhabitants of the pueblos clear the portions allotted to them, cutting away the brush that has accumulated. When a tree has fallen across the track, those who next pass that way make a fresh track through the forest around the fallen mass as best they may, for they rarely carry with them axes, or have time to spend an hour or two in clearing the road. Those who follow enlarge the track with their cutlasses, and thus one is continually coming on narrow and difficult turns.

The principal road in Maynas is that leading from Tarapoto to Moyobamba, and thence to Chachapoyas. As far as Moyobamba it is just practicable for horsemen, who, however, have to pass some dangerous places on foot, but laden beasts cannot traverse it. From Moyobamba to Chachapoyas it is said to offer still greater natural obstacles, but to be kept in better order, so that mules can be used if carrying a single burden of five arrobas (160 pounds). Thence to the coast there is a good broad road on which mules can pass carrying ten arrobas, divided into two equal portions one on each side.

From Tarapoto to Tabalosos—two short days' journey—the road is good enough to allow mules to pass, and the latter part of it (from Lamas to Tabalosos) is especially well kept, which is due to the Cura of Lamas having often to traverse it, and as the people hold him in great respect they take care that he shall find everything as smooth as possible. All the brush is kept down and no stumps are left sticking out.

But from the first stream beyond Tabalosos the road is in a deplorable state, and the natural obstructions are very great. To avoid a ravine on the Cachi-yacu, a steep ridge (the Andarra) has to be crossed, in many parts by climbing high natural steps which are very dangerous on horseback. On the other side of the Andarra the channel of the Cachi-yacu has been followed for about an hour, sometimes on one side, sometimes on the other, and here and there a cliff has to be scaled by the aid of roots spreading over it. The crossings of the river are the worst, for the water is always turbid, and runs rapidly over and amongst slippery rocks, so that on stepping into the water one rarely sees what one is going to tread on. The water is always knee-deep, and sometimes more; indeed, if the stream be swollen it is quite impassable, and travellers have to wait till it abates. The whole number of these crossings is twelve, and after leaving it a tributary stream of scarcely less size has to be crossed thrice in ten minutes.

Many attempts have been made to find a way which shall avoid the gorge of the Cachi-yacu, but hitherto without success. Beyond this there is a long painful ascent to a spring of clear water called Potrero, where the traveller begins to emerge on the grassy plateaux and declivities of the Campana.

In imitation of the tambos or houses of rest and refreshment placed by the Incas along their great roads, the modern Peruvians have erected sleeping-places wherever the pueblos are at too great a distance to be reached in one day. To these also they give the name of tambos, but they are as inferior to the ancient ones as are the modern roads to the solid structures of the Incas. They consist of a roof supported on four bare poles, without walls, but when large and well-made such shelters answer their purpose tolerably well. Of course they have no permanent occupants, and the only thing a traveller can calculate on finding when he reaches a tambo is fire, which is rarely allowed to become extinguished, as it is the custom for those who have last occupied it to leave their fire well heaped up with rotten logs. A slight channel is made round the tambo to carry off rain-water, and the soil taken out serves to heighten the flooring, which, being spread with palm-leaves or with fern, the traveller extends thereon his mattress or his blanket, and wrapped up in his poncho and another blanket, may calculate on passing the night without suffering from mosquitoes, though a snake may creep to his side for warmth, or he may be disturbed by the invasion of a jaguar, especially if he has allowed his fire to get low. Tambos are always placed near good water, and as every traveller carries his coffee-pot and provisions, he has it in his power to enjoy one of the greatest of earthly pleasures—a cup of good coffee after a long and fatiguing walk or ride. Pans for cooking can rarely be carried, but meat and plantains can easily be roasted.

The inhabitants of Tarapoto have often good broad roads to their farms and cane-mills, especially when several of these lie in the same direction. A great obstacle to the use of these, and indeed of all other roads, is in the swelling of streams and the improvidence of the people in making no bridges; and though the waters generally fall as rapidly as they have risen, several hours must sometimes be passed on the banks, at great inconvenience or loss, awaiting their abatement.

[Besides this main western route to Moyobamba, two other roads or mule-tracks lead out of Tarapoto to the south and east. That to Juan Guerra has been referred to in Spruce's letter to Teasdale describing his journey to Tarapoto. Another goes nearly due east till, after crossing the rivers Shillicaio and Aguashiyacu, with their intervening hills and ridges, it sends a branch south-westwards, and then again eastward to Chapaja on the Huallaga river, while the main route continues over a high shoulder of Mount Guayrapurima to Chasuta, at the lower entrance of the pongo of that river. Along all these roads Spruce collected assiduously, but he also made numerous expeditions to the mountains which surround Tarapoto, especially on the north, east, and west, as well as along the banks, up the valleys, and through the gorges of the numerous streams and rivers that issue from them into the pampa of Tarapoto. If the difficulties along the beaten tracks were often great, it may be imagined what they were when he had to penetrate these almost untrodden mountains and valleys, densely covered with virgin forest, and for the most part rarely or never visited by any of the inhabitants of the surrounding country. Owing to the almost complete absence of any account of these various journeys, I can only give a bare enumeration of them, with a few scattered notes on some of their features where such exist.

During the first month of his residence (June to July 1855) we have only the note—"Collecting near Tarapoto." This no doubt means within the limits of a day's walk, which would take him over nearly the whole surface of the pampa. From various notes and scattered remarks we learn that although this pampa had been more or less completely cleared of its original virgin forest, and cultivated for more than a hundred years, yet strips and patches of the original vegetation remained along the steep banks of the numerous rivers and a few other precipitous or rocky portions, while considerable tracts had reverted to second-growth woods, mostly of shrubs and low trees, thus furnishing work for the plant-collector at the flowering seasons of the various kinds of plants. We accordingly find a similar note for the month of September, then in January 1856, again in July and in September 1856, and in November of the same year.[1]

After the first month he began the more difficult excursions—to the pongo of the Shillicaio, to the river Aguashiyacu, and to Mount Guayrapurima. This latter mountain he visited twice afterwards—in January and in June 1856, staying some days, or perhaps even weeks, each time. Of the second of these excursions there are a few notes.

This mountain, whose highest summits lie about 12 to 15 miles due east of Tarapoto, sends out spurs to the Huallaga, while to the north-west it extends till it mingles with the more prominent mountains north of the town. It consists of many steep ridges, which from some aspects give it a serrated appearance, while from Tarapoto it has a pyramidal outline with much-broken sides. It is penetrated by deep and almost impassable ravines and valleys.

The meaning of the name is "Where the wind blows," and Spruce says that on the high ridges (over one of which the road to Chasuta passes) the wind seems to be almost constant, and so strong as in precipitous parts of the track to be dangerous. They blow always from the north, and where Spruce slept, a few hundred feet below the

[1] [Among the miscellaneous "notes" on the vegetation we find this very interesting remark: "Going out of Tarapoto in different directions, although the soil may be the same, there is much difference in the vegetation." This accounts for the large amount of time he devoted to this pampa, and it is also instructive as showing that differences of conditions quite imperceptible to us determine the presence or absence of certain species at certain localities, and no doubt in some cases their absolute extinction or preservation. Of course the same phenomenon occurs everywhere around us, as every botanist knows, but they sometimes forget what a striking proof such facts afford of the severity of the struggle for existence, even under what appear the most normal and favourable conditions, and the rigidity of the "natural selection" that determines the result.—Ed.]

ridge on the eastern side, he heard it blowing all night. On the top of the narrow ridge of crumbling sandstone covered with a dwarf herbaceous and shrubby vegetation, it is hardly possible to walk on account of its violence. Spruce here remarks: "The descent on the east side of this col, towards Chasuta, is very abrupt; the trees are mostly low; they, like the rocks and the ground, are densely clad with Hepaticæ (especially Mastigobryum, Lepidozia, and Plagiochila), among which grew several ferns, especially some interesting arborescent species of small size. In places where the road has been cut or worn down, so as to form deep hollows, the walls (red sandy clay) are clad with mosses and ferns, especially a pretty little Lindsæa and three species of Trichomanes."

Later, in the Journal of his voyage from Tarapoto to Ecuador, he speaks of this descent from the ridge of Guayrapurima to a clear stream called Caraná, as being "the richest bit of fern ground I had seen in the world"; while, after another hour's journey and a steep descent, he reached the Yacu-catina, which he describes as "a most picturesque rivulet with a magnificent fern and forest vegetation."

His next expedition was to Chapaja, on the banks of the Huallaga, in October 1855; but of this there are no notes.

Early in November he took a two days' expedition "to the head of the Cumbasa river and Mount Canela-uesha, on the way to the stream Cainarache, down which canoes pass to Yurimaguas."

In November and December 1855, he took his

Fig. 4.—The Mountains north of Tarapoto (3000–5000 feet). Mount Pelado on the right. (R. Spruce, Aug. 1856.)

first expedition to the Campana Mountain and Lirio-pampa, already described at some length.

In February 1856, he made an excursion to the head of the Puca-yacu, on the western slopes of Mount Guayrapurima.

In March he went to the Upper gorges of the Shillicaio river.

In May he went to the top of Cerro Pelado, and to the upper gorges of the Aguashiyacu, Uchulla-yacu, etc. There are a few notes on Mount Pelado, which consists of bare sharp ridges running about S.E. and N.W., the N.E. side being very precipitous but sloping more gradually towards the plain of Tarapoto. The rocks are covered with lichens, a few ferns, some rigid-leaved Liliaceæ, and a few dwarf shrubs. From the S.E. the ridges dip abruptly to deep ravines, which form the sources of the streams of the pampa, as well as of some tributaries of the Huallaga. Lower down the slopes are clad with low forest which is densely mossy. The summit of all the ridges is a white, friable, coarse-grained sandstone, in thin layers, inclined at a very high angle. The Cumbasa rises to the north of this group of mountains, and many of the deep ravines above mentioned are some of its tributaries.

(The accompanying beautiful drawing of the mountains north of Tarapoto is the only one of large size which was carefully shaded by Spruce himself. With the one exception of the immediate rough bit of foreground, it has been photographed from the drawing as he left it fifty years ago. The curiously ridged mountain to the right exactly corresponds to his description of it above given; and we can well understand the difficulties of the ascent of such a mountain through many miles of tropical forest, among deep ravines and impassable gorges, along a track used only by Indians crossing the mountains to a good fishing stream which flows directly into the Huallaga, as described at p. 40.)

In July 1856, Spruce went for a month to Lamas and Tabalosos, making the latter place his head-quarters for the exploration of the eastern slopes of the Campana Mountains, where, at about 4000 feet elevation, is a natural pasture called Potrelo, "around which is low forest with many interesting flowering plants, palms, tree-ferns, ferns, and mosses." The position of Tabalosos is picturesque, being situated in the midst of mountains. On the opposite side of the Mayo (to the N. and N.E.) there is a very bold and lofty peak, at no great distance, whose rocky slope seems to be nearly perpendicular. Those who go from Yurimaguas to Moyobamba by way of Balsapuerto have this peak on their left. The inhabitants are nearly all Indians, with very few half-breeds. Hardly any speak Spanish. They grow large quantities of vegetables, and are much employed as carriers on the route from Tarapoto to Moyobamba.

(The drawing here given of the rude clock-tower of Tabalosos shows this remarkable mountain immediately to the left of it, and nearly in the centre of the picture, while the Indian ringing the two very small bells gives life and character to the scene.)

From the summit of the Pingulla mountain there is a splendid view of the whole lower course of the Mayo, with Lamas, Tarapoto, and all its surrounding mountains, to Chapaja on the Huallaga river.

Fig. 5.—View from Tabalosos, looking across the River Mayo. (R. Spruce, July 1856.)

The ridges and peaks are of white sandstone, as are those of the Andarra farther up the river. Both are very bare of vegetation, being burnt almost every year and overrun with the common fern *Pteris caudata*. The ascent to Potrela up the rocky valley of the Cuchi-yacu is, however, through luxuriant forest especially rich in ferns and mosses.

To conclude this sketch of the Tarapoto district investigated by Spruce, I will give a few passages translated from his "Précis d'un Voyage" published in the *Revue Bryologique* for 1886:—

"The first thing that strikes the eye of the botanist at Tarapoto is the abundance of ferns. These plants are by preference, as we know, either maritime or subalpine. On the hills of Brazil a tolerably large number of species are found, but in the interior of the continent and in the great plain of the Amazon valley, although ferns are not wanting, yet the species are never numerous and several of them repeat themselves at every step even up to the roots of the Andes. One may therefore judge of the riches of the Eastern Cordillera of Peru in ferns by the fact that there, within a circle less than fifty miles in diameter, the author found 250 species of ferns and their allies, of which many were new, especially among the tree-ferns."

Among the most interesting plants in this region, next to the ferns, may be named the Rubiaceæ, of which Spruce collected 98 species. A small number of these were already known through the researches of Ruiz and Pavon, Poeppig and Matthews, but the majority were new. The "Précis" then continues:—

"Some genera of mosses, absent in the plains, began to appear in the lower forest zone of the Andes. For example, those splendid mosses of the genera Phyllogonium, Rhacophilum, and Hypopterygium, all of which, by their primary leaves arranged in double rows, and in the latter-named genera accompanied by stipulated folioles, appeared at first sight to be Hepaticæ rather than true mosses. Among other mosses which are met with in the Andes of Peru, but which are never found in the plain, are Helicophyllum, Disticophyllum, Cryphæa, Pterobryum, Entodon, Fabronia, etc. The Tortulæ, represented along the banks of the Amazon, but very rarely, by the single *T. agraria*, begin to be less scarce; also the genus Bryum, of which the *B. coronatum* and a barren form of *B. argenteum* are the only species found on the Amazon.

"With regard to the Hepaticæ, while the Lejeuneæ are almost as abundant as upon the banks of the Amazon, and still show the same preference for the living leaves of trees, the Frullaniæ, of the subgenus Thyopsiella (which are related to our *F. tamarisci*), appear there for the first time. Among other genera of the Eastern Andes which are never seen in the plains may be named Porella, Herberta, Mytilopsis, Adelanthus, Leioscyphus, Jungermannia, Scalia, Marchantia, Dendroceros, and Anthoceros. Lepidozia, which is represented in the plain by a microscopic species (and that found only once!), is met with in the mountains of Tarapoto in the form of large and elegant species."

On examining Spruce's descriptive catalogue of the plants which he collected, and which are numbered consecutively, I find that there are

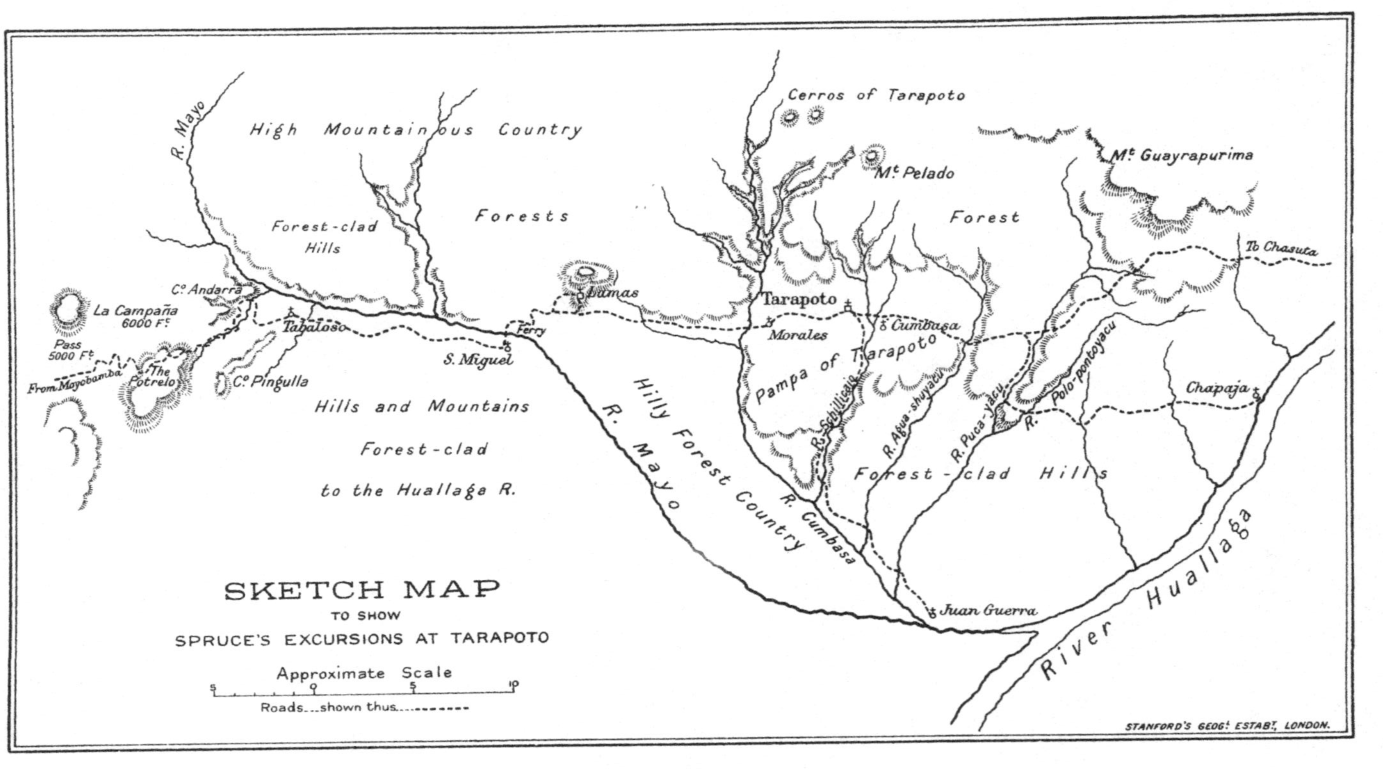

London: Macmillan & Co., Ltd.

1094 species of flowering plants and ferns, to which must be added several hundred species of mosses and Hepaticæ—his favourite groups—which here for the first time formed an important part of the vegetation. It must be remembered that this by no means affords any near approach to the whole flora of the Maynensian Andes (as he termed the district of which Tarapoto formed the centre), because, both by inclination and necessity, he limited his collections as much as possible to species which he had not met with before, and especially to such as he believed to be unknown to European botanists. We know from his Journals that often he could not possibly collect all he saw, especially among the forest trees, and that he was accustomed often to leave ungathered many new species in favour of others which he believed to constitute new genera. These Tarapoto plants were the result of about eighteen months' collecting; for, although he resided there a year and three-quarters, at least three months were lost by illness and in the preparations for his journey to the Ecuadorean Andes.]

CHAPTER XVII

IN SMALL CANOES FROM TARAPOTO TO CANELOS: 500 MILES ON THE HUÁLLAGA, MARAÑON, PASTASA, AND BOMBONASA RIVERS

(*March* 23 *to June* 14, 1857)

[THIS journey up the little-known Pastasa and Bombonasa rivers in small canoes for a distance of perhaps 500 miles, following the curves of the rivers, was a very painful and tedious one, owing to the whole country being almost depopulated, and provisions not to be obtained. It occupied nearly three months, of which Spruce kept a very full account in his Journal, and as the whole route is almost unknown to English naturalists, I have selected all the more interesting portions (about one-half) for presentation here. It is full of details which may be useful to future travellers, and contains a good deal of curious information as well as several rather strange occurrences. Some German botanists who descended the rivers from Canelos in 1894 found the villages rather better peopled on account of the increasing rubber-trade, but otherwise just as Spruce described them.]

ABSTRACT OF JOURNAL

(BY THE EDITOR)

[As stated in the letter to Mr. Bentham of March 14, Spruce arranged to make the difficult and costly as well as dangerous journey from Tarapoto to Baños in Ecuador in the company of two merchants of the former place, Don Ignacio Morey and Don Victoriano Marrieta. Each party had its own canoe with a crew of seven Indians, and Spruce was accompanied by a youth of twenty years, named Hermogenes Arrebalo, probably an Indian, as his servant. I cannot find either in the letters or journals any further reference to his assistant at Tarapoto, the young Englishman, Charles Nelson, and we are left in darkness as to where Spruce first met with him or why Nelson did not accompany him to Ecuador.

On this journey the travellers first went overland to Chasuta, occupying two days, and the latter portion of this route was so full of obstructions and mud-holes, the weather being continually wet and stormy, that in order not to lose his shoes Spruce was obliged to walk barefoot and arrived at Chasuta both lamed and suffering from fever.

The canoes in which they descended the river were entirely open, in order to pass the falls more safely, and the travellers were therefore exposed to the rains, which were almost continuous, while the passage of the cataracts was difficult, and the boats narrowly escaped being swamped. This incident, with one of its rather singular results, is well described in his first letter to Mr. Bentham from Baños, and is as follows:—]

I arrived here on the 1st of July, after a voyage of exactly a hundred days from Tarapoto. Such a journey! I can hardly bear to think of it, much less to write at length of what I saw and suffered. In a postscript to my last letter written at Yurimaguas, I mentioned that my canoe had been nearly swallowed up in a whirlpool in the pongo of the Huallaga. That the peril had not been slight you may have some idea from the following circumstance. I had with me a large handsome dog whom I had reared from a pup. There was not such another dog in all Maynas, and latterly he made my house respected by the drunken cholos, who, instead of pestering me as formerly, took care to give us a wide offing. In one of my last walks about Tarapoto, he pulled me down a fine deer. When my canoe was caught in the whirlpool, the horrid roar of the waters, which drowned our voices, and the waves, which splashed over us, so frightened the dog that he went mad! From that hour he would drink no water, and after the first day would take no food. Six days I kept him by my side, at great personal risk, hoping to cure him. When we went on shore in the villages he ran straight off, uttering the most unearthly sounds, and putting to flight dogs, pigs, and cows, sometimes biting them severely. At length he began to snap at the people in the canoe, and being worn almost to a skeleton, I saw all hope of saving him was vain, and was obliged to shoot him.

Respecting the voyage, I may say in brief that from the first day to the last my progress has been impeded by swollen rivers and steeping rains. . . . Join to this a monotonous river whose flat shores rarely rose 2 feet above water, almost destitute of settled inhabitants (we once passed fifteen days without meeting a soul), and you will have some idea how heart-sickening such a journey must have been. From embarking on the Huallaga till entering the mouth of the Bombonasa (which is now the frontier of Ecuador), we had the usual accompaniment of mosquitoes by day and zancudos by night.

[A few details of this portion of the voyage will now be given from the Journals and letters. At La Laguna, near the mouth of the Marañon, the travellers stayed three days in order to get fresh men and make toldos (thatched cabins) in the stern of the canoes. They also had to collect provisions for the voyage up the Pastasa to Andoas at the mouth of the Bombonasa river, the greater part of the shores of those rivers being without inhabitants. At La Laguna Spruce witnessed a curious example of voluntary flagellation which he thus describes :—]

The Indians of La Laguna have a custom of scourging themselves in the Holy Week. We were setting out a fortnight before Easter, so that there would be no opportunity of performing this at the proper time, our Indians therefore determined on undergoing their "penitencia" on the Saturday evening. For this purpose whips of a most barbarous description had been prepared —cylinders of pitch six inches long were stuck full of bits of broken glass, projecting about half an inch. About four in the afternoon the penitents began to promenade the streets, giving themselves smart blows on the naked shoulders with thongs of thick skin of cow-fish, that the blood might flow more freely on application of scarifiers, which was done by themselves in like manner when all assembled in a large house. At the Oracion they sallied forth to the church, walking by twos—their backs one mass of gore and their white trousers (their only garment) soaked and dripping with blood. I have never seen a more horrible sight. They unceasingly applied the cow-skin straps, making the blood spurt in all directions and sprinkling my clothes, though I took care to keep at a respectful distance. In the church a little below the altar was extended a mat, and on the mat a crucifix laid on a cushion, with a cup by its side to receive contributions of penitents. As the latter advanced in their turn they knelt down and kissed the crucifix, beating themselves with redoubled energy. At the same moment their wives or mothers, who walked by their side, dropped each an egg into the cup. Whilst this was doing, the Sacristan chanted a Miserere. Each Indian, after kissing the crucifix, walked out of the church, in the order he entered, nor suspended the flagellation until reaching his own house. The value of an act of penitence like this may be estimated by the fact that every one of the penitents was intoxicated. They believed, however, that it would ensure their safe return from the perilous voyage, or, at any rate, should they be killed by the Infidels, their souls would be immediately received into glory. Many white men would have kept their beds for a month after such a punishment, but our penitents sat down to their oars before noon on Monday (the next day but one) without showing any inconvenience from their wounds. They have an idea that the beating after the application of the scarifiers drives out the coagulated blood from the wounds and prevents any formation of pus.

[On April 6 they left La Laguna, and on the 7th entered the Marañon, and though the distance up that river to the mouth of the Pastasa is only about 25 miles, they did not reach the latter till the 11th. On the afternoon of the 7th they came upon a small village of six huts, where the remnant of the pueblo of Santander on the Pastasa had established themselves. Here they learnt that five men of San Antonio (a village just above the mouth of the Pastasa) went into the forest to cut palm-leaves, and never returned, but remnants of their clothes had been found, showing that they had been murdered by the savage Huambisas.

On the morning of the 9th the travellers came to the deserted pueblo of Shiruri, half a day below the mouth of the Pastasa. There were about a dozen cane houses standing on level ground scarce a foot above the highest floods. Spruce thus describes what he found :—]

The exodus appears to have been very hasty, for pans and tinajos of all sizes are left scattered about, and even several arrobas of rice in pots and baskets. The neat beds made of stems of bamboo opened out into sheets and laid side by side are mostly in their places, but the termites are everywhere and will speedily complete the destruction of everything vegetable. The ground is fertile, and the colonists had made their plantations of plantain, sugar-cane, yucas, etc., not omitting several sorts of the necessary Capsicum and the flowers used by women for adorning their hair (cockscombs, African marigolds, etc.), nor the verbena which is a panacea for every disease. A few Crescentias had been planted and in another year would have begun to yield cuyas. What a picture of disappointed hopes is suggested by the view of such desolation! With what lamentation must the poor women have deserted the spot where they had just completed preparations for rearing their young families, and had calculated on growing old amidst plenty and tranquillity!.

[Thenceforward when sleeping on shore Spruce and his companions took turns to keep watch during the night, allowing the Indians to sleep. The latter, however, usually stuck their lances and bows and arrows at the head of their mosquito nets, so as to be ready in case of an attack. The Journal continues :—]

Just above the point where we got into the main channel were three houses in the midst of large platanales on the left bank, probably remains of the new pueblo of Santander, though our Indians refused to tell us. It is impossible to get from them any information about places and distances, as they are afraid we should want to go ashore at the deserted pueblos, where the Infieles might be in ambush to fall on them. Even where we have cut plantains in deserted chacras (which are frequent along the shores, though generally hidden by a strip of forest and not reaching the river margin as on the Marañon), it has been necessary to go ashore ourselves first with our firearms.

A little before sunset we reached the upper point of an island, clad with a willow-like Composite, and rapidly becoming covered with water. Here we made fast, intending to pass the night, but shortly the Indians took alarm at seeing how easily an enemy could approach our encampment concealed by bushes which, although growing pretty close, admitted an easy passage; so we moved off to the middle of the river, here very broad and shallow, with several prostrate dead trees sticking out which the rising waters had not yet liberated, though they were beginning to move them. I was not sorry for the change, for zancudos were very numerous and fierce on the island, though not entirely wanting on the river. The nocturnal zancudo is a small slender gnat with spotted wings—rest of body a uniform black. It is called birotillo (the little dart) because its puncture is so cruel, often leaving pain and swelling. When the days are dull we have them in the canoes at all hours, and the small mosquitoes are as abundant as on the Marañon. My skin has been in a very sensitive state since the journey from Tarapoto to Chasuta, and some of the mosquito wounds are beginning to ulcerate. In the woods I have made acquaintance with a minute and very active tick, which sucks a little here and there, and does not, like the other species, hang on to one place till it gets full; its bites cause an intolerable itching, and if one scratches, ulcers ensue.

April 12 —Don Victoriano's dog, which had

been ill for several days, was now unable to stand, and excessively bloated. . . .

We took it on shore where we made our breakfast this morning, and, as it was evidently in a dying state, before we re-embarked its master put an end to its sufferings by a couple of pistol-shots. Thus our two handsome dogs, on whose services as sentinels we had so much calculated, had been left as food for beasts and birds of prey—my poor "Sultan" in the forests of the Huallaga, and Don Victoriano's "Muchacho" in those of the Pastasa!

At sunset we reached the ancient pueblo of Santander on the left shore. Standing on a steep bank of red earth, it reminds me, by its position, of Barraroa on the river Negro. I invited our Indians to go there to sleep, but they shook their heads and could not even be induced to take that side of the river. There are still two large houses standing—possibly church and convent.

[During the next fortnight the journey was wearisome and monotonous, with almost continuous rains, rarely any dry land to sleep on, and not a single village or settlement of any kind. The only break to the monotony of the succeeding days was an occasional success in procuring game, such as curassows or wild ducks, once an armadillo, and once by great good fortune a tapir. Only once they met a solitary canoe with a young Indian man and woman who said they came from Andoas. On reaching that place they learnt that the man was the son of the chief, and that he was running away with the girl to somewhere on the Marañon.

On the evening of the 25th, to their great delight, they saw a fire on shore, and found a small farm where three men and two women were cutting palm-leaves and preparing the fibre to make hammocks for the Governor of Andoas. The Journal now continues :—]

April 26 (*Sunday*). — Starting at four this morning, about seven we reached a playa where we found three families of inhabitants of Pinches encamped. We bought of them part of a very large tapir they had killed the previous night, and some pieces of baked agouti in very fine condition. Here we breakfasted, and then proceeded; but our men were completely at a loss in the broad shallow river, and were continually running us aground, so that we did not reach the village till 3½ P.M. Pinches Nuevo stands on the left bank on a barranco 20 feet above high-water mark. It is reached by rude steps cut in the cliff, which is of tenacious red earth, without the least mixture of stones or gravel. There are but some ten houses, including church and cabildo (guests' house), all of Caña brava, or of strips of palm-stems, roofed with palm-thatch. Very few inhabitants were present, and we had some difficulty in procuring five heads of plantains and a basket of yucas, especially as their chacras are new and they still bring the greater part of their plantains from the site of the old pueblo. The inhabitants are ill-looking, and some are affected with caracha (leprosy). They are the remnant of a nation of Pinches Indians, and still speak a peculiar language, though all understand the Quichua.

April 27.—Navigation now gets more difficult, hardly anywhere is there sufficient water to float our canoes. Beaches appear in different places from last year, and our guides can hardly pick their way.

Several times the men have had to leap into the water and drag the canoes by hand a good distance over the shallow bed before finding again sufficient water to float us. . . .

[Early on April 29 they reached the much-desired Andoas, situated on the left bank of the Pastasa, where they had to engage fresh crews to take them up the Bombonasa river to Canelos. The village stands on a low ridge, on each side of which is a little stream, the mouths of which are about a quarter of a mile apart. The soil is loamy and very fertile. Spruce was only able to take one short walk in the forest during his five days' stay here, and noted that while the trees seemed mostly familiar to him, the shrubby and herbaceous plants were nearly all new. The following rather characteristic incident is noted in the Journal:—]

At Andoas it was necessary that some one should sleep in the canoes, to take care of their cargoes, and I and Don Ignacio, as being most interested, undertook to do it, although we must thus deny ourselves the pleasure of sleeping under a roof, which the rest of our party took advantage of. Our salt fish was stowed in the fore-part of the canoe and covered over with palm-leaves, on which were laid logs of wood, so that the fish could not easily be got at by the dogs who visited the canoes every night in a troop. Nevertheless, they found out some part not so well secured as the rest where they one night introduced their muzzles and gnawed at the fish, and on the following night I lay awake until I heard them at work, and then seized my gun and rushed out of the cabin; but they made off too quickly for me and disappeared over the top of the steeply-sloping bank. One dog, however, turned round when he reached the top and barked at me. I fired (with shot) at his legs, intending only to wound him, but his shattered legs failing him, he rolled howling down the bank into the river and was drowned. His body was retained in an eddy a little lower down, and there it was found by the women when they went to fetch water at daybreak. The Governor had told me to shoot those pilfering dogs, for they were vagabonds who had no owners; but this one chanced to belong to an old woman, who made an outcry about it, and the Governor told me that if I did not succeed in pacifying her we might have some difficulty in getting our complement of mariners, so I sent for her and asked her how much she wanted for her dog. She said ten needles! I was glad to give her an entire packet of the best I had, with which she went away content, having therewith enough to buy three dogs such as the one she had lost.

Andoas differs from Pinches only in size, as it contains some twenty houses and about sixty married couples, but the aspect is equally miserable. The walls of the houses are of wild cane or palm, while the church is of bamboo stems opened out into boards, and in a very dilapidated state. The church divides the town into two nearly equal portions or partidos, that to the south or down the river being inhabited by Indians of the Andoas nation, and that to the north by Indians of the Shimigái tribe. . . . In external appearance the two tribes inhabiting the village of Andoas show no difference. The men are of lowish stature, not robust, mouth wide, but lips not disproportionately thick, nose straight or slightly Roman, forehead lowish, rather receding, and with the bump of locality universally strikingly developed. Their hair is cut off straight just over the eyes, and allowed to hang down long behind, usually reaching the middle of the back. They streak their faces daily with anatto, and sometimes pour the juice of jagua over their bodies, but this is not done (as by the inhabitants of Tarapoto) to hide spotted skins, as they are quite free from caracha.

The characteristic dress is a sort of poncho called a cueshma, which is a long narrow rectangular piece of cloth (coarse cotton, the manufacture of Anito or Tarapoto) with a slit in the middle through which the head is passed; as it is narrow it covers the body before and behind to below the knees, but not at the sides, so that the arms are free. The legs are encased in breeches of the same material, tight, but not fastened at the knees. . . . A few of them who have been down to the Amazon wear shirt and trousers. The women are none of them pretty, though there are some countenances not unpleasing. They cut their hair like the men, and as the latter are of slender make the two sexes can scarcely be distinguished at a distance. Generally a pollena constituted the article of dress of the women, the body from the waist upwards being naked, but they hang a profusion of beads (white, red, and blue) round their necks, and sometimes use armlets of the same. . . .

The forests on the opposite side of the river abound in animals, and those who go in search of the tapir rarely fail in killing one. Don Ignacio and I paid two men—to one three yards of English calico, to the other a Rondin—to seek us each a tapir. They brought us two fine animals with quite as much flesh on them as a Tarapoto cow, and we had charqui (dried meat) made of them for the voyage. The weapon used in chasing tapir is a lance with large well-tempered iron head, brought from Quito or Riobamba. The dogs used in tracking the animal are a small breed with little triangular heads and curled bushy tails—colour usually iron-grey or fawn colour. One of our hunters went alone with his dogs; the other took two companions. . . . Like most Indians who have been brought to "Christianity," they have no manufactures of any kind. Their canoes, hammocks, blowing-canes, matiris, etc., are all bought from the "Infieles"!

The present Governor of Andoas is Don Benito Sumaita, a native of Moyobamba, who treated us as kindly as his means would allow, and aided us much in procuring men and food for the voyage. He is subject to the recently-created and probably not very permanent Prefectura del Alto Marañon. The head-quarters of the Prefect are at Jeberos or Xeberos, on the Aipena river which enters the Huallaga near its mouth, which, though much larger than Andoas, is quite as miserable a place. Don Benito has been two years in this banishment alone amongst the Indians save his son, a little boy of ten years; and he told us he slept more securely the few nights we were there than he had ever done before in Andoas, for he knew not on what night the Infieles might break into the village and murder him in his bed. He was almost in despair, poor fellow, for he has no salary from the Government, and has not even received pay for cargoes of wax and other products of the country which he had taken or remitted to his superiors in Jeberos at their request.

May 5 (*Tuesday*).—This day at noon we got off from Andoas. Our crews were eight men to each canoe. Eighteen bunches of plantains were embarked in each, for we calculated on fourteen days to Sara-yacu (about 100 miles farther up the Bombonasa), and the existence of plantains on the route was uncertain. Besides plantains, we took a great store of yucas, sweet potatoes, and pine-apples; and the Indians so filled the canoe with their pots of masuto (fermented yucas), beds, etc., that they had not room to work. . . .

May 6 (*Wednesday*).— . . . This morning at three we got off and shortly afterwards entered the mouth of the Bombonasa, which was about 60 yards wide, winding, muddy because nearly full, with vegetation exactly the same as on the Pastasa, where the shore was flat—grasses (*Panicum amplex*, etc., Gynerium, and other genera and species with Cecropias, Ingas, etc.). On the steep loamy banks there were ferns, especially a Mertensia, and the forest trees of Pastasa, with *Iriartea ventricosa*, and a stout tall palm near the *Œnocarpus Patawa*. In some respects it reminds me of the Casiquiari towards the upper mouth. The muddy, shallow water—winding considerably—the dense, intricate vegetation of the shores where low—are the same, but the Bombonasa is much smaller.

. . . .

May 7 (*Thursday*).—The river went down nearly 1½ feet in the night.

. . . .

Several small streams of black water were passed to-day. There was no perceptible current in them, and when the river is fuller it evidently enters some way up them. . . . The river winds much, and reminds me of the Upper Pacimoni. This morning we passed one reach due S. (*i.e.* where the course of the river is N.), and towards evening we made much easting.

. . . .

May 9.—When our Indians have been an hour or two on their way in the morning they proceed to take their chicha. From the mass of crushed and fermented yucas which they keep in a monstrous jar in the prow, they take out handfuls and mix with water to a drinkable consistency. The drinking-vessels used are wide shallow basins varnished and painted, whose use is general amongst the Indians of Maynas. Each Indian will drink one of these full twice or thrice—equivalent to about half a gallon. In the process they occupy at least half an hour, and are as merry and noisy (but not so quarrelsome) as a lot of navvies over their beer. At the same time they make their toilet, which consists in carefully combing out their hair with cane combs of their own manufacture, then tucking up the back hair with a liana passed round the head, while the narrow strip of long hair at the sides is allowed to hang down over the ears, and that on the forehead has been cut short, as already mentioned. After this comes the painting. Each man carries in his bag a slender bamboo tube, a little larger than one's finger, filled with anatto or chica; from this he extracts a portion with a small stick, and with the point of his forefinger makes three broad red streaks from ear to ear, one below the eyes, another along the base of the nose, and the third below the mouth. This done he no doubt considers himself dressed for the day, and holds his head a full inch higher.

. . . .

May 11 (*Monday*).—After a gloomy but dry night, we got under way this morning at 3 o'clock, the river having abated 4 feet. The history of to-day varies little from that of preceding days. The same winding turbid river—in no place more than 80 yards wide, and sometimes narrowing to 40 yards, when the current is stemmed with difficulty.

. . . .

May 14.—The banks now begin to be picturesque: cliffs clad with ferns and mosses, a Helicomia with distichous leaves and pendent scarlet and yellowish spikes; a Calliandra like that at the Pongo of the Huallaga, etc.; tiny cascades falling over the cliffs.

We breakfasted at the mouth of the Puca-yacu, the most considerable stream we had seen entering the Bombonasa. It comes in on the left bank with a strong current—water muddy, reddish. Above the mouth of this the water of the Bombonasa is sensibly clearer, depositing very little earthy matter when allowed to stand; it is whitish, like the Upper Orinoco.

May 15.—Yesterday at $5\frac{1}{2}$ P.M. we reached Palisada-Zipishko, and remained all night on an island, where there was the broadest beach we had seen on the Bombonasa. Pebbles begin to be larger and more numerous; they are chiefly quartz and a compact blue stone. . . .

Coasting along a low shore, our men spied a small white alligator basking in the sun by the margin, and killed him with their lances. His stomach was distended by some food he had taken, and on piercing it, a snake's tail protruded. I laid hold on it and drew out the snake, which was closely coiled up; it was still alive (!), though so much crushed below the head as to be unable to move away. It was a terrestrial species, not venomous—yellow with black spots on the back. The body thick, passing abruptly into a short slender tail—full 3 feet long, and its destroyer no more. Thus we go on preying on each other to the end of the chapter. This poor snake, while watching for frogs among the moist stones and roots, little dreamt he was about to serve for an alligator's meal; nor the alligator, while devouring it, that he himself would soon be eaten up by Indians.

. . . . . .

May 16 (*Saturday*).— . . . The aspect of the river is unchanged, save that there is more rarely any low shore. We have passed some strong currents to-day, but the water is fortunately low. Beaches are now covered with large pebbles, and where we breakfasted it was like a mosaic pavement, stones of so many colours formed our seats and table.

May 17 (*Sunday*).—Near 8 o'clock A.M. we spied a tapir a little ahead, making his way up-stream. On perceiving our approach he took to shore, where from a narrow margin rose a steep barranco, which he was unable to ascend; he therefore again entered the water and attempted to pass down-stream. At this moment we poured in shots upon him from musket and pistol, which, however, did not disable him, and he dived out of sight, but on coming up near one of the canoes, an Indian planted a lance in his breast. Several Indians then leaped into the water, which was scarcely breast-high, and speedily dispatched him. When swimming he had only his head above water, and his mouth wide open displayed a formidable set of teeth. At the first reach we went ashore and cut him up; he was a fat, well-grown male; few of the shots had gone much beyond his thick skin. The finest pieces were salted down and the rest partly consumed on the spot and partly roasted for the morrow.

. . . . .

May 19 (*Tuesday*).— . . . We stopped to breakfast at 10½ A.M., in the mouth of a stream called Sara-yacu, which enters on the left bank. It is of considerable size, with clear water and pebbly bed. Here was a house and chacra with several people. . . . In the canoes moored here I saw several batéas (wooden dishes) for gold-washing; they were made of some light wood, and were about 1½ feet in diameter—either in the form of a meniscus or of a very low cone—and two projecting pieces had been left on the margin for handles. The gold found here is only in small quantity and in very minute fragments. My companions washed two or three pans of gravel, and in each found three or four grains of gold; but it would be necessary to go to a considerable depth to wash with any chance of success, for the gravel is very loose and wet, so that the fragments of gold sink into it by their weight.

At $4\frac{1}{2}$ P.M. we reached the pueblo of Sara-yacu, on the left bank. It stands on a steep ridge 15 feet above the high-water mark of the river and distant perhaps 200 yards. On each side of it and at a short distance is a deep ravine with a rivulet; at the mouth the streams are barely 30 yards apart, but the space between them widens higher up. The track leading up to the pueblo has in one place a steep slope on each side, with barely room for one person to pass another. A barricade across this strait would render it defensible by two men against a hundred. This position has no doubt been selected for the pueblo with an eye to its defence from attacks of Infieles, and it is far stronger than that of Andoas, though there is some similarity.

. . . . . .

May 20 (*Wednesday*).—Our Indians from Andoas should have returned home from Sara-yacu, but as we found there neither Governor nor Curaca, we persuaded them to go on with us to Puca-yacu, where the Governor was at present residing, and so paid then each 2 varas of Tocuyo for the additional labour—all save two who could not be persuaded to go farther. This day was passed dully enough in the port of Sara-yacu, waiting till the Indians should stuff themselves with masuto; enlivened only by disputes about the payment to Puca-yacu, such as are unavoidable in all traffic with Indians.

May 21 (*Thursday*).— . . . We left at an early hour, and the slight rise of the waters gave us more depth in the rapids, so that we got on capitally, and at 4 P.M. reached the port of Puca-yacu. Here we found that the village was nearly a mile from the river and elevated 250 feet above it, the ascent being very steep and slippery.[1] We climbed up to pay our respects to the Governor, and then returned to sleep in the port, I and Don Ignacio in our canoes, and the rest on a narrow beach scarcely elevated 2 feet above the water. The beach was margined by a bank of earth 6 feet high, densely clad with overhanging trees and bamboos, and then after a narrow strip of nearly level ground rose a gentle acclivity. As we supped at sundown, thunder was heard at no great distance, and the heavens gradually became entirely obscured by a dense mantle of clouds. The Indians, who had gone up to the pueblo to take chica, now rejoined us and also prepared to pass the night on the beach. We had scarcely resigned ourselves to sleep, at about 9 o'clock, when the storm burst over us, and the river almost simultaneously began to rise; speedily the beach was overflowed, the Indians leaped into the canoes; the waters continued to rise with great rapidity, coming in on us every few minutes in a roaring surge which broke under the canoes in whirlpools, and dashed them against each other. The lianas by which the canoes were tied had to be moved every now and then higher up the trees, and finally broke. The Indians held on by the branches, and fortunately found two contiguous lianas of Bignonia, which having cut below, they fastened to the prow of each canoe, their upper part being securely entwined

[1] [By barometrical observation, Spruce found the altitude of Puca-yacu village to be 425 metres = 1394 feet.—ED.]

in the branches overhead. Here we held on, the Indians using all their efforts to prevent the canoes from being smashed by blows from each other or from the floating trees which now began to career past us like mad bulls. So dense was the gloom that we could see nothing, while we were deafened by the pelting rain, the roaring flood, and the crashing of the branches of the floating trees, as they rolled over or dashed against each other; but each lightning-flash revealed to us all the horrors of our position. Assuredly I had slight hopes of living to see the day, and I shall for ever feel grateful to those Indians who, without any orders from us, stood through all the rain and storm of that fearful night, relaxing not a moment in their efforts to save our canoes from being carried away by the flood, or dashed to pieces by swinging against each other, or against the floating timber. As the waters rose higher, the stern of my canoe got entangled in overhanging prickly bamboos, which threatened to swamp it, and which we with some difficulty cut away. Every hour thus passed seemed an age, and the coming of day scarcely ameliorated our position, for the flood did not abate until 10 o'clock. About an hour before this, the river began to fall a little, and as soon as the rain passed we got the cargoes out and carried up to the Governor's house. It was past noon ere we got breakfast—wearied to death, and myself in a high fever, which happily passed off in the following night.

The river is only 40 yards broad in that place (indeed before the flood there had not been more than 25 yards of water, nowhere 3 feet deep), and the rise during the night had been 18 feet. I have not yet mentioned that our companion Don Victoriano and the two muchachos, when the rising waters drove them from the beach, thinking that it was merely a brief thunder-shower which had caught us, gathered up their beds and climbed the barranco, where they set up two palm mats belonging to the canoe, and sheltered themselves under them as well as they could; but scarcely had they accommodated themselves here when the flood reached them and burst on them so unexpectedly that several articles which were loose, trousers, handkerchiefs, etc., were swept away. They retired in all haste, and in the dense gloom, ignorant of whither they were going, the only guide to their position being the roar of the river. They wished to enter the canoes, and called out at the top of their voices, which were drowned by the loud conflict of the elements, and the cries of the Indians in the canoes were all unheard by them. Thus they wandered about all night, the flood continually obliging them to retreat farther inland, and when day broke it found them half dead with cold, and their clothes and bodies torn and wounded by prickly bamboos and palms. To reach the canoe they had to wade with the water to their waists. As we were unloading the canoes, the barranco by which we had at first been moored fell into the river with several large trees on it; another peril which we happily escaped by having had to move lower down.

Puca-yacu consists of but eight houses besides the convent and church; they are in the same style as those of Andoas, and there is no cultivation near them, though most have an odd tree of

Wingo, another of Anatto, and some roots of the twining Bignonia (Carajaru) planted by the door. The Governor resides in the convent, which is remarkable for having an upper story, the flooring of which is of bamboo planks resting on rafters of Tarapoto palm. The ground floor is scarcely made any use of, for the kitchen is a low shed standing a few yards apart; but the upper story is divided along the middle by a bamboo partition, the northern half being open at the sides, so as to form a wide veranda, where the family pass the day; and the southern half is divided into two dormitories, where they keep their household gods and pass the night. The whole is very light and cleanly, with superabundance of ventilation; but we have not yet experienced any high winds, the force of the squalls being broken by higher ground across a valley to north and north-east. We live with the Governor, who has given up one of the dormitories to us.

From the village there is a track in a northerly direction which continues all the way to the river Napo. At half an hour from the village it crosses a stream called Baha-yacu, whose mouth is a very little below the port; there are a few chacras on it, and the gold-washings are said to be the best of any of Bombonasa. The banks are steep and fall in with every flood. The water runs over beds of indurated clay, such as most of the rock on the Bombonasa; though easily broken by the foot, it resists remarkably the action of water. Pebbles of quartz and blocks of compact blue stone are evidently alluvial deposits.

In something under half a day the track brings

us to the head of the river Rutuno, a considerable stream whose mouth we had seen below Sara-yacu. All the way along it there are tambos of inhabitants of Sara-yacu, Puca-yacu, and Canelos, who go there to wash gold.

After the Rutuno the head of the river Tigre is passed; this river holds its course nearly midway between Pastasa and Napo, and falls into the Marañon.

A large stream, the Villano, is next passed; this runs into the river Curaráy, whose junction with the Napo is not far from the mouth of the latter.

From the Villano we come to its tributary, the Giguino, on which there is a largish pueblo of Zaparos.

Next to this is another tributary of the Villano—Callána-yacu, and then we come to

Ananga-yacu, which runs direct into the Curaráy.

The Curaráy itself is now reached. On this also are several Zaparos.

The Noshúro, to which we now come, has a pueblo of Zaparos; it runs direct into the Napo, as does also the Washka-yacu.

Passing these, we reach the Napo, at a small pueblo called Aguana, not far from Santa Rosa; whence there is a route over the Cordillera to Quito which is impassable from June to September on account of the streams being swollen by the melting of the snows on Cotopaxi, as also by the depth of snow on the highest point of the pass.

The Governor, Don Gabriel Cordena, is an elderly man of about fifty, with quiet and very devout manners. He has been twelve years on the Bombonasa, but his native place is Quito. Canelos, Puca-yacu, and Sara-yacu are all under his rule, and he divides his residence equally among them. It should be observed that his title is Lieutenant-Governor, the Curaca of each pueblo being considered its real governor. The labour of the Indians is entirely voluntary, nor is there any tariff of prices strictly adhered to. In consequence, the Indians are sufficiently impertinent and difficult to treat with. The pueblo of Puca-yacu contains some nine men accustomed to carry cargoes to the Sierra; and after more than a week's delay, Don Ignacio and Don Victoriano have with much difficulty persuaded five of these to accompany them; the rest excuse themselves from pretended sickness or some other motive, so that I, who need seven cargueros, am still waiting to see if I can induce the Indians of Sara-yacu to accompany me, as they are much more numerous than those of Puca-yacu. The Indians of Canelos are away at their tambos on the Rutuno, etc., with licence of absence for three months, which does not expire till June 20.

. . . .

Don Gabicho (as he is familiarly called) presented himself to us with shirt outside trousers (Amazon fashion), so that it stood for jacket as well, and his head adorned with a broad-brimmed hat of tamshe, similar to those woven by the Indians of Maynas of the same material; well ventilated but affording no protection against rain; so, to render it waterproof, he had stuck it all over with the feathers of small birds, the points all directed to the brim. I have rarely seen a gayer or stranger head-gear.

Puca-yacu is a colony of the still considerable pueblo of Canelos. It contains also four or five Jibaros, who are married to women of Canelos. The Governor has in his house a Jibaro girl whose history is singular. It seems that among those Indians when a man of note dies it is the custom to put his wives to death, in order that their spirits may accompany him, as they did while in the body. An old chief died two years ago, leaving four wives, whereof one was scarcely nine years of age. This poor creature, knowing that they would seek to kill her, fled into the woods, and though pursued, succeeded in reaching Sara-yacu, where the Governor then was, and placed herself under his protection. Her "friends" have since reclaimed her, but the Governor refuses to give her up, and she still remains with him, and is an excellent servant to his wife. She has been baptized by the name of Magdalena, the Governor and his wife standing sponsors. She looks little like a widow, with her slender, girlish figure and smart chitty face. The Jibaro Indians still abound on the Pastasa (above the mouth of Bombonasa) and on its upper tributaries. There is a settlement of them, commonly called the Jibaria, at three days from Canelos, near the river Pindu, on the route to Baños. . . .

There is a magnificent view looking west from the plateau of Puca-yacu, but I saw it only once, for about a couple of hours, in all its entirety. It takes in an angle of about 60°, bounded left and right by forest on adjacent elevations. At my feet stretched the valley of the Bombonasa, taking upwards a north-westerly direction, its waters not visible, and audible only when swollen by rains. Beyond the Bombonasa stretched the same sort of boldly undulating plain I had remarked from Andoas upwards, till reaching one long low ridge, perhaps a little higher than Puca-yacu, of remarkably equable height and direction (north to south); this is the water-shed between the Bombonasa and Pastasa, and the latter river flows along its western foot; a little north of west from Puca-yacu, the course of the Pastasa is indicated by a deep gorge stretching west from behind the ridge. This gorge has on each side lofty rugged mountains (5000 to 6000 feet), spurs of the Cordillera; one of those on the right is called Abitagua, and the track from Canelos to Baños passes over its summit. All this was frequently visible, but it was only when the mist rolled away from the plain a little after sunrise that the lofty Cordillera beyond lay in cloudless majesty. To the extreme left (south), at no very great distance, rose Sangahy (or the Volcan of Macas, as it is often called), remarkable for its exactly conical outline, for the snow lying on it in longish stripes, and for the cloud of smoke almost constantly hovering over it. A good way to the right is the much loftier mountain called Los Altares, its truncated summit jagged with eight peaks of nearly equal elevation and clad with an unbroken covering of snow, which glittered like crystal in the sun's rays, and made me think how pure must be the offering on "altars" to whose height no mortal must hope to attain. Not far to the right of Los Altares, and of equal altitude, is Tunguragua, a bluff irregular peak with rounded apex capped with snow, which also descends in streaks far down its sides. To the right of Tunguragua, and over the summit of Mount Abitagua, appeared lofty blue hills, here and there painted with white; till on the extreme right was dimly visible a snowy cone of exactly the same form as Sangahy but much more distant and loftier; this was Cotopaxi, perhaps the most formidable volcano on the surface of our globe. Far behind Tunguragua, and peeping over its left shoulder, was distinctly visible, though in the far distance, a paraboloidal mass of unbroken snow; this was the summit of Chimborazo, so long considered the monarch of the Andes, and though latterly certain peaks in Bolivia have dethroned it, for ever immortalised by its connection in men's memory with such names as Humboldt and La Condamine. Thus to right and left of the view I had a volcano. Cotopaxi I never saw clearly but once, but Sangahy was often visible when the rest of the Cordillera was veiled in clouds, and on clear nights we could distinctly see it vomiting forth flame every few minutes. The first night I passed at Puca-yacu I was startled by an explosion like that of distant cannon, and not to be mistaken for thunder. It came from Sangahy, and scarcely a day passed afterwards without my hearing the same sound once or oftener; my ignorance of its origin at first amused the people of Puca-yacu, to whom it was a familiar sound.

[During his twenty days' delay at Puca-yacu, besides making notes on the general botanical features of the district and collecting all the new Mosses and Hepatics he could find, Spruce also made, as he states in his *Précis d'un Voyage*, "a collection of the beautiful Coleoptera (beetles) which were to be found there in great abundance." No doubt these were obtained in some of the newly-cleared plantations of the natives on the road to the Napo river, which he explored for some distance.]

June 10 (*Wednesday*).—This day at 8 A.M. I got off from Puca-yacu, where I had been waiting three weeks. My companions had started on the last day of May, and after their departure the Governor went to Sara-yacu and with much trouble found cargueros for me, as they had been frightened at the large size of my trunks when I passed up. I again lightened them as much as I could by selling and giving to Don Gabriel and his family everything not absolutely necessary, and for one trunk in which I had deposited my drugs, barometer, and some other valuables I paid two cargueros. The pay to each was 3 D. 2 Rs., with three varas of bretaña (English calico), and to one who carried a long but not heavy trunk I paid 4 D. and a red handkerchief. They arrived at Puca-yacu on Monday, but Tuesday being very rainy we could not get off; the canoes, however, were put in readiness for the following morning. There were four of them, one lent me by the Governor and the rest furnished by the Indians themselves, and intended to be left in Canelos till their return. We started, sixteen in number, for each of the seven cargueros took with him a boy or young woman to carry his food. The canoes are small, light, flat-bottomed, not capable of carrying more than two of my trunks. . . .

[*June* 12.—Reaching Canelos in the morning,

Spruce found there only two Indians, from whom he was able to buy some fowls and other provisions to complete what was needful for the long journey through the forests. Here all the elaborate packing of the baggage by the Indian carriers had to be done, and the straps carefully arranged in a peculiar manner, so as to be suitable for a route where they are liable to be entangled by creepers overhead and other difficulties. Then there was food for the whole party of sixteen persons to be carried by the boys and girls brought by the Indians themselves, so that they were not ready till late the next day. Then a heavy storm came on which caused the actual start to be put off till the morning of the 14th, at which date the Journal continues the story in the next chapter.

The region described by Spruce in the last three chapters is characterised by the presence of the singular plant usually called the Vegetable Ivory palm, but which is now considered to form a distinct natural order intermediate between true palms and Cycads. Its very hard albuminous seeds, nearly the size of hen's eggs, are contained in compound fruits as large as a man's head, which are concealed among the leaves close to the ground. These seeds are largely exported and used to make buttons, umbrella handles, and other small objects. The plants occur thinly scattered from the mouth of the Napo to Tarapoto and the Forest of Canelos on the lower slopes of the mountains up to about 2500 feet, and on the river-banks.

Spruce only once collected ripe fruits, and then unfortunately lost them, as he describes in his

Memoir on the Equatorial American Palms. I here quote the incident:—

"On my voyage up the Huallaga in May 1855, I gathered one morning some fully formed fruits of Yarina, and as they were infested by stinging ants, I laid them near the fire, where our breakfast was being cooked, to disperse the ants, and then plunged into the forest in quest of other objects. During my absence the Indians, not knowing I wanted to preserve the fruits, struck their cutlasses into them, and finding the seeds still tender enough to be eaten, munched them all up and thus destroyed my specimens. I never again saw the Yarina in good condition, except when I and my attendants were already laden with specimens of other plants."

Two species very closely allied (*Phytelephas macrocarpa* and *P. microcarpa*) are spread over the Eastern Andes, and Spruce described another species (*P. equatorialis*) from the Western Andes of Ecuador, which differs in having a trunk sometimes reaching 20 feet high. The leaves, of a fine deep green colour, are from 30 to 40 feet long. The plate here given is from a photograph taken on the river Ucayáli.]

Fig. 6.—Vegetable Ivory Palm (*Phytelephas microcarpa*).

CHAPTER XVIII

THROUGH THE MONTAÑA OF CANELOS TO BAÑOS

(*June* 14 *to July* 1, 1857)

[THE Journal of this portion of Spruce's travels is so full and interesting, and the district passed through is in many respects so remarkable, that I have no hesitation in printing the account of it almost entire. In the half-century that has elapsed since it was written no other English traveller has, I believe, passed over it. Two German botanists made the return journey from Baños to Canelos in April 1894, when they had better weather than Spruce; but they describe the forest between the Topo and Canelos as being quite uninhabited, and the track so seldom traversed and so ill-defined that even the guides lose their way!]

CANELOS TO BAÑOS

(*Journal*)

June 14, 1857.—It was about 8 A.M. when we got off. We had a steep slippery descent to the Bombonasa, which was crossed with difficulty and risk, as the turbid, swollen waters careered violently among and over rocks and stones. We crossed near where it is joined by a large stream (Tinguisa), by the side of which our course lay for above an hour, sometimes crossing it, sometimes plodding among stones and mud on its margin. At length we turned away to the right and began to ascend to a ridge, which gradually runs higher and sharper, like many such in the Andes, whence they are called cuchillas (knives). It separates the valley of the Tinguisa from that of the Bombonasa. As we ascended it, we had often on our left a steep bare barranco of sand-rock and pebbly alluvium, quite like what I had remarked along the Bombonasa. At 2 P.M. we had come out on high ground, nearly level, but still with steep declivities left and right—where a cool wind was blowing. Though so early, our men declared that there we must pass the night, because it was the accustomed stopping-place on the first day from Canelos, and they set to work to clear the ground and to collect materials for ranchos. Here, as in most other places on the way, we occupied four ranchos, one for myself and my servant, and the other three for the cargueros, who generally chose a site a little retired—say, thirty paces or more from our rancho. The ranchos were merely a fall-to roof, resting on the ground, and were erected in this way. Two stout sticks about 9 feet long were stuck sloping into the ground, about 4 to 6 feet apart; across these were tied palm-fronds, after the fashion of large tiles, till the roof had reached the required width, and it was then secured at an angle of about 45° by a forked stick stuck in front of each of the two whereon the roof was framed. The palm-fronds used were those of two species of Iriartea and of *Wettinia Maynensis*. Of the Iriartea, the fronds were split along the middle and the two halves placed alongside, with the point of one to the base of the other; but of the Wettinia, the pinnæ of one side were doubled over so as to fall between those of the other side, and as they are remarkably canaliculate-concave, a series of alternate convex and concave surface was thus obtained, resembling remarkably well the tiled roof of a house. Several entire fronds with their pinnæ in the natural position were fastened along the top of the roof, so as to throw the rain both ways. On the ground beneath other palm-leaves were extended, and on these were placed our beds and boxes. The fire was made midway, under the ridge of the roof. A stick set up on each side, to sustain a cord stretching across the fire, was essential for hanging up our wet garments through the night to dry and smoke. Two of the cargueros were considered my personal attendants on the way, viz. the one who carried my bed, the necessary changes of linen, and other things likely to be needed, in a waterproof bag; and the one who carried the provisions in a saparo, a nearly cylindrical basket 3 feet long and 2 feet in diameter, covered by a lid made of an outer and an inner framework woven of the liana Tamshe, with two or three layers of leaves of Vijao securely packed between them so that no rain could enter. The duty of these men was to erect my rancho, and collect me firewood sufficient to burn through the night. When we had got our house set up and the necessary fuel and water brought to it, my first care was to prepare coffee —the greatest consolation a traveller can have after a day's work in the wet forest. After coffee a salt fowl was boiled and plantains roasted for supper. Then, wrapped in my blanket and stretched on my mattress, with my feet near to a good fire, I prepared to pass the night, and I may say that however much I might have suffered through the day, I generally slept tolerably well and rarely suffered from cold.

June 15.—We had heavy thunder-showers from 2 to 4 A.M., and wet dripped from the roof on to the foot of my bed. The day was cold and drizzling throughout. Our course was still mostly along the top of the ridge, gradually ascending, rarely descending a little to pass slight rivulets. About noon we reached the highest part at a place called the "Ventanas" (windows), where the track ran along the edge of a steep barranco to the right, down which we looked into a tremendously deep valley, whose bottom was obscured by rolling mist, though we distinctly heard the murmuring of the nascent Bombonasa along it. Travellers and cargoes arrived pretty well soaked at the end of this day's journey, and the same was the case through nearly all the rest of the way. The ground to-day was mostly gravelly.

June 16.—Again heavy showers before daylight which left the forest soaking wet for our journey. There was a little sun till 9 o'clock, then came on showers, which, with very short intervals, lasted till 4 P.M. Our cargueros were accustomed to breakfast at daybreak, I and my muchacho at the same hour made our coffee and cooked a fowl to be eaten on the way by some stream of cool water, whenever hunger should invite us. On reaching the first stream from our sleeping-place, the women prepared large draughts of masato for the men, as they said, to give them force, and the process was repeated once or twice during the day. They had also generally their marked resting-places, where they made long halts after carrying their loads an hour or an hour and a half together. On reaching one of these, the women used to cut palm-leaves and spread them on the ground, and the men, after depositing their loads, threw themselves on the leaves at full length. This day they had made very long halts, so that although we went along very slowly, and I often delayed to pluck a moss from the branches, we had got far ahead of them. The day was wearing away, and the clouds and rain made the forest so gloomy that night seemed nearer than it actually was. We waited a good while at a place that seemed convenient for the ranchos, till I began to shiver with cold, and I actually turned back to see what had become of them. The Indians from the first had been complaining, *more suo*, of the heavy cargoes, then of the rain and the wet forest, and of the long distance they had to go. They might at any instant leave their cargoes and return to Canelos, without giving us a hint. Such a thing had happened many and many a time. Even these very Indians on their last journey towards the Sierra—conducting the Padre and his cargoes—left him and his goods at the Rio Verde, a day's journey from Baños. The night is generally chosen for these elopements, and when day breaks the unfortunate traveller finds himself alone. Fortunately, my misgivings in this instance were without foundation, and after I had gone back a good distance I heard the voices of my people advancing, and conducted them to the site I had chosen for our resting-place.

The road had been gently descending for most of the day and was not so gravelly as yesterday, while much sloppy ground had to be passed.

June 17.—A shower at 3 A.M. At daybreak rain again came on and continued without intermission till near noon, when we set off. We had gone for scarcely two hours when we reached the large stream called Púyu, a tributary of the Pastasa, and found it so swollen that there was no hope of crossing it; we must therefore again set to and construct ranchos, and there await the river subsiding. My chagrin at this delay was somewhat lessened by the circumstance of finding myself in the most mossy place I had yet seen anywhere. Even the topmost twigs and the very leaves were shaggy with mosses, and from the branches overhanging the river depended festoons of several feet in length, composed chiefly of Bryopterides and *Phyllogium fulgens*, in beautiful fruit. Throughout the journey, whenever rains, swollen streams, and grumbling Indians combined to overwhelm me with chagrin, I found reason to thank heaven which had enabled me to forget for the moment all my troubles in the contemplation of a simple moss. We had hoped to reach the Jibaros settlement this day. The chacras were said to be near, and two of our men swam across the river Púyu and before nightfall returned with plantains.

June 18.—Slight showers before daybreak, but the river had sufficiently abated to allow of our passing it, and at 6 A.M. we started. On the opposite side we were not long in coming on large plantations of yucas, plantains, yams, etc., and about nine we reached a house where we found an old man and several women. Here we remained an hour, and I bought a cock of the old man, though I must needs shoot it with my gun, as it was wild and would not allow itself to be caught, he said. After a short chase among the wet yucas, I brought it down and we bore it off in triumph. It took us two hours more to reach the centre of the settlement, where are the Curaca's and two other houses. The way was very muddy, and in that short distance traversed by above twenty streams, with steep slippery descents to them. It was noon as we reached the Curaca's house. We had had drizzling rain for some time this morning, which with the heavy rain of yesterday and the soft muddy nature of the earth had put the track in very bad order and we reached our halting-place in pitiable plight. A good many years ago, it seems, some missionary had induced these Jibaros to become Christians, and to erect a church and convent, after the fashion of those of Canelos and Puca-yacu, but they have long ago renounced Christianity and the church has fallen to decay. The convent was still tenantable and we took possession of it—that is, I and my servants, for the Indians installed themselves in the Curaca's house. The Curaca was absent in the forest and did not return till evening, when I bought a couple of fowls and some plantains of him. His name is Huêléca—a young man of middle stature, slender in body, but with remarkably muscular arms and legs. Compared with our "Christian" Indians from Sara-yacu, we found him a person of gentlemanly manners and with none of the craving selfishness of those people. I had therefore quite a pleasure in offering him such little presents as I had kept in store for that purpose. His wife was a tall young woman with pleasing features, and they had four small children, all ill of catarrhal fever. The Curaca and every one about him were complaining of illness, especially of rheumatic pains, which was not to be wondered at from the wet and mud among which they live at this season.[1] In dry weather the site must be rather pleasant; the ground is highish, rising from the Púyu, which furnishes water, though it is a good ten minutes' walk to the river and back. When the sky is clear, Mount Tunguragua, with its cope of snow, and the lower wooded ridges in front of it are seen very distinctly.

The afternoon of the day we arrived was nearly fair, though cloudy and cool; but at two of the following morning it came on to rain heavily and continued without intermission till midnight.

Next day (20th) drizzling rain from sunrise till nightfall. The sloppy ground, the soaked forest, and the unceasing rain kept us close prisoners. My Indians had been occupied in preparing chicha for the remainder of the journey; this task was completed, but the weather and the road were so dreadful that we could not think of starting. They declared they were quite out of heart, and they

[1] Shortly after I passed by the Jibaria, Huêléca removed with his family to Sara-yacu, to consult some noted medicine-man; there his wife and one of his children died, and I have since learnt that he has burnt down his house and the convent, and that he has removed to some other part of the forest where the whites never pass, for to their contamination he believes that he owes his bereavement.

absolutely refused to stir a step further unless I would lighten my cargoes. They had received their pay beforehand and I was therefore completely in their hands. I had brought from Tarapoto a boxful of drying paper, and on our way up the rivers I had dried a sprig or two of everything accessible, and especially of Cryptogami, by placing them in paper under my mattress in the canoe. At Puca-yacu, fearful of increasing the weight of my cargoes, I limited my collections to mosses. The only way of lightening my cargoes was to throw away all the paper not occupied by plants, and then divide the remainder of the effects nearly equally among my five boxes. This I did—with a heavy heart—for I knew I should have much difficulty in replacing the paper when I got out into the Sierra. The savages made a bonfire of my precious drying-paper and danced round it!

Sunday the 21st.—The sun shone out in the morning, and we were gratified by the day holding out dry and hot. We waited, however, till the following morning to give time for the forest to dry a little. Early on the 22nd we resumed our journey. I had gathered small quantities of many interesting mosses in the Jibaria, chiefly on logs in the platanal by the convent, and on trees in the forest by the Púyu; of these I made small bundles, putting alternate layers of Mosses and Hepaticæ so that there might be no confusion of fallen lids and calyptras, and dried them in the sun and by the fire. The same plan I followed through the remainder of the journey, depositing such mosses as I could snatch from the branches in a bag hung at my side, when we halted for the night tying them up in bundles, and then hanging them up through the night *to smoke* along with our soaked garments.

Monday was also happily a sunny day. The way was mostly along level ground, often through beds of tall prickly bamboos, and lodales (muddy places), the mud being, as might be supposed, congenial to the bamboos, and often hiding fallen prickly branches of the latter which wounded our feet. I wore throughout the journey a pair of india-rubber shoes which I had fortunately bought off the feet of a wandering German I met in La Laguna. They were slippery in the descents, where I required to step cautiously in them, and they were easily pierced by thorns and stumps, but they were uninjured by mud and wet, and so long as I kept in movement my feet were never cold in them, even when they filled with water. In fording the streams I kept them on my feet; on reaching the opposite bank I slipped them off and poured the water out, then in an instant slipped them on again and resumed my march without experiencing the least inconvenience. We had got off about seven, and it was near ten o'clock when we reached another Jibaro hut, and the last of the pueblo of Pindo. Here we rested awhile, and my Indians partook of chicha which was offered them. I considered myself fortunate in buying a couple of fowls and the leg of a tapir. Shortly after we crossed the Pindo, a considerable stream with a broad white beach strewn with blocks and much resembling the Cumbasa below Tarapoto. This stream receives the Púyu (which also we crossed this day, quite near the Jibaria), and the two united are navigable for small canoes to the

Pastasa, which is at no great distance. We were gradually approaching the Pastasa, and we slept at night on a plain where the rushing of its waters was distinctly audible.

June 23.—About 10 A.M. we reached Allpa-yacu, a stream of clear cool water about the size of the Pindo. This also was low and we got across it without accident. There were steep cliffs of gravel on the east bank just above the ford. Our way to-day was almost entirely across a plain, bounded on the left by a very steep alluvial cliff (which gives the name of Barrancas to the site), at whose foot ran the Pastasa. There is a great contrast between the aspect of this river here—leaping and foaming over rocks with a din which throughout the rest of our journey we heard more or less distinctly—and in the lower part of its course, where it spreads out into a broad placid river. The track in places ran along the very edge of the cliff, and the projecting bushes menaced thrusting us over. At about 2 P.M., on the top of a low hill, we came to a rancho, but as our Indians were still disposed to proceed we determined to sleep at a more advanced post. From this place we descended into a deep ravine, and crossed a narrow clear stream with some peril, as the ford was over slippery stones on a steep declivity. To our right the water came down from a lofty hill in a cascade. To climb out of the ravine we had to use hands as well as feet, but a winding path might be easily made, for the soft sandstone admits of being cut by a spade. We slept about half-way down the descent of the other side of the mountain, but were wetted by a shower ere we could get our ranchos put up.

June 24.—This morning in less than an hour we reached a narrow but rather deep rocky stream, remarkably like so many others in the Montaña of Canelos for its crystalline water. We crossed it near its junction with the Pastasa, on the banks of which and above its mouth rise lofty cliffs from the river's edge, to avoid which it is necessary to climb over the most formidable mountain on the whole route, named Abitagua, and perhaps 6000 feet high. It was near midday when we reached the summit. At something more than half-way up is a puesto (resting-place) called Masato, whence there is a view down the valley of the Pastasa, extending, it is said, in clear weather even to the Marañon. I could distinguish the water of the river Pastasa apparently a little below Andoas, but beyond this the sky was too hazy to make out anything. From Masato upwards the ascent is painful—steep, rugged bits alternating with flats of mud, sometimes over the knees. On the top is a long narrow plain, where the intervals between the trees are occupied by loose mud. At the western extremity of the plain is a small open dryish space where a cross has been erected. From this site the heights of Patati and Guayrapata in the Sierra are visible, as are also the much nearer ridges running from Llanganati between the Topo and the Shuña. From the cross there is a steep short descent, and then another long muddy level, about midway of which, and a little to the right of the track, there is a hollow filled with clear cold water—in fact, it may be called a lagoon, though there are mounds here and there on it with trees, true Vaccinia, etc., on them. Perhaps never a day passes without rain on this mountain, and its summit is nearly always enveloped in mist, which looks as if it were permanently hung up in the trees. The trunks and branches of the latter, and often even the uppermost leaves, are densely enveloped in mosses. Various species of Plagiochila, Mastigobryum, Phyllogonium, Bryopteris, etc., hang from the branches to the length of 1 to 3 feet, and in such thick bunches that when saturated with rain they often break off even green branches by their weight. I have been told by the cargueros of Baños that when they pass with cargoes through the most mossy parts of the Montaña after much rain has fallen they step with constant dread of being crushed by some ruptured branch. I examined hastily such mossy branches as had fallen across our path, and often found on them a Holomitrium and a Bryum, which I never got in any other situation.

We had fortunately fine weather until reaching the cross of Abitagua; after passing this we had smart rain all the way down. The descent was long and rugged and took us two hours and a half. At the base was a stream of beautiful water quite like that on the eastern side. On a hill of small elevation, called Casha-urcu ("Prickly Hill," because of the ground being strewed with thorny twigs of bamboos), rising from the opposite bank of the stream, we drew up for the night.

June 25.—We had heavy rain from midnight; when day broke we prepared for the journey, hoping that the rain would pass, but in vain, for it abated not till two in the afternoon, when it was too late to start. This was a most dismal day, and filled us with anxious thoughts for the passage of the Shuña and Topo, which rivers the Indians began to predict would be swollen. They, however, were consoled by meeting near our ranchos a band of large monkeys, several of which they brought down with their blowing-canes.

June 26.—Rain again from midnight, but about nine in the morning it abated so much as to allow us to get under way. Road dreadful, what with mud, fallen trees, and dangerous passes, of which two in particular, along declivities where in places there was nothing to get hold of, are not to be thought of without a shudder. In three hours we reached the Shuña, a larger stream than any we had previously passed; it comes from the north-east in a steep rocky course, and can only be forded after long-continued dry weather, and even then with danger. Now we found it much swollen, but as the tops of the rocks on which it is customary to rest the bridge were out of water, though we had to wade in 3 feet of water to get to them, we set to work to get materials for the bridge. These were merely three long poles, not of the straightest, laid from rock to rock and lashed together with lianas. An Indian posted on each rock held up the opposite ends of a fourth pole to a convenient height to serve for a hand-rail, by means of which one could cross the narrow slippery bridge with some degree of security. We all got safely across the Shuña, but it had again come on to rain, and we bent our steps towards the Topo with misgivings that we should find it altogether impassable. On the west side of the Shuña there is a steep cliff, perhaps 150 feet high, of dangerous ascent. In some parts of it on projecting ledges poles are set up with notches cut in them wherein to step, but they were very slippery, and in clambering up them I trusted more to my hands than to my feet. Beyond this the ground is nearly level to the Topo, which we reached in an hour more. Here our worst fears were realised. The Topo, as far as we could see up it, and downwards to its junction with the Pastasa, was one mass of foam, and the thunder of its waters against the rocks made the very ground shake to some distance from the bank. The Topo is perhaps the largest tributary of the Pastasa on the north side; its course is much shorter than that of the Bombonasa, but more water seems to come down it. Its source is in the snowy mountain Llanganati—the fabulous El Dorado of the Quitensians. . . . This mountain and its offshoots occupy nearly all the space between the head of the Napo and the Rio de Patate, both which rivers rise in Cotopaxi. . . . The Topo is never low enough to be fordable on foot, and though numerous explorations of its banks have been made for some leagues up, no place has been found practicable for a bridge save the accustomed one, which is about 200 yards above its junction with the Pastasa. Here, on each side of the river, which is perhaps 100 feet wide, stands a large rock, nearly flat-topped, and rising some 12 feet above high water; they are rather difficult of access, but can be clambered up. . . .

In the middle of the river, and in a line with these two rocks, is a smaller one of equal height, to which bridges could be thrown, and a third short bridge to the right bank of the river (where is a narrow channel, sometimes dry), between the large rock and the actual margin, rendered the crossing of the river complete. Ordinary floods did not reach these bridges, but after long and heavy rains they were carried away, the rocks supporting them being laid deep under water. Yet they sometimes lasted so many months that the bamboos began to decay, and have given way under people who incautiously attempted to pass them. In one of these high floods, some eight years ago, the intermediate rock was toppled over, and as it now lies it is so much lower than the others that it no longer serves to support the bridges. From this cause, the Topo has now to be passed by four bridges, thrown from the sides to three rocks in the water, about 20 yards higher up than the ancient site. These rocks are all smallish and uneven-topped, and the middle one is so low that a very slight flood suffices to render it inaccessible. When we reached the margin, this rock was barely visible at long intervals, and then came surging waves which laid it 1 to 2 feet under water, and would have swept away instantly the poles attempted to be laid on it. The Indians declared that until this stone should be left uncovered there was no hope of getting across; we therefore cast about to make the preparations necessary for passing the night in this place. So many travellers have been detained here by the swollen Topo, that the narrow isthmus between the Topo and the Shuña has been ransacked of everything available for food or shelter. Not a palmito is now to be met with, nor even a palm-leaf wherewith to thatch a rancho. Our Indians therefore made the roofs by tying long slender sticks across each other, so as to form small squares, and then overlaying them with such large leaves of terrestrial and epiphytal Aroideæ (chiefly species of Anthurium) as they could meet with. Roofs so constructed are not proof against heavy rains, and the leaves soon begin to shrivel and rot. Our huts being put up, we cooked our humble supper and lay down to sleep. At 9 P.M. heavy rain came on and continued without intermission till daybreak (5 A.M.) of the 27th. When we looked out in the morning we saw that the river had risen still higher, and there was no hope of getting across this day. Our provisions began to run low. The Indians had drank their last chicha, and they had all along kept robbing me of such eatable things as I could not keep under lock and key, so that my stock of salt fowls was reduced to three, and I had only besides a few dried plantains in a tin secured by a padlock; with their usual carelessness for the morrow, they had already eaten up the large monkeys killed at Casha-urcu, and all their provision consisted of a few baked plantains.

The day continued gloomy, but no more rain fell. I sallied forth along the river-bank to see if I could meet with anything eatable. Rude granite blocks, often with quartz veins, and here and there small masses of pure quartz, were so heaped up as not to be passed without difficulty and danger. Among them grew scattered plants of a small Cardamine, of which I gathered all I could find to eat as salad. I then struck into the forest and anxiously scrutinised all the trees and the ground beneath them, in the hope of meeting some edible fruit; but it was not the proper season, and I could only find a single tree of a Miconia (Melastomaceæ) about 20 feet high, with small insipid black berries about the size of swan-shot. This I decided to cut down the following day, should we be unable to get away, and boil up the berries with about a handful of sugar which I had still left. Neither I with my gun nor the Indians with their blowing-canes could meet a single living thing save toads.

At about four in the afternoon the sun shone out among the clouds, and though the river fell not, there seemed some chance of its abating before morning; so, that all might be in readiness for this desirable contingency, I set the Indians to work to get out the bamboos and lianas required for the bridges. About a quarter of a mile back from our ranchos, and on moist rising ground, are large beds of bamboos affording abundant materials for bridging the Topo. The old stems are so inwoven to one another and to adjacent trees, by means of their arched thorny branches, that, though cut off below, it is impossible to get them down. On this account, stems of a year's growth are chosen; these are as tall as the older ones, but have no branches, only spiniform pungent branch-buds at each joint, which must be lopped off, or they would wound the hands and feet. About 40 feet of the stems is available for the bridges; above this height they are generally so much thinner as to be easily broken off. When cut down and trimmed, each man drags one to the river's brink, which is no easy task over ground where there are so many obstructions; and in the bamboo-flats so many dead thorny branches are strewed that the feet do not fail to be sorely wounded. When a dozen bamboos had been dragged out the Indians fell tired and could not be induced to fetch the four more which were needed to make the bridges sufficiently strong, so we had but three instead of four for each bridge.

At nightfall the river seemed to be falling slightly, and we retired to rest not without hope of seeing it passable when day broke; but after midnight heavy showers came on and continued till near 5 A.M. (June 28), so that the morning light showed us the river as much swollen as ever. The sun looked out on the wet forest for a brief interval and then was hidden by clouds, which speedily overspread the whole heaven, so that we could not doubt more rain was coming. The Indians had had long consultations amongst themselves the previous day, the purport of which I could not doubt was the expediency of deserting me and returning to their homes. I also had proposed to them that two or three of their party should return to the Jibaria, and from thence bring plantains for the rest, as I had been told by the Governor of Canelos that such a thing was sometimes done. But they shook their heads and said that if one went they must all go, that they were weary and famished, and that the women would die if they returned not soon to their own country; so that I plainly perceived if I once sent them away I should see their faces no more. On the 28th, however, they began to talk openly of the necessity of returning, seeing, as they said, that before the river could abate we must all perish of hunger. And in truth our state seemed desperate, our provisions altogether would not suffice for more than a couple of meals, say to keep body and soul together for two days. Of the painful thoughts that passed through my mind at this critical juncture my rough notes contain no record, and writing now, after six months have elapsed, I shrink from recalling them. The conclusion of my cogitations was to remain by my effects till death or help should arrive; and my lad, who promised not to desert me, was of the same opinion. We calculated that we should be able to keep alive for a week, and in that time perhaps some trader might come from the Sierra on his way to Canelos. The Indians also were loath to turn back for this reason that they had received their pay in money, with which they hoped to buy great store of calico in the Sierra, where it would cost them but a real the vara, whereas if they took the money back to Sara-yacu they must give four reals the vara for the same sort of calico to some trader who should by a rare chance go thither. I called a council by the river-side, in order to consult on the possibility of throwing the second bridge to a rock a few yards higher up the stream than the one that was under water, but so much higher out of the water than the first stone that the bridge resting on it must necessarily slope considerably, and so far apart that it was doubtful if the bamboos would span the distance. I had proposed the same thing to them yesterday, when they had declared it impossible, but now they seemed to think that if the bamboos would only reach the upper rock the plan was feasible. There was no time to be lost, for heavy rain was coming, and it was probable the river would speedily rise, so to work at once we went. Though the crossing these frail bridges is a ticklish operation, it may well be supposed that the fixing them is far more perilous. A bamboo was placed resting towards the base on a stone by the margin; its point was then elevated considerably by two or three men weighing down the end by their united force; in this position it was swung round till it hung over the rock on which it was intended to rest, when the point was gradually lowered till the bamboo lay as it was required. By the same process a second bamboo was placed alongside the first, and then a man at the imminent risk of his life crawls along them till reaching the rock whereon they rest. He carries a liana rope attached to the root-end of a third bamboo, which he now, with some help from those on shore, draws after him and places alongside the other two; the bridge is thus stronger than if all the points were laid the same way. Finally, the bamboos were lashed tightly together by lianas at about every 2 feet, and stones laid on them at each end to keep them firmer. So deafening was the roar of the waters that all these operations were carried on through the medium of signs. A movement by the hand to imitate chopping was the signal that a knife or cutlass was wanted, and the hands twirled round one another asked for a roll of liana. The first bridge was short and completed without difficulty, but when they came to throw the bamboos to the second rock, which, as I have said, was much more distant and higher out of the water, it was found that their points merely reached the sloping side of the rock and not to its summit, and that the surging waves every now and then washed over them. Four bamboos were laid side by side before any of the Indians would venture to pass to the other extremity, though one of them was afterwards drawn away to enter into the composition of the third bridge. They were at length securely lashed together, and then the third bridge was completed with more facility, being somewhat shorter though sloping from a high to a low rock. The fourth and last bridge was short and speedily constructed. It was near noon when the bridges were ready for crossing. It had been raining heavily for some time, and the river already began to show signs of a further rise ; our safety depended therefore on getting over as speedily as possible. And now became evident what I had all along feared, namely, that the second bridge was so long and so weak, and bent so much when a man went over it, that a very little addition to his weight would plainly either cause it to break or the farther end to slip off the rock whereon it rested but too insecurely. To get across my heavy boxes would be plainly impossible; the Indians indeed flatly refused to risk themselves on the bridge under the weight of any one of my boxes.

It was doubtful if an additional bamboo would make the bridge strong enough, and there was now no time to get one out. I had therefore no alternative but to leave my goods where they were, and trust to be able to send from Baños to fetch them away. With some difficulty I got across my bed and a change of linen and what little money I had, and left my boxes as well protected as I could from the moisture both above and beneath.

We were a good while in all getting across, for

we must pass one by one with slow and cautious steps, where one slip might be fatal. Though the bamboos were scarcely so thick as one's leg and completely wetted, the natural asperity of their cuticle rendered passing along them less insecure than I had feared; but the longest bridge bent so low when we reached the middle that beyond this it was like climbing a hill, and in this part a surging wave wetted me to the knees, but I stood firm and allowed it to pass. The river was obviously rising and our bridge must soon be swept away.

Those who have escaped from death by hunger or drowning may understand what a load was taken off my heart when we had all got safely across the Topo, although I had been obliged to abandon so many things which to me were more valuable than money. On the following day we might hope to reach the Rio Verde, where is a hacienda for the fabrication of cane-brandy, and the first habitation on the skirts of the Montaña. The rain came down heavier than ever, and the forest was like a marsh, but we dashed on as quickly as we could. The track lay mostly along nearly level ground, with a high cliff to our left, and the Pastasa roaring along its base. In one part we had to wade for nearly a mile though fetid mud in which grew beds of gigantic horse-tails 18 feet high, and nearly as thick as the wrist at the base. At length we came to where we had to descend to the beach of the Pastasa, or "Arenal" as it is called. Here it might truly be said "C'est le premier pas qui coûte," for the descent began by a ladder—merely a notched pole down a rock which overhung the very Pastasa at a height of 150 feet above it; and it may well be supposed how each as he descended the pole clung to it like grim death. We all got safely down to the beach, where we could get along more pleasantly.

When the two Spaniards left me at Puca-yacu I sent by them a tin box asking them to return it full of bread from the Sierra, when they should send back their cargueros. I had hoped to meet the bread about the Jibaria, but I afterwards learnt that my companions had had a long disastrous journey through the Montaña, and that the swollen Topo kept them waiting three days. However, when we got down to the Arenal, we saw some Indians advancing and recognised them for our friends of Puca-yacu. They brought my bread, which thus came very opportunely, and I immediately shared out a loaf to each of my hungry companions, reserving enough for other two rations.

The Indians of Puca-yacu, on learning the state of the Topo, did not delay a minute, but started off at the top of their speed. I afterwards learnt that when they reached the Topo the bridges were beginning to move, that they crossed with some peril, and that immediately afterwards the longest bridge was carried away. We continued along the margin of the Pastasa till the sun began to get low, indeed the rain did not clear away so as to allow us to see his face until 2 o'clock, and at about 4 P.M. came on a rancho thatched with leaves of Arrow-reed, where we drew up to pass the night.

We were still a good way from the end of the Arenal. Whilst my supper was preparing I had leisure to examine it a little. The gorge of the Pastasa, though still bounded on the north side by the same high cliff as we had seen from Barrancos upwards, opens out here to a considerable width, and here and there the river forms islands. The broad sandy beach, strewed in some parts with gravel and in others with angular blocks, bears marks of having been at some epoch permanently under water, but much of it lies now above the limit of the highest floods, and is in some parts covered by a dense but not intricate vegetation, among which the Laurel is the most conspicuous plant. I was also much struck by a Diosmeous shrub with sarmentose pinnate branches, and small flowers of which the petals persist after flowering and become distended by a purple-black fluid which I afterwards found to be the universal substitute for ink at Baños. On the sand grew a pink-flowered Polygala 9 inches high, and some other herbs, but especially *Melilotus officinalis*, which must have been brought down from the mountains; and amongst the under-shrubs a bushy digitate-leaved Lupin was very frequent. These plants were all new to me, but along with them, and especially in places which the floods still reach, grew abundance of *Gynerium saccharinum* with the same tall Gymnogramme and the same Composite tree as were so abundant on the beaches of the Mayo and Cumbasa near Tarapoto. They were accompanied by an Equisetum, resembling *E. fluviatile*, and distinct from the tall species mentioned above.

June 29.—The night was fortunately dry, and at daybreak I had our last fowl cooked and the remainder of the plantains distributed among the Indians, besides a loaf of bread to each. At sunrise we got off, and about the same hour rain came on, and continued till noon. Though not very heavy, it had the accustomed effect of putting the forest in weeping plight. The track, instead of improving as we approached the residences of civilised people, was this day decidedly worse than ever, and the natural obstructions were multiplied almost tenfold. At 8 o'clock we reached the terminus of the beach, above which the Pastasa ran close to the barranco, so that we could no longer follow its banks.

And now commenced a series of ascents and descents, of which I counted eight from Mapoto to Rio Verde. Of these, the first two ridges were the highest and most fatiguing. Beyond these was a narrow sloppy plain at whose further side we had to pass a long puddle-hole called Runa-cocha, in which are laid slender poles from one projecting stone or tree-stump to another, and as they were now covered by water it was difficult to step on them. I had, in fact, the pleasure of slipping off them into the water nearly up to my waist. As the Indians travelled now without cargo, they got much ahead of me, and I know not how long they had been at the Rio Verde when I came out there, at 3 P.M., very much wayworn. What a pleasure it was to see again a white man's habitation, with plots of cultivated land! The hacienda has only been recently established, and the dwelling-house, which has an upper story, was unfinished; but there was a cane-mill worked by water-power, and from twenty to thirty people at work cutting cane in the adjacent cane-piece, distilling brandy, etc.

The Rio Verde is very little less than the Topo, and, like it, is unfordable. We crossed it by two stout poles laid from rock to rock at a part where the river was confined to a narrow gorge. Immediately below, it opened out into a deep basin where the water was so clear and green that one sees the name of "Verde" has not been given to the river without reason. Its course is down a steep valley from north to south, and at its mouth it falls over the barranco of the Pastasa in an unbroken cascade of perhaps 200 feet high.

We had obviously been ascending all day, and when we came out on the open ground of the Rio Verde, a cold, penetrating wind was blowing. Here we found that the common plantain would no longer bear the climate, though the small species called Guineo was still flourishing. Oranges and sugar-cane did not attain the size they did on the Amazon. On the other hand, productions of cooler climates began to make their appearance, such as potatoes and zanahovias, which seem a sort of parsnip. These are planted in far too small quantity to suffice for the consumption of the people employed in the hacienda, who being from the Sierra, their food consists chiefly of potatoes, pea-meal, and barley-meal. I was therefore disappointed in my expectation of finding materials for a plentiful refection for all my party, and with much difficulty bought a few potatoes and zanahovias, and a small quantity of barley-meal, besides a couple of bottles of aguardiénte for the Indians, who esteemed it much more than the food.

June 30.—Although at the Rio Verde I slept under the shelter of a good roof, I suffered more from cold than I had done in the forest, for a cold wind came through the unfinished flooring and walls of the upper story, where I had made my bed.

From Rio Verde to Baños, a distance of some 15 English miles, the road runs near the Pastasa, but only in two places, in each for near a mile, along the actual beach; in other parts it passes over elevated pampas, or makes detours over hills to avoid steep cliffs, especially at the cataract of Agoyan. For the first hour from Rio Verde we were on elevated, nearly level ground, called Quillu-túru or yellow mud. As to the mud, well does it deserve to be signalised by such a name, though the actual tint is as often black as yellow. In no part of the Montaña had we harder toil in tramping through the mud than here; in other respects the road was a tolerably good mule-track, not very wide, but kept clear of rubbish; and after passing Quillu-túru it was mostly sound and often gravelly. At nearly two hours from Rio Verde we came to a hacienda on a beach by the Pastasa, called the Playa de Antombos, where the mistress, a very hospitable lady, must needs have us enter and take some refreshment. Here we learnt that the late rains had been equally heavy in the Sierra, and that on the preceding day the Pastasa had swollen so much as to break the bridge of Agoyan, though this is 40 feet above the river at low water. She had yesterday sent a lad to the town with aguardiénte and counselled us to await his return, as if he did not come it was a sign that the bridge was impassable. Here was another delay, and it seemed as if my progress must be arrested by swollen rivers up to the very last day, as it had been almost from the first. The lad did not arrive until near evening, all too late for us to start again for Baños, although he reported that the bridge might still be passed by one person at a time without much risk.

Here I found a stronger and cooler wind than even at the Rio Verde, and as I had with me no garment proper for the cool region, I was glad to purchase of a carpenter at Antombos a new poncho, of two thicknesses, of scarlet bazeta or baize. This was a welcome addition to my blankets at night, and afterwards served me much in riding about on the cold high lands.

July 1.—Our kind hostess gave us an early breakfast, and then we began our last day's journey. On reaching the bridge of Agoyan, we found it to consist of three or four trunks of trees laid across from cliff to cliff (for the river here foams between steep black walls of trachyte), and covered with branches of Retama (*Spartium junceum*) and earth. Of the trunks only one remained unbroken, and we crossed it with cautious steps and slow, but without accident. We were still a league from the village of Baños, but a short way beyond the bridge we reached a farm called Ulva, where the owner (a widow lady) was so good as to lend me a horse, mounted on which I arrived at Baños early in the afternoon. Following the recommendation of the lady of Antombos, I sought out the Teniente parroquial, and requested him to procure me a lodging. He accordingly put me into an unoccupied house in the playa, one of the only two tiled houses in the village. See me, therefore, at the end of my travel of 102 days (counting from my departure from Tarapoto on the 22nd of March), but by no means at the end of my "travayle."

As I arrived at the Hacienda del Rio Verde—

the first habitation of civilised men—on June 29, the journey up to that point had lasted just 100 days.

[As a conclusion to this chapter it will be well to give here the short account of the Forest of Canelos—geographical, historical, and botanical—contained in the *Précis d'un Voyage*, which is of much general interest, as it is now, probably, in exactly the same condition as when Spruce traversed it, if not, from the point of view of the traveller, even worse. The translation follows the original in being written in the third person.]

The Montaña de Canelos has not any fixed limits. It extends between the parallels of 1° and 2° S. latitude, and the meridians of 77° to 78½° west of London, exceeding these limits in a few places. Within this space are included the sources of several tributaries of the Pastasa and the Napo, and a part of the upper course of these rivers themselves. It is bounded on the west by the volcanoes Cotopaxi, Llanganati, and Tunguragua; and on the east it slopes imperceptibly down to the plain of the Amazon, towards the middle of the course of the Bombonasa.[1] It will be understood that, with the exception of the little plantations made by the Indians, the whole of this district is primeval forest. It was in this forest of Canelos and on the banks of the Curaray and the Napo, that Gonzalo Pizarro wandered for more than two years, searching for cities as rich as those of Peru, which he imagined must exist there; hoping besides to discover that great river, which, uniting all the rivers of the Cordillera, ran from west to east, to empty

[1] Spruce spells this word either with or without the "m."

itself into the Atlantic Ocean—an honour of which he was robbed by his lieutenant, Orellana. He had left Quito in December 1539 with 350 Spaniards and 4000 Indians, and he returned with only 80 Spaniards, having lost all the Indians either by death or flight.

Two hundred and thirty years later, Madame Godin des Odonais, wife of one of the fellow-labourers of M. de la Condamine, wishing to join her husband at Cayenne, chose the route of the Amazon. Leaving Riobamba, a town of the Andes of Quito, towards the end of the year 1769, she arrived at Canelos without any accident. There she found the village deserted on account of an epidemic of smallpox. The Indians of the Sierra, who had until then carried the effects of Madame Godin, fearing the infection, immediately retraced their steps. There then remained with her only her two brothers and six persons of her suite, all unaccustomed to navigation. Finding no boat at Canelos, they constructed a kind of raft, but not knowing how to manage it, on the second day it was upset and they lost almost all their effects, including the provisions. Attempting then to follow on foot the course of the Bombonasa, they lost themselves in the forest, and after having wandered about for several days, they succumbed one by one to hunger and fatigue, so that soon Madame Godin alone remained alive. Moved more by the necessity for separating herself from the sad spectacle of her dead brothers than by the hope of saving herself, she pursued her way in the forest, and happily she was able to find some tinamou eggs and wild fruits, which sufficed for her sustenance.

On the morning of the tenth day after the death of her companions she arrived at the bank of the river, just at the moment when two Indians were embarking in a boat. These good people succoured her and conducted her to Andoas, whence she could continue her journey to La Laguna, and from there descend into the valley of the Amazon as far as Cayenne, where her husband was expecting her. During the time that she had wandered lost in the forest of Canelos, her hair had become perfectly white; and to the end of her life she could never speak, nor even think, of those terrible days without a shudder. Every time that the author recalls the calamities with which this poor lady was overwhelmed, he feels that his own sufferings in the same region were but very inconsiderable.

But to treat now of the vegetation. He does not think that he is mistaken when he claims for the forest of Canelos the honour of being the richest cryptogamic locality on the surface of the globe. The trees even, in certain parts, seem to serve no other purpose than to support ferns, mosses, and lichens. The epiphytic ferns, which are the most abundant, are principally Hymenophylleæ and Polypodium (in the widest acceptation of the term). Among the ferns growing upon the ground there are some that attain a height which is almost gigantic: they belong to the genera Marattia, Hypolepis, Litobrochia, etc.; but the really arborescent species come behind those of Tarapoto in variety. Among the mosses, the genera Hookeria and Lepidopilum occupy the first place, and he was able to enrich them with several new species.

Among the species already known may be mentioned the fine *Hookeria pendula*, discovered by Humboldt and Bonpland in New Granada, and the *Hemiragis aurea* (Lam.), Brid., which adorned the trunks of trees with its great clusters. . . .

The most precious of the Hepaticæ are often, as we know, very minute; in order to find them a scrupulous search made without haste is necessary. In spite of that, he found some novelties, and among them an unpublished genus, the *Myrio-colea irrorata*, represented on Plate xxii. of his book,[1] which is perhaps the most interesting that he has ever found. It was growing on bushes watered by the stream of the Topo, and it is the only agreeable souvenir he preserves of that river. All the Hepaticæ gathered in the valley of the Pastasa at a height from 5500 down to 1000 feet, that is to say, from the cataract of Agoyan downwards, belong to the forest of Canelos, and, as will be seen from his book, they are very numerous.

. . . . . .

Baños lies just at the foot of Mount Tunguragua, and upon its wooded sides there was plenty to occupy the author, but he did not cease thinking of the beautiful ferns he had seen on the other side of the Topo, and as soon as paper arrived from Guayaquil he made preparations for again penetrating into the forest. With four cargueros, his servant, and provisions for twelve days, he took the Canelos road on the 6th of October. But the rains had not yet abated on the eastern side of the Cordillera, and when he arrived at the Topo he found crossing impracticable. Two nights he waited on

[1] *Hepaticæ Amazonicæ et Andinæ*, 1885.

the banks; the day was stormy, but the second night it did not rain, and he saw with joy on the morning of the third day that the water had decreased. He made then no delay in having the four bridges thrown across, and took care to make them very solid, lashed together so as to make one single continuous bridge, hoping to find it there on his return. But although he remained only three nights on the other side of the Topo, on returning to the banks on the fourth day, towards sunset, the bridge was there no longer, having been carried away the previous night by terrible storms which had lasted for twelve hours, inundating the travellers' rancho and putting out their fire, so that at daybreak they found themselves soaked with wet, sitting upon their baggage, and with their feet in water. Fortunately, during the day, the Topo had sufficiently abated for them to discover the tops of the rocks; so the bamboos were felled and arranged upon the rocks, and they were able to cross in the last rays of twilight. He learnt when too late that it was only during the months of December, January, and February that one might hope to find the rivers of the forest of Canelos low enough to be crossed easily and without danger. But he was content to have been able to devote an entire day to Mount Abitagua, besides collecting interesting plants all along the road; and he returned to Baños, having enriched his collection with a considerable number of very beautiful specimens.

[Returning to the Journal, the following short note on the few plants observed during his journey may appropriately be given here:—]

NOTE ON THE VEGETATION OF THE MONTAÑA OF CANELOS

The circumstances under which I travelled prevented me paying any attention to the phænogamous plants, nor did I throughout the journey see any large tree in flower, save two or three times a species of Laurel. After the first two days from Canelos, I was much struck by the abundance and variety of the ferns and mosses: every day I saw ferns new to me. The scarcity of tree-ferns was notable, since around Tarapoto, at the same altitude, I had seen such abundance and variety of them.

Between Alapoto and Rio Verde I first came on a tree-fern growing gregariously; it was a species of Cyathea, with a stout trunk, and I cannot distinguish it from a Tarapoto species.

Among the stemless species was a handsome Marattia, and I was much struck by twining species of several genera. In an excursion since made (October) as far as Mount Abitagua, I have, however, been able to gather several of these ferns.

Among the mosses what I most remarked was the great abundance of Hookeriæ, which was indeed equally notable on the Upper Bombonasa.

The most abundant palm, as far as Mount Abitagua, was *Iriartea ventricosa*, and up to this point extends the Wettinia, but west of the Abitagua it entirely disappears.

In descending the western side of that mountain I first saw the noble Wax palm, *Iriartea andicita*, which is said to exist in some abundance on the ridges running down south from Llanganati. Between the Topo and Rio Verde there is a good deal of a slender inclined Chouta (Euterpe), rarely exceeding 15 feet, which affords a delicious palmito. On the opposite side of the Pastasa, in this part, rise steep hills clad to the very summit with *Iriartea ventricosa*. I have nowhere seen so dense a palm-vegetation, save in the Mauritia swamps of the Amazon.

Among inundated rocks at the Topo, Púyu, etc., I noticed the same small bushy Cupheæ as on Bombonasa.

CHAPTER XIX

BOTANICAL EXCURSIONS AND ASCENTS IN THE ECUADOREAN ANDES

(*July* 1, 1857, *to December* 31, 1858)

[WITH the exception of the first six months spent at Baños (and several shorter visits to it afterwards), the town of Ambato was Spruce's head-quarters during his three years' continuous exploration of the Andes of Ecuador. During the year and a half comprised in this chapter there is nothing in the shape of a Journal, but the letters to his friends, Messrs. Teasdale, Bentham, and Sir William Hooker, furnish materials for a fairly complete account of his life and work during this period. His explorations covered a large extent of the surrounding mountains and forests, and as he was often away from Ambato for weeks or months at a time, I shall now commence each chapter with the full "List of Botanical Excursions," which is sufficiently detailed to enable the reader, by the help of the map, to follow his wanderings, and thus to better understand the references he makes in the letters to the places he has visited. It will be seen that he made a stay of some weeks at Quito in order to explore the neighbouring mountain

Pichincha; and also made several short visits to Riobamba, where his fellow-botanist, Dr. James Taylor, was then living.]

LIST OF BOTANICAL EXCURSIONS

Baños

1857.

July 2-31. Collecting at and around Baños, especially on adjacent wooded slopes of the volcano Tunguragua. Many fine ferns and mosses during this and the following months.

Aug. 1-31. Around Baños. Also excursion to Mount Guayrapata (on the way to Ambato), and to the lake of Cotaló; and down the Pastasa as far as Antombos.

Sept. 1-30. On lower slopes of Tunguragua. On cliffs and beaches of Pastasa, etc.

Oct. Excursion into the Forest of Canelos

„ 6. From Baños to the farm of Antombos.

„ 7. To the Rio Verde.

„ 8. To the beach of Mapoto, on the Pastasa.

„ 9. To the river Topo.

„ 10. Delayed by the swollen Topo.

„ 11. Across the Topo to the place called Terromotillo.

„ 12. To Casha-urcu at western foot of Mount Abitagua.

„ 13. Up Mount Abitagua and returned to Casha-urcu.

„ 14. Crossed the Topo.

„ 15. To Mapoto.

„ 16. To Rio Verde.

„ 17. To Baños.

„ 18-31. At Baños and collecting in the district near it.

Nov. 1-30. Ascent of Tunguragua from the farm of Juivis, along the course of the great lava stream of 1773. Excursion to Mounts Guayrapata and Mulmúl, and to the village of Huambato. Excursion down the Pastasa to the cataract of Agoyan, to the Rio Blanco, etc.

Dec. 1-31. Collecting around Baños. Ascent of Mount Tunguragua by way of the chacras of Pondóa and the forest beyond, where there are numerous tree-ferns, palms, laurels, Weinmannias, etc., at a height of 8000 to 11,000 feet.

1858.

Jan. 1-15. About foot of Tunguragua, some fine trees.

AMBATO

1858.

Jan. 16-31. Removed to Ambato. Until end of the month packing the Baños collections.

Feb. 13. Excursion from Ambato to Riobamba to visit Dr. Taylor, and thence to Penipe, Mount Paila-urcu, the cataract Huandisagua on western side of Tunguragua.

March 10. This excursion lasted from February 13 to March 10. (This and all future long excursions were made on horseback.)

" 11. From this date until April 20 chiefly engaged in examining and packing ferns, mosses, etc.

BAÑOS

April 20-May 22. From this date to May 22 at Baños, and excursions from thence down the Pastasa to Antombos and the Rio Verde; also up Tunguragua by Juivis and the Alisal, etc. (on the line of the eruption of 1773). Many fine ferns, Hepaticæ, etc.

AMBATO

" 23-31. At Ambato.

June. At Ambato, and excursions thence to Mount Guayrapata and to Tisaleo (and Mount Carguairazo), chiefly in quest of mosses.

July. At Ambato. Excursion to the villages Quisapincha and Pasa, and the neighbouring cool but wooded valleys running down from the northern shoulder of Chimborazo. Many fine mosses gathered amidst fogs and rains.

" 27. Journey to Tacunga (on way to Quito).

" 28. Journey to Romenilo at foot of Mount Rumiñahui.

" 29. Arrive at Quito.

QUITO

" 30-Aug. 23. Exploring the vicinity of Quito and slopes of Pichincha up to 11,000 feet.

" 24-27. These four days collecting at Nono, on the northern declivity of Pichincha.

" 31. Until end of month preparing collections.

Sept. 1-14. At and near Quito, including ascent of Pichincha by the farm of Lloa.

" 15-17. Return journey to Ambato, by Machache and Tacunga.

AMBATO

1858.
Sept. 18-22. Packing Quito collections.
,, 23. Ambato to Riobamba.
,, 28. Riobamba to Cajabamba.
,, 29. Cajabamba to Pangor.
,, 30. To a hill-top in forest of Pallatanga.
Oct. 1. To village of Pallatanga.
,, 12. Collecting around Pallatanga.
,, 13. From Pallatanga to forest below Pangor.
,, 14. To the Hacienda de las Monjas.
,, 15. Across the Paramo de Naba to Riobamba.
,, 16-20. At Riobamba.
,, 21. Return to Ambato by the Paramo de Sanancajas.
,, 22. Ambato.
,, 23. To Patate and back to Ambato.
,, 24. Back to Riobamba by way of the Paramo de Sabañan.

RIOBAMBA

,, 25.
,, 31. Collecting near Riobamba, at Guano, etc.
Nov. 1-8. Collecting near Riobamba.
,, 9-14. At the farm of Titaicun on the slope of a ridge (not rising to snow-line) above the village of Chambo.
,, 15-30. At Riobamba.
Dec. 1-2. Excursion to the Paramo del Puyal, a prolongation of the southern shoulder of Chimborazo.
3-8. Collecting near Riobamba.

AT THE FARM OF TAMAUTE, AT THE CONFLUENCE OF THE RIVERS GUANO AND CHAMBO

,, 9-31. Excursions around Tamaute: On the banks of the Chambo—on the salt-plain of Elen—to Guanando—to the head of the river Guano (at the foot of Chimborazo), etc.

[The following letter to Mr. Bentham, written two months after his arrival at Baños, gives an interesting summary of the whole journey to that place, with so many vivid personal touches that, despite of some repetition, I will here give the latter part of it. It also carries on the narrative to a later period, and gives some account of the little-known village of Baños, and of the surrounding mountains, with their more interesting botanical features.]

To Mr. George Bentham

BAÑOS, ECUADOR, *Sept.* 1, 1857.

The last part of the journey, namely, the overland part of it, was by far the worst.

. . . . . .

Road there is none, but only the merest semblance of a track, such as the tapir makes to its feeding- and drinking-places; often carried along the face of precipices, where had it not been for projecting roots on which to lay hold, the passage would have been impossible. No one ever opens the road—no fallen trees have been cleared away—no overhanging branches cut off. From Canelos the rains set in with greater severity than ever—the dripping forest, through which I had to push my way, soaking my garments so that towards evening my arms and shoulders were quite benumbed—and the mud, which even on the tops of the hills was often over the knees—made our progress very slow and painful.

The Indians were little accustomed to carry burdens—some of them had never been out before—and though I had made the loads as light as possible, they grumbled much and often threatened to leave me. I had brought from Tarapoto a trunk full of paper for drying my plants, but when we reached the Jibaro settlement, where unceasing rains kept us three days, I found it absolutely necessary to throw all the paper away if I did not wish to be deserted by the Indians. . . . At length we reached the cataracts of the Topo, which have to be crossed by throwing over them four bamboo bridges—from one side to rocks in the middle and thence to the opposite side. As far as we could see up and down it, it was one mass of foam, with here and there black rocks standing out, and so much swollen that one of the rocks used as a support for the bridges was completely under water. Here we waited two days and nights in the vain expectation of seeing the waters subside; and finding ourselves on the point to perish of hunger, we with great risk threw bridges across at a place some way higher up. One of the middle bridges was so long (at least 40 feet), that the three slender bamboos of which it consisted almost broke under the weight of a man, even unloaded, and it was found impossible to get my boxes across. I crossed myself and got over my bed and a change of clothes, and the last of my party had scarcely got over when the waters rose and swept the bridge away. In three days more we reached Baños, and my first care was to seek out and pay fresh cargueros to fetch my baggage from the Topo. Eleven days they waited ere the river went down, and twice I had to send them out supplies of provisions. My goods had been left under a rude rancho, thatched with Anthurium leaves (for there were no palms near); but when the men found them the leaves had fallen on them and there rotted; the leather covering of the trunks was half rotten and *full of maggots*; yet fortunately the contents were very slightly injured. You can perhaps fancy my sorrowful position in

Baños awaiting the fate of my goods. After so long a voyage I was much fallen in flesh, and my thin face nearly hidden by a beard of three months' growth. The cold at Baños I found almost insufferable—thermometer sometimes as low as $48\frac{1}{2}°$ at daybreak, and at its maximum not passing 64°—rains still continuing. I was attacked by catarrh, with a cough so violent as often to bring up blood from both nose and mouth. Perhaps I should never see again my books, journals, instruments, my Peruvian mosses, and other things which no money could replace—all perhaps rotting on the shores of the Topo. There was not a book in all Baños, save breviaries and "doctrinas." The weather scarcely allowed me to get out, or I might have put off sad thoughts by the sight of new plants. I had no drying-paper, but I found some coarse calico, and with this began to dry the mosses and ferns I found on the dilapidated walls of my garden. I had also to lay out the mosses I had snatched up as we came along from Canelos, and which by chance had been brought along with my bed, and this occupation diverted my thoughts from my painful situation. The Cryptogamic vegetation of some parts of the Montaña of Canelos is wonderful. There is one mountain, called Abitagua, which though not more perhaps than 5000 feet high, is continually enveloped in mists and rains. The trees on it, even to the topmost leaves, are so thickly encased in mosses that a recognisable specimen of them would be scarcely procurable, if indeed they ever flower, which must be very rarely. I gathered a tuft of everything I saw in fruit and stowed it in a pouch by my side. In the evening I made them into bundles—putting alternate layers of Musci and Hepaticæ, and hung them up to "smoke" through the night, along with my soaked garments. Even gathered in this hasty way, I have a great many fine things; of Hookeria alone there seem to be not fewer than fifteen species. I saw also great numbers of new ferns, but could not take them, save two or three of the minute ones that I had not seen elsewhere.

Having perforce to remain at Baños till my goods were got out from the Topo, and finding it favourably situated for exploring Mount Tunguragua, which, like much ground in the neighbourhood, scarcely any botanist has visited, I determined to make it my station for some months. It is a poor little place, much subject to earthquakes and violent winds, and not abundant in provisions. Bread is brought from Ambato and other places where the climate is more suitable for the growth of wheat. Baños is some 5500 feet above the sea according to my barometer. Its position much reminds me of that of Argélez in the Basses-Pyrénées, though the valley is narrower and the schistose grassy hills that bound it seem much higher than those of Argélez. In the gorge of the Pastasa we have still oranges and the sugar-cane, and on the hills that rise from it barley, beans, and potatoes. The volcanic cone of Tunguragua is perhaps the highest in the world; it is quite isolated from the rest of the Cordillera, and on its eastern side is joined by a narrow col to the wooded hills which subside into the great Amazonian plain; taken from the valley of Baños, its height cannot be much less than 15,000 feet. It has

Fig. 7.—Tunguragua, from the North (towards Ambato).
El Altar just visible on the right.

much more wood on it, at the same altitude, than Chimborazo or Cotopaxi, or any other of the lofty mountains I have seen. Its ascent begins from my very door, but to get up to the snow-line and make any collection there, would occupy at least two or three days. When the rainy season, or, as it is called here, "tiempo de las nevadas," is fairly over, I hope to attack Tunguragua in earnest. The snow has been very low down, even into the forest, but is now beginning to subside. I was at first much hindered in my operations for want of paper—at Ambato I could get only white letter-paper, very dear—but I have now, through Mr. Mocatta's kindness, got a stock of paper from Guayaquil, and you may consider me in constant work, though the rainy weather interferes rather with collecting and drying. Ferns and mosses are in full bloom and in great abundance—flowers still rather scarce. It must be from having been so long among lofty trees that all herbaceous vegetation has a weedy look to me, yet I have felt great pleasure in renewing my acquaintance with several European genera among the humble plants; such are Ranunculus, Geranium, Alchemilla, Hypericum, Cerastium, Stellaria, Silene, Cardamine, Centunculus, Tillæa, Hydrocotyle, Viola, Veronica, Valeriana, Medicago, Cytisus, and several others. Species of these genera grow along with several characteristic Peruvians—Fuchsias, Calceolarias, and most abundant of all a pretty Labiate shrub (Gardoquia sp.) with copious reddish tubiform flowers. The arborescent vegetation, especially towards its upper limit, is what most interests me; but very few trees are in flower as yet, and amongst them only one European genus (Alnus). Amongst the trees hitherto gathered are an Erythrina, a Pithecolobium, three Polygaleæ (Monninæ), a hexandrous Myrtus, a Proteacea (Roupala, sp. n.), a Verbenacea, a Petiveriacea, a Cratægus (or something nearly allied), etc. etc. A curious tree on wooded hills, at 6000 to 9000 feet, most resembles Polemoniaceæ in its characters, but has nothing of the habit of that order. I believe most of the trees will be undescribed. A Rutaceous shrub with long sarmentose pinnate branches, called Shangshi, has the peculiarity that the petals, at first smaller than the sepals, persist and become three times larger, being at the time so much distended by a dark purple fluid—the universal substitute for ink at Baños—as to simulate the valves of a berried capsule. It is so abundant that it must surely have been previously gathered, yet I can find no description of it. Another sarmentose shrub, growing some 15 feet high, is a species of Cremolobus, which seems to me to have as good a claim to be considered a Capparid as a Crucifer. On mossy declivities about the base of Tunguragua are several Ericeæ, Vacciniaceæ, and small-flowered Orchideæ.

A fortnight ago I went to explore a wooded hill called Guayra-páta (*i.e.* Windy Height) about 9000 feet high, a few hours farther up the Pastasa, towards Chimborazo. I slept at a small hamlet called Cotaló, 8000 feet high and terribly cold, because situated on a plateau, exposed directly to the winds that blow up the valley. At Cotaló there is a small lake choked with weeds of the same genera as I might have found in an English lake (Myriophyllum, Lemna, and Callitriche). Guayrapata is almost as mossy as Abitagua, and much more flowery.

. . . .

[In a letter to his friend Mr. Teasdale, written a few days later, there are some details which are additional to those given to Mr. Bentham. After describing the journey to Baños in much the same terms, he proceeds :—]

September 14, 1857.

Baños is a poor little place of about a thousand souls ; and it takes its name from certain hot springs that well out at the foot of a cataract of very cold water, falling from an offshoot of Tunguragua. The patron saint—"Nuestra Señora de las Aguas Santas"—is a very miraculous saint, and "romeros" (*i.e.* pilgrims) come to adore at her shrine from far-away towns. In large troops they come—bathe nine days in the hot wells, assist at nine masses, rosarios, and processions, get drunk every night of the nine—*all* in honour of the virgin—and then, after these "actas de devocion," as they are called, return to their homes rejoicing, having fulfilled some previously-made promise to the saint, and feeling secure of her protection for the future.

Baños is nearly 6000 feet above the sea, and nestles under Tunguragua in the gorge of the Pastasa, where the deep narrow valley widens out a little at the estuary of a small river (the Baccún) which rushes down from the volcano. In the village we have oranges, bananas, and sugar-cane, and on the hills close by barley, beans, and potatoes. Wheat is grown farther up the Andes, and bread is brought to Baños from Ambato, Pillaro, and other towns. It is rather dear, and when I arrived from the forest—half-famished, and my thin face nearly hidden under a beard of three months' growth—I could easily demolish sixpenny-worth a day. Beef and mutton can mostly be had at 2½d. the pound. In fact, good solid eatables are not scarce, and as my troubles had not taken away my appetite, I assure you I have gone deep into them. I have now got up my strength again, and I don't think I was ever so stout in my life as I am at this moment. At first I suffered much from the cold. Think of the thermometer 48½° at sunrise, when even at 70° on the Amazon I (and everybody else) used to shiver with cold; but I am gradually becoming inured to it.

We are still (September 14) in the "tiempo de las nevadas"—the snowy time on the summits of the Eastern Cordillera—and the snow has been very low down, even into the forests on the flanks of Tunguragua.

Earthquakes begin to find a place in my Journal. We had one here on the 7th of August, a little before seven in the morning. I was sipping my chocolate when it came on—at first with a gentle undulatory movement, then with a brisk shaking, and then gradually subsiding, its whole duration being about three-quarters of a minute. I was trying to ascertain its direction and the number of vibrations per second (about four), when a piece of plaster fell from the wall at my back, whereat I snatched up my chocolate and walked to the door, thinking it quite as safe to continue my observations outside. Whilst the shock lasted, the ground, the trees, and the houses oscillated to and fro, in a way to quite upset one's notions of the earth's stability. I cannot walk abroad in any direction without seeing evidence of former earthquakes, far more violent than this one, and some of them of not very ancient date.

A short time after this earthquake, I was talking about it to a neighbour, when he remarked, "It is seven years ago since we had an earthquake that did any damage, and then only a single house was destroyed, and it stood exactly where yours does, which was built on its ruins." This was startling, but I was reassured when I learnt that the house overthrown was built of adobes, and therefore easily thrown down, whereas the new one was of wooden pillars and wattles, the interspace being filled with earth, and both inside and outside plastered and whitewashed; and that the pillars, being of "helechos" (trunks of tree-ferns), were so tough as to sway backwards and forwards without ever breaking. All the other houses in the village had the uprights also of tree-fern.

[On January 16, 1858, Spruce removed to the town of Ambato, situated on the highroad from Guayaquil to Quito, and about midway between the two cities. This town continued to be his head-quarters for two and half years, when he finally quitted the higher Andes.

A series of extracts (made by Spruce himself) from letters to his friend Teasdale carry on the narrative of his more general observations and experiences during the year 1858. In this period he visited Riobamba and Quito, as well as Baños, several times, and made numerous excursions to the mountains and valley around, as noted in his "List of Excursions" given at the beginning of this chapter. The letters to Mr. Bentham and Sir William Hooker treat chiefly of botanical and other matters connected with his scientific work, and these are arranged in order of date so as to form as far as possible a consecutive narrative.]

To Mr. John Teasdale

AMBATO, NEAR QUITO, *March* 13, 1858.

. . . I came hither from Baños two months ago. My labours there were brought to an abrupt termination in consequence of having filled all my paper and the non-arrival of a further supply I had sent for to Guayaquil. Ambato is conveniently situated for communicating both with the coast and the capital; but the low grounds at the western foot of the Andes are now inundated, and the road will not be passable for beasts of burden until July, up to which time Ambato will be my head-quarters. As soon as I shall have dispatched my collections to the coast I shall probably move on to Quito; for, although Ambato is the prettiest town in Ecuador, and the most abundantly supplied with all sorts of provisions, it is a miserable place for a botanist. It stands on a plateau, half-way down the slope of a deep narrow valley, at the bottom of which runs the river Ambato, a considerable stream, coming from Chimborazo. There is a broad green band of gardens, orchards, and plots of lucerne on each side of the river, but outside the valley the eye rests on little else than hills and rolling plains of sand; streaked here and there

Fig. 8.—Ambato, from the surrounding Gardens.
Carguairazo and Chimborazo in distance.

with long lines of Cactus and American Aloe, the fences of the country, against which the wind piles up the sand like snow-drifts. It is true that most of this sandy country produces scanty crops of barley, peas, and lupines, and that, where it is accessible to irrigation, it is rendered even very fertile; but at a distance it often looks quite naked. On account of the sand, and of the violent wind that gets up as the sun approaches mid-heaven, it is only in the early morning one can go out on foot, and then not with much pleasure, for although Ambato has such a coquettish appearance, and has been built entirely anew since the great earthquake of 1797, notions of cleanliness are so lax that it is necessary to proceed with cautious steps and slow to avoid the "quisquilia" that are copiously strewn about and salute the olfactory organs with an odour by no means "sweeter than smell of sweetest fennel" (vide *Paradise Lost*). At early dawn it is difficult to avoid stumbling over the "bodies" squatting down at the street sides, and even in the principal square, like so many toads, and it is not uncommon for a decent-looking woman in that position to look up in your face as you pass her and give you the "Buenos dias, Señor!" with an air of the most unconscious innocence. At 10 o'clock—or sometimes not until noon—the wind gets up from its sleep, and from that time till about sunset blows over these high bleak grounds with the fury of a hurricane, raising up the fine sand, which obscures the landscape as it were with volumes of mist, and penetrates the narrowest chinks in doors and windows. Few people, except the native Indians, stir beyond the precincts of the town on foot; and it is customary for both men and women, when riding on horseback, to protect the face by a gauze veil from the sand, the scorching sun, and the cold piercing wind. After being exposed for some hours to this adverse combination without such protection, the eyes become bloodshot, the skin peels off the face, and the nose becomes red and swollen, in which state it is emphatically styled a "nariz tostada" (toasted nose). I suppose I may have cast the skin of my nose not fewer than ten times since I came to Ambato. From this brief sketch of the climate you will not be surprised to hear that acute catarrhal complaints, influenza, spitting of blood, etc., are frequent; but they are very rarely fatal, and I have not yet seen a single case of pulmonary consumption; and the climate on the whole must be considered conducive to longevity. A countryman of ours, Dr. Jervis, nephew of the first Earl St. Vincent, died two or three years ago at Cuenca, at the age of a hundred and fifteen. As fires are used here only for cooking, the natives have no calefacients beyond food, clothing, and solar heat, and the latter is often considerable, although the thermometer in the shade scarcely ever passes 65°. Very old people are sometimes put into a basket of cotton and set in the sun, with a wide-brimmed hat on their head. Then they remind me of newly-hatched goslings I have seen similarly treated.

. . . .

It is but two days since I returned from Riobamba, about 40 miles away to the south, where I remained about four weeks, on a visit to my countryman Dr. James Taylor, who has been in

South America near upon thirty years. He was formerly medical attendant to ex-President Flores, and lecturer on Anatomy in the University of Quito; but he married several years ago a young Peruvian in Cuenca, the widow of one of Bolivar's generals, and he has since then resided in Cuenca and Riobamba. He has but one child living—a boy of about fifteen. Dr. Taylor is a native of Cumberland, and has had a good education when young; he has still Greek enough to read and enjoy Anacreon; and what is much better, he is a very kind-hearted, honourable man, which can't be said of many Englishmen I have met in South America.

. . . .

I found it a rather fatiguing day's ride to Riobamba. Instead of starting at five in the morning, as we ought to have done, it was ten when we got off, in consequence of a delay in bringing in the horses. The first 17 miles was mostly over loose sand, where the horses sometimes sank to the knees. This brought us to Mocha, a small village, some 1500 feet higher up than Ambato, and with a very chilly climate. The chief industry of its scanty population is the keeping of horses and mules for hire to Quito and Guayaquil. From Mocha there is a steep descent to a stream, and then we begin to reascend towards Chimborazo. The ground becomes firmer, and grassy, and at about two-thirds of the ascent a road branches off to the right, which leads to Guayaquil. It crosses the southern shoulder of Chimborazo, at a height of over 14,000 feet. We keep straight on; and up, up, up, till we come out on an elevated grassy plain (the Paramo de Sanancajas), which stretches along the eastern base of the mountain for about 8 miles, and at a height of 11,000 to 12,000 feet. Here the icy cope of Chimborazo seemed so near that one might have touched it by stretching out the hand—an illusion caused by the transparency of the atmosphere. The temperature was pleasant, for the bright sun tempered the cool breeze, and there was no sand. But as I returned, a few weeks afterwards, I crossed the paramo in a piercingly cold misty rain, and when I reached Mocha I scarcely knew whether I had any hands or feet. If you have been up Teesdale as far as the Weel, you have seen in that chilly treeless solitude something very like the paramos of the Andes. The Weel itself is not unlike the small lagoons scattered about in hollows on Sanancajas. They are often to be seen covered with small wild-ducks that no one cares to disturb. Herds of shaggy wild cattle roam over the paramo, and pick up a scanty subsistence from the sedgy herbage.

You may have read of the paraméro—the deadly-cold wind, charged with frost, that sometimes blows over the paramos, and withers every living thing it meets. A person has told me that when a boy he was once crossing the highest point of Sanancajas, towards Guayaquil, along with his father, when they saw a man sitting by the wayside and apparently grinning at them with all his might. "See," said the boy, "how that man is laughing at us!" "Silence, my son," replied the father, "or say a prayer for the repose of his soul—the man is dead!"

I have had to face a paraméro, but never of this

Fig. 9.—Chimborazo, from the Paramo of Sanancajas.

intense kind. Its approach is indicated by the wind beginning to whistle shrilly in the distance among the dead grass-stalks. When he hears that ominous sound, the horseman takes a pull at his flask, draws his wraps close around him, and his hat down over his eyes; and his horse too seems to nerve himself for the encounter of the withering blast—carries his head low, and throws forward his shaggy mane.[1] It seems to be the first shock of the cold blast that kills. If a man can sustain it unscathed, he generally escapes with his life. Horses are much more rarely frozen to death than men. Indeed, the amount of cold and wet these mountain horses will bear is surprising; but they are to the manner born, and have never known the luxury of sleeping under cover.

The descent from the southern side of the paramo of Sanancajas is along a ravine, worn deep into the black turfy soil and subjacent volcanic alluvium by the rains and melting snows from Chimborazo. One of my two horses carried my trunks, and got along so slowly that night closed over us as we reached San Andres, a village nearly 9 miles from Riobamba. We would fain have remained there for the night, but there had been a bull-fight that day in the plaza, and the houses were so thronged with noisy, drunken men, that we saw

[1] I have been reminded by this sound on the paramos of the Andes of our bleak Yorkshire moors and moor-pastures, where the wintry wind whistles through the "windlestraws"—the dead flower-stalks of Bent-grass and Dog's-tail grass (*Agrostis canina* and *Cynosurus cristatus*). In the Pyrénées, the strings of Eolus's harp are the slender stalks and rigid pungent leaves of *Festuca Eskia*—the "Esquisse" of the shepherds—which grows on bleak mountain sides at great elevations. In the Andes the whistling grasses are chiefly *Festuca Tolucensis* and *Stipa Javava*, whose thread-like leaves and stalks are most apt for the wind to play upon.

no chance of obtaining a supper and a bed, and we had no alternative but to hold on our way to Riobamba. Having crossed the plaza, we entered a dark narrow street, some way down which we heard several men uttering angry shouts. On nearing them, my horse reared straight up against the wall—alarmed at the sight of the dead bull the men were dragging along, and which the gloom had hindered me from seeing. I gave him the lash and he cleared the obstruction at a bound, but his rider narrowly escaped being spilt. Beyond San Andres we had stony descents and ascents; a drizzling rain came on, which made the night more dark, and we had to leave it entirely to the horses to pick out the way. As I returned by the same route, with daylight, I was horrified to see that for a space of nearly two miles we had skirted the edge of a precipice, where a single false step would have hurled us to destruction.

Riobamba has about as many inhabitants as Ambato (8000), but it covers more ground, because the streets are wider; and it is less neatly built. It is of equally modern date, and stands three leagues away to eastward of the ruins of ancient Riobamba (overthrown in 1797)—in the midst of a flat sandy desert, where the winds have full play, and raise up whirls of sand that look at a distance like waterspouts at sea. An open aqueduct from the paramos of Chimborazo, 15 miles away, supplies the town with water, which by the time it reaches Riobamba has got so fouled as to be undrinkable until it has been passed through a filtering-jar (called an "estiladera") that answers its object admirably.

Fig. 10.—Riobamba and the Eastern Cordillera. Sangáy in eruption.

It is a striking sight to look over the great square at Riobamba on a market-day, and see it crowded with Indians and rustics in dresses of the gayest colours; while the shops that surround the square have their glittering and gaudy wares hung outside, or spread out on mats on the wide pavement; and at the back, Chimborazo towers high into the sky—its snow shining in the sun like polished silver—and seems to touch the very houses of the town at its base, although half a day's journey away.

Several snowy peaks, besides Chimborazo, are visible from Riobamba, the chief being El Altar, La Candelaria, Sangáy, and Tunguragua. The last I call *my* mountain, because I explored its flanks for seven months, from Baños. I made a desperate attempt to get in at the south-western side of it, from Riobamba, and devoted several days to it, but paid dearly for my presumption. My aim was to ascend by a magnificent cataract, called Guandisagua, which comes out from under the snowy cope of Tunguragua, and falls at three leaps into the warm valley of Capil, where flourish Seville oranges, alligator-pears, and sugar-cane—a total height of some 8000 feet (15,700–7500). What with alternately wading in the cold snow-water and climbing up cliffs under a hot sun, I had to keep my bed for four days afterwards, with rheumatic pains from head to foot.

. . . I met with agreeable society in Ambato which I had not reckoned on. The Hon. Philo White, American Minister to Ecuador, resides here, with his wife, nine months in the year. They find the climate suits them better than that of

Quito, where they are obliged to be during the three months the Congress is sitting. Mr. White is a man of middle age. In his younger days he has been United States Consul at Hamburg and other ports in the north of Europe, and he has travelled also in England and France. Afterwards he was for some years Navy Agent on the coast of Peru and Chile; so that he is a man with more cosmopolitan sympathies and fewer local prejudices than many of his countrymen. Like many diplomatic gentlemen, he is apt to run into long-winded dissertations, not remarkable for either depth or brilliancy; and, at the same time, he is a very amiable, sound-hearted man. Mrs. White is a very friendly, chatty lady, who gets all her dresses out from New York, in the latest style of fashion, to the admiration and envy of the belles of Ambato. I often step into Mr. White's of an evening, just as I used to do into yours, when in England. We have, however, no chess-playing, and, instead, we rail against the people of the country—after the fashion of foreigners in all countries—and I listen patiently to Mr. White's lectures on political aspects and complications.

To Mr. George Bentham

AMBATO, *March* 16, 1858.

. . . . . .

As I mentioned in my last letter, my labours at Baños were terminated sooner than I wished in consequence of having filled all my paper; and this was the more provoking because just at that time there were more trees in flower than at any other.

Fig. 11.—Chimborazo and Carguairazo, from near Riobamba.

In my last excursion to Tunguragua I was obliged to leave several things because I had no paper to put them in. I have just returned from Riobamba, after a stay there of nearly a month with Dr. Taylor. Like Ambato, it stands in the midst of sandy deserts where hardly any vegetation is visible save the fences of Agave and Cactus and the common weeds that grow in their shelter. I made a desperate attempt to get in at the southern side of Tunguragua, where there is a magnificent waterfall (Huandiságua) which comes down from the very snow at three leaps into the warm valley of Capíl—full 8000 feet; for this purpose I moved to Penipe, about four leagues east of Riobamba, and from thence reached the cataract in an excursion of fourteen hours. But what with alternately wading in the cold snow-water and climbing up cliffs in a burning sun, I was confined to bed for four days afterwards with fever and rheumatic pains from head to foot. The worst was that, so dried up was the forest with the protracted summer, I did not get a single plant in good state. The weather is still dry, and until the rains come there will be no herborising; but I am occupied in arranging and packing my Baños collections, which I hope to dispatch to Guayaquil in June. In May I ought to revisit Baños to procure plants of two fine Orchises I found on Tunguragua.

I should be very glad to return to England, as you recommend me, to distribute my mosses, but I am fearful of again falling into delicate health if I go there. I have, besides, no funds beyond what are in your hands; these would soon be exhausted, and poverty is such a positive crime in England, that to be there without either money or lucrative employment is a contingency not to be reflected on without dread. On the other hand, I already feel myself unequal to the painful mountain ascents, exposed at the same time to a burning sun and a piercingly cold wind. The eastern slopes of the Andes no doubt contain much fine ground, but for want of roads they can scarcely be explored, except by one to whom the pecuniary value of his collections would be no object, and who could go to any amount of expense. I have often wished I could get some consular appointment here, were it only of £150 a year; but I have no powerful friends, without which a familiarity with the country, the inhabitants, and the languages go for little. A person is much wanted to watch over the interests of Europeans on the Upper Amazon, but I can hardly suggest a station for him which would not be liable to some objection, and an itinerating consul is something I have never heard of, though it would really be very useful here. The Brazilians have a vice-consul in Moyobamba. The French have a vice-consul in Santarem and another in the Barra do Rio Negro.

To Sir William Hooker

AMBATO, *March* 24, 1858.

. . . Several friendly letters have passed between Dr. Jameson and myself, but I have not yet had the pleasure of meeting him. The upper part of the Rio Napo (where is the Indian village of Archidona), which Jameson has lately explored, is nearly parallel to the upper part of the Pastasa (and at no great distance from it), which has been my hunting-ground for the last seven months. Though so near, it would seem that a great proportion of the plants are distinct. . . .

At the foot of Tunguragua, in dripping situations, I found a small Polypodium creeping on branches which has the fronds deeply sinuated so as to resemble a narrow oak-leaf. All the fronds are fertile, and I take it to be nothing more than a variety of a Marginaria with linear lanceolate fronds that grows near.

A short time ago I found in a strip of forest by the Pastasa, about 10 miles below Baños, a very strange little fern with compound fleshy fronds, looking not unlike one of the small Asplenia, but completely different from that genus and its allies. The sori, immersed in the margin of the frond, recall those of some Davalliæ, in which genus, however, the structure of the receptacles seems essentially distinct. I enclose a small specimen, and if the fern be really new and you would like to describe it, I will send you the largest plant I have, which is about three times the size of this one. Unfortunately, I could find the fern on only a single tree, though I spent two hours in searching the neighbouring trees, and my stock of it is rather small.[1]

From my letters to Mr. Bentham you will have learnt how much I suffered in the Montaña of Canelos, on my way hither. This name is popularly given to the forest from near Baños, where the natural pastures begin, at the actual foot of the Cordillera, to Canelos on the Bombonasa. It is the finest ground for Cryptogamia I have ever seen, but when I passed through it with Indians I was obliged to lighten my cargoes by giving and throwing away whatever I could best spare, so that I could bring no plants along save a few mosses. . . . One striking feature among the ferns was the number of sarmentose, or even actually climbing, species of various genera. On the Bombonasa a true Selaginella climbs into the trees to the height of 30 feet, and the twining caudex sends off fronds 4 feet long; in some places it forms impenetrable thickets. A handsome Marattia was a great acquisition to me, as I had not previously seen that genus growing. Two small Asplenia, looking quite like Hymenophylla, crept along the branches of shrubs by shady rivulets. But the most remarkable plant in the forest of Canelos is a gigantic Equisetum, 20 feet high, and the stem nearly as thick as the wrist! . . . It extends for a distance of a mile on a plain bordering the Pastasa, but elevated some 200 feet above it, where at every few steps one

[1] It is *Davallia Lindeni*, Hook. Sp., Fil. I, p. 193, and has been found at Caraccas and also in the Organ Mountains of Brazil.

sinks over the knees in black, white, and red mud. A wood of young larches may give you an idea of its appearance. I have never seen anything which so much astonished me. I could almost fancy myself in some primeval forest of Calamites, and if some gigantic Saurian had suddenly appeared, crushing its way among the succulent stems, my surprise could hardly have been increased. I could find no fruit, so that whether it be terminal, as in *E. giganteum*, or radical, as in *E. fluviatile*, is still doubtful, and for this reason I took no specimens at the time, though I shall make a point of gathering it in any state.

Mount Tunguragua is nearly as fine a locality for ferns as the forest of Canelos, but great difficulties attend its ascent. First, there is the actual height, for Baños is but 5500 feet high, and from thence to the snow-line (15,000 feet) is a great way to climb. Then there is the want of water, for between Baños and Puela, that is, for about five leagues along the northern base of the mountain, all the ravines are dry. The streams that formerly traversed them all became submerged when the great earthquake of 1797 took place, and now run in subterranean courses, coming out on the actual margin of the Pastasa, sometimes in considerable volume. But the greatest obstruction to the ascent is the dense, untracked, quasi-Amazonian forest, to penetrate which the knife is needed at every step, and which extends to a height that I have not yet exactly ascertained. I could not have believed, unless I had seen it, that at 11,000 feet elevation on Tunguragua there are laurels 70 feet high and 12 feet round.

I trust my collections will not disappoint your expectations; they do not, however, quite come up to mine, for I have suffered much here from the cold, and especially from the sudden alternations of burning heat to frosty cold, and I have consequently been unable to do so much field-work as I could wish. Since entering the Ecuador I have gathered forty-five species of Polypodium (including Goniophlebium, etc.), all, with two or three exceptions, different from what I gathered in Peru. They include some very pretty things, especially in Polypodium proper. I have also some pretty Asplenia; but the species of this genus and Diplazium give me more difficulty than any other—to know what are species and what varieties.

. . . .

[The next letter to Mr. Bentham contains, among other valuable botanical matter, an exceedingly interesting estimate of the probable number of species of plants now living in the great Amazon valley, founded on his own observations. It is far beyond what other botanists have supposed to be likely, but no one had ever before given the same close attention to the species of forest-trees over so large an area as Spruce had done.]

To Mr. George Bentham

AMBATO, *June* 22, 1858.

I have just completed packing up three cases of plants to be dispatched to you. . . .

There are a few specimens of a Balanophorea which I have included in the general collection, and a single specimen (being all I could find) of a plant allied to Rafflesiaceæ, which please give to Dr. Hooker. The latter grew on the root of a tree in the forest on Mount Tunguragua; when fresh the involucre was dull purple and the florets violet—it has shrunk about half in drying. I only *guess* at its affinities, for I did not wish to injure the specimen by examining it.

The Phanerogamic collection is not so interesting as I could wish. As I mentioned in a previous letter, I was prevented from gathering many interesting trees about Baños by having filled all my paper. I have lately revisited Baños and spent a month there, but the weather was very gloomy and rainy, and there were scarcely any flowers. In consequence of this I found it impracticable to procure plants of the fine Orchids I have found on Tunguragua. Nor did I find a single moss that I had not gathered during my previous residence there—so eagerly, it seems, I had searched for them—though I got twenty-one ferns and a few Hepaticæ which had previously escaped me.

. . . . . .

What a fine chance there is now for your friend Dr. Caapanema, or for any other wealthy and scientific Brazilian not afraid of heat, rains, and mosquitoes, to explore the Amazon and its tributaries in a small steamer, where everything necessary could be carried, and their collections preserved and stowed away!

I have lately been calculating the number of

species that yet remain to be discovered in the great Amazonian forest, from the cataracts of the Orinoco to the mountains of Matto Grosso; taking the fact that by moving away a degree of latitude or longitude I found about half the plants different as a basis, and considering what very narrow strips have up to this day been actually explored, and *that* often very inadequately, by Humboldt, Martius, myself, and others, there should still remain some 50,000 or even 80,000 species undiscovered. To any one but me and yourself this estimation will appear most extravagant, for even Martius (if I recollect rightly) emits an opinion that the forests of the Amazon contain but few species. But allowing even a greater repetition of species than I have ever encountered, there cannot remain less than at least half of the above number of species yet to be discovered.

At the highest point I reached on the Uaupés, the Jaguaraté caxoeira, I spent about a fortnight, in the midst of heavy rains, when (according to my constant experience) very few forest trees open their flowers. But when the time came for my return to Panuré (for I had to give up the boat and Indians by a certain day) the weather cleared up, and as we shot down among the rocks which there obstruct the course of the river, on a sunny morning, I well recollect how the banks of the river had become clad with flowers, as it were by some sudden magic, and how I said to myself, as I scanned the lofty trees with wistful and disappointed eyes, "There goes a new Dipteryx—there goes a new Qualea—there goes a new 'the Lord knows what'!" until I could no longer bear the sight, and covering up my face with my hands, I resigned myself to the sorrowful reflection that I must leave all these fine things "to waste their sweetness on the desert air." From that point upwards one may safely assume that nearly everything was new, and I have no doubt that the tract of country lying eastward from Pasto and Popayan, where are the head-waters of the Japurá, Uaupés, and Guaviare—probably nearly conterminous—offers as rich a field for a botanist as any in South America. But I have made inquiries as to the possibility of reaching it, and I find that it will be necessary to cross paramos of the most rugged and inhospitable character, and afterwards risk oneself among wild and fierce Indians, so that I fear its exploration must be left to some one younger and more vigorous than myself.

If I remain in this country and do not make Quito my head-quarters, I suppose I must go to Loja, where the climate is more temperate and the flora no doubt magnificent. People who travel that way all speak with admiration of the abundance and beauty of the flowers.[1]

. . . . . .

To Sir William Hooker

QUITO, *Aug.* 15, 1858.

. . . . . .

The house in which I reside is on the very slope of Pichincha, and is actually the last house in Quito,

[1] The route along the Cordillera to Loja is now little traversed, and is very difficult and expensive. The so-called "road" has no mending (or marring) save what it gets from the rains and the hoofs of the mules. There is, in fact, not a single road in the Ecuador.

ascending by the stream which runs through the city. Jameson's house is about 150 yards lower down, and poor Hall lived on the opposite side of the stream. Dr. Jameson is, however, at the present moment in Guayaquil. . . . I had the pleasure of spending a day with him in Ambato, on his way down. He is a tall ruddy Scot, and although on the shady side of sixty years, may very well reach a hundred and fifty, for he shows no signs of age yet. People who are naturally robust and live temperately *do* reach very advanced age in these mountains. Our countryman Dr. Jervis died lately at Cuenca, aged a hundred and fifteen years; and here is Mr. Cope, turned of eighty-five, trotting about as nimbly as a young man. . . .

The weather is extraordinarily dry just now, for Quito, and vegetation is much burnt up. Before I put myself in the doctor's hands I contrived to scramble some way up Pichincha, and to gather a few mosses; although I had already gathered the greater part of what it produces in other parts of the Cordillera. In my garden I have *Brachymenium Jamesoni*, Tayl., *Tortula denticulata*, Mitt., and some other mosses, and there are many more pretty things by the stream close at hand. When I came out on the Cordillera last year, one of the first mosses I recognised was the curious Orthotricheid moss, *Streptopogon erythrodontus*, Wils., which grows perched on twigs in bushy places, just as *Orthotrichum affini* and *striatum* do in England. Another of my first findings was your *Didymodon gracilis*, Bridel. *Grimmia longirostris*, Hook., I have gathered abundantly on Chimborazo, the original locality. I have little difficulty in recognising your and Humboldt's mosses, but many of Taylor's I cannot satisfactorily identify. Besides the incomplete analysis, there is a laxness in the use of terms relating to form in his descriptions which makes me almost in every case feel uncertain whether I have got his plant or not. I have three claimants for his *Nickera gracillima* (Pichincha), and as I find that only one of them grows on Pichincha, I have no doubt of its being the species intended, though it is the one least like his description of the three.

I am glad that Mr. Mitten is working up the Indian mosses, as I hope we shall thus be able to ascertain whether it be really the fact that, while so many South American Hepaticæ are identical with Indian species, scarcely any mosses are. I feel sure I have many European species among my Andine collections. I was surprised to see on Chimborazo dense tufts of *Hypnum Schreberi* growing among the heather just as they do in England.

. . . .

To Mr. George Bentham

QUITO, *Aug.* 17, 1858.

I have lately had the pleasure of receiving your letter of June 1, containing the names of my Venezuelan plants. My notes on these are in Ambato; still, I recognise the greater part of them. . . . Nearly every plant I gathered at the highest point of the Guainia which I explored proves to be new; and this increases my regret that I could not, on account of illness, follow out my original project of ascending that river as far as the cataracts. . . . I have a new genus allied to Henriquezia in the forest of Canelos, but when I shall be able to go and gather it I cannot tell. It is an immense tree, with leaves three together, and with large yellow flowers 6 inches long—five equal stamens—but a much longer corolla-tube than Henriquezia.

I am satisfied to find that the small collection from Maypures has arrived in an identifiable state. When I opened it out at San Fernando, after my long illness, both plants and paper were one mass of mould. By little and little, as my trembling hands and dim eyes allowed me, I brushed away the mould and transferred the plants to other paper. When I reached San Carlos the process had to be renewed, so that I had reasonable doubts of their preserving any of their original semblance when they reached England.

Whatever steps you think necessary to take for lessening your labour in the distribution of my plants will meet my cheerful acquiescence. . . . I am deeply thankful to you for having bestowed so much of your valuable time on my plants; but, on my part, I can truly say that I have had no greater stimulus in collecting than to think, whenever I have gathered a new or strange plant, "This will surely please Mr. Bentham."

I wrote to you in June last, on the occasion of dispatching to you three cases of plants from Ambato. . . .

I am sorry the collection does not contain more trees; but the number of species of trees is actually much fewer in cold regions than in warm, and I miss much here the excellent climbers (Indians, I mean, not lianas) I used to have on the Uaupés and Rio Negro. However, if I remain in the country, not many of the trees shall escape me. There are a great many arborescent Compositæ, of which I have as yet taken very few—do you think I ought to gather them all?

My object in visiting Quito was partly to get my few books bound and a few clothes made, for in so many years in the forest all I had with me had got into a very dilapidated state; as also to get such substitutes as I could for the warm under-clothing sent out by yourself and Mr. Pritchett and lost on the way. I hope also to herborise a little on the western side of the Cordillera, but I have been seriously ill, and am still in so much pain that I cannot write for more than a few minutes together.

. . . . .

RIOBAMBA, *Nov.* 2, 1858.

. . . Since I last wrote to you I have several times done the 40 miles from Ambato to Riobamba in one day, and the distance begins to seem much less than at first. But my back is just now aching considerably from having ridden 112 miles in three days, for the most part along steep and dangerous declivities. I left Quito in September and came straight on to Riobamba, and then 60 miles farther, in a south-westerly direction, crossing the summit of the Cordillera at an elevation of 12,500 feet, and then descending to the valley of Pallatanga at 5000 to 6000 feet. This pass, called the Paramo de Naba, is far lower than that over the shoulder of Chimborazo (14,000 feet) on the way from Quito and Ambato to Guayaquil. I scarcely suffered from the cold on Naba, although I was buffeted by a hail-storm; and I gathered there some very interesting plants, including the beautiful *Gentiana cernua* found by Humboldt and Bonpland on Chimborazo. It is great pity that these fine Andine Gentians have proved so difficult to cultivate in England. Anderson, the famous nurseryman of Edinburgh, has succeeded in raising a great many plants of the Andes, from seeds sent to him by Professor Jameson of Quito, but I am told that none of the Gentians have survived. It is difficult to imitate the conditions of their growth; for some of them endure frost nearly every night of their lives, yet so light is the pressure of air upon them that the frost injures them not; yet they die when frozen in the dense atmosphere of the plains. I have seen epiphytal Orchids—Oncidiums, Odontoglossums, etc.—growing in the Andes at 10,000 feet, where they must frequently endure frost; yet these are precisely the kinds that have been found (hitherto) most difficult to cultivate in England.

The greatest height to which I have yet climbed was 13,000 feet, on the volcano Pichincha, near Quito. It is practicable to ride up to the very edge of the crater (15,000 feet), and it was my intention to do so, but my guide mistook the way, and we got entangled in thickets at about 11,000 feet, where we had to dismount and cut a way for the horses to pass, and finally to leave them tied to bushes and continue the ascent on foot. I had only lately emerged from the sick-room, and got very much fatigued with two hours of steep, rugged climbing. At the highest point we reached, we lay down to rest on the grass, and I had lain a few minutes with my eyes closed when I suddenly felt as it were a flag waved over my face, and looking up saw an immense condor sailing over us at only a few feet distance. My companion sprang to his feet with a shriek, and prepared to defend himself with his staff. "He thinks we are dead," said he, "and if we had lain a moment longer we should have felt his beak and claws in our faces!" The condor was immediately joined by two others of his species, but being baulked of their prey, they rose in slowly widening circles, and at length appeared only specks on the bright heaven. This incident was additional confirmation to me that the vulture tribe hunt by sight and not by scent. The condor is a magnificent bird, but yet looks very much like a turkey-buzzard on a large scale, and has not the noble aspect of the golden eagle and the royal eagle of the Amazon.

Fig. 12.—Quito, on south-east Slope of Pichincha.

. . . Here, on the eastern side of the Cordillera, summer fairly began last month (October), but its continuity has been interrupted for some days by a succession of terrific thunderstorms, one of which has caused a break of two hours in the writing of this letter. Three days ago, two women were struck dead by lightning while gathering sticks on the plain outside the town; and yesterday six people were killed and a wheat-stack burnt down at a village a little south from us.

. . . Matters are in a very unsettled state here, and preparations for war with Peru resound on every hand. Recruiting—forced contributions of money and horses—people hiding in the forests and mountains to avoid being torn from their families—scarcity and dearness of provisions—such are some of the precursors of the contest. And the war—if it actually comes—will be something like what you have read of in India; yet nobody knows what it is to be about! These Spanish Republics are not unlikely to squabble among one another until —like the Kilkenny cats—there is nothing left of them but their *tails*; and then Jonathan will step in and make an easy prey of their mangled *carcasses* (Hibernicè loquitur).

. . . How often have I regretted that England did not possess the magnificent Amazon valley instead of India! If that booby James, instead of putting Raleigh in prison and finally cutting off his head, had persevered in supplying him with ships, money, and men until he had formed a permanent establishment on one of the great American rivers, I have no doubt but that the whole American continent would have been at this moment in the hands of the English race. It should be noted that this consummation has also been hindered by our unbroken alliance with the most beggarly nation in Europe (Portugal)—the nation which most hates the English, because they have most interfered with her staple trade—the traffic in human flesh!

[Among a quantity of loose notes, headed "Quitonian Andes," the following, on the "Bridge of Baños," seems worth quoting:—]

"The Pastasa runs in a tortuous course, about 40 feet broad, between perpendicular walls 150 feet high, sometimes much excavated at the base, the water foaming against blocks and down cascades into deep caverns, whence it issues in a savage whirl. Across this chasm the frail bridge is thrown, and is higher at its northern side. The adjacent rocky ground seems as if it had been shaken into irregular rather small fragments, not separated but as if the original mass of rock had been crushed without much displacement. The ground rises abruptly to a great height on the left, but lower on the right; and a col stretches on one side towards the other, looking as if it might formerly have been the barrier of a lake.

The view down the Pastasa as one descends the hill towards the bridge is savagely sublime. A dense grey curtain of Tillandsia—sometimes 30 feet deep—hangs from the cliffs and adjacent trees, contrasting with the black trachytic rock over which it hangs."

[The bridge here referred to was probably of similar construction to that at Agoyan (described at p. 163), which was passed on the route from

Canelos, and which consisted of a few trunks of trees covered with bushes and earth.

In the following "note" of a visit to Penipe—a hill-village near Riobamba—a much more perilous kind of bridge, common in the Andes, is described.]

"The distance from Riobamba is about 4 leagues. The road leads a little to the south of Guano; at near half-way it passes some low flats inundated in winter, or interspersed with small lagoons, now (February) mostly left dry, and covered with a whitish saline deposit. In places where moisture is preserved there are beds of tall Cyperacea (*Scirpus validus*), of which mats are plaited. After passing this the road ascends gradually to a considerable elevation (about 1500 feet above Riobamba), whence there is a splendid view of the western side of Tunguragua, which is its most striking aspect. The top of the ridge reached, there is a long descent to the river Pastasa, with a narrow plateau about midway, along which the road runs for some distance parallel to the river. At last there is a steep winding descent to the hanging bridge of Penipe, which is formed by cables made of roots of Agave, 4 inches thick, stretched as in an ordinary suspension bridge; and the roadway consists of sticks tied across the cables. These sticks should be flattened and touching each other, but many of them are left in their original rotundity, and they are sometimes wide enough apart for a foot to slip between. The bridge sways to and fro when the wind is high, and oscillates fearfully as one passes over it. It has also become lower on one side, and several sticks are slipping away on that side. A rope is stretched on each side at some height above the bridge; its original intention seems to have been for a hand-rail, but the bridge has so sunk away from them that they can only be reached here and there even by a very tall man."

[These are the only references I can find to the dangerous bridges in the Andes, which Spruce must have so frequently had to cross in his numerous excursions.]

MAP SHEWING
SPRUCE'S PRINCIPAL ROUTES IN THE ANDES
By RED LINES

N.B.—The routes in the southern half of this map are better shown on the Map of the "Red Bark Region." The discrepancy in the position of some places on these two maps is due to the fact that the latter was made nearly fifty years ago, and the country is still very imperfectly surveyed.

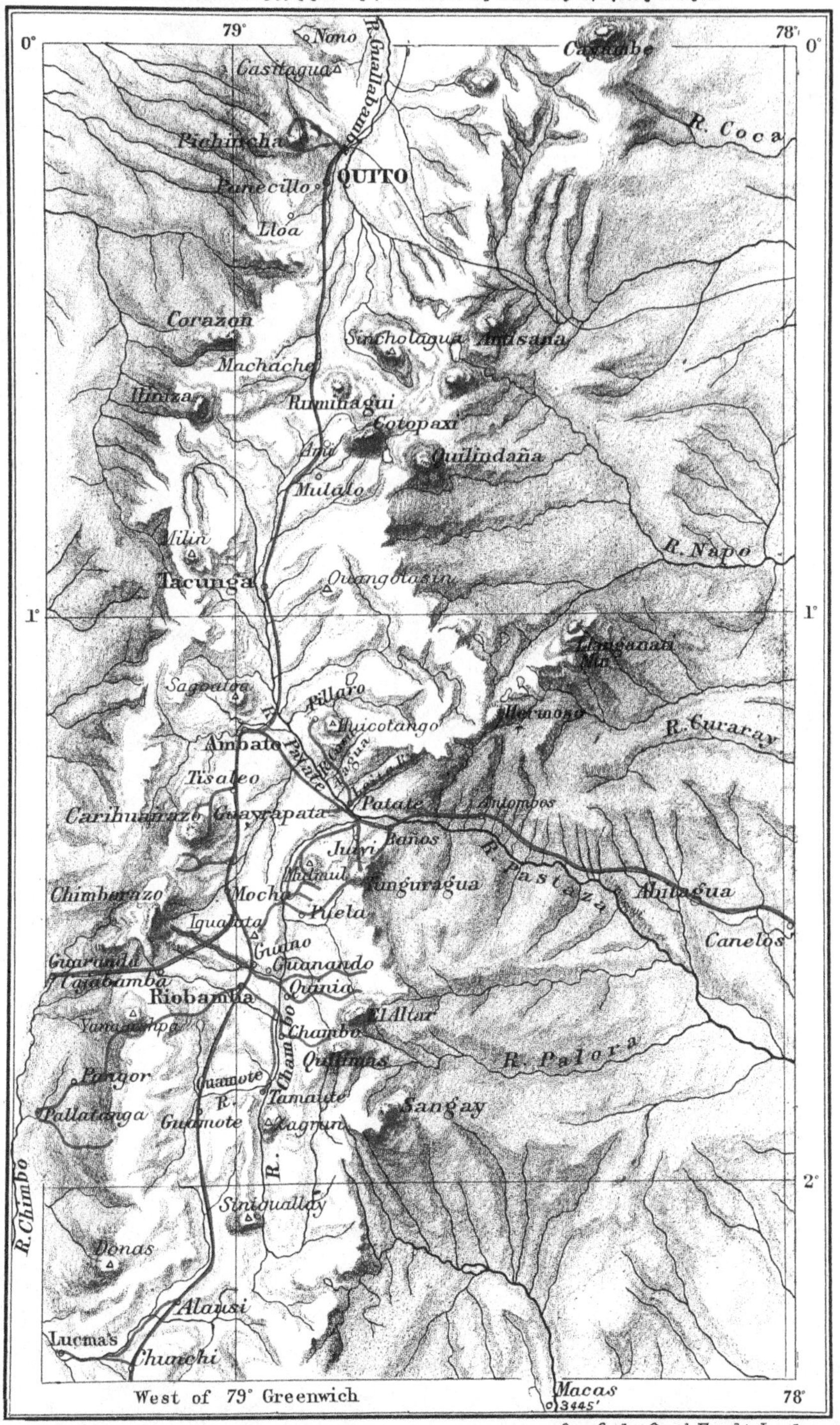

CHAPTER XX

AMBATO AND THE CINCHONA FORESTS OF ALAUSÍ, ON THE WESTERN SLOPES OF MOUNT AZUÁY

(*January to December* 1859)

LIST OF BOTANICAL EXCURSIONS

At Farm of Tamaute

1859.		
Jan.	1.	Descended to Baños.

At Village of Puela

"	2.	Ascended from Baños to Puela, a village at the north-western foot of Tunguragua (height 7000 feet) and near the confluence of the rivers Puela and Chambo, and remained there till the 16th, making in that time three ascents of Tunguragua by the ravine Mapa-yacu.
"	16.	From Puela to Tamaute. Collecting at Tamaute until end of month, including an ascent of Mount Condorasco (a shoulder of El Altar) by way of Quinia.
Feb.	2.	From Tamaute to Riobamba.
"	3.	To Ambato.

Ambato

"	4-7.	At Ambato.
"	8.	To Baños.
"	9-11.	Collecting near Baños.
"	12.	To Ambato.
"	28.	Until end of month examining and packing collections.

1859.	
March–April.	Throughout these months examining and packing ferns and mosses, and making short excursions near Ambato.
May.	At Ambato; and excursion of four days to the valley of Leito.
June.	At Ambato: the first half of the month confined to the house by fever.
July 1-21.	At Ambato: excursion to Guayrapáta, etc.

EXPEDITION TO VALLEY OF ALAUSÍ

,, 22.	From Ambato to Riobamba.
,, 26.	From Riobamba to farm of Miraflores near Guamóte.
,, 27.	To Alausí.
,, 28.	Down valley to Chunchi.
,, 30.	To farm of Guataxí, where I established myself.
Aug. 2.	Down valley to Lucmas.
,, 4.	Down valley to Puma-cocha, where the Red Bark grows (4000 feet).
,, 5.	At Puma-cocha.
,, 6.	At Lucmas.
,, 7.	Back to Guataxí.
,, 8-22.	Collecting on the hills and by the streams of Guataxí.
,, 23-26.	In the forest of Llalla, on the western slope of Mount Azuáy. Altitude 9000 feet.
,, 27-31.	Returned to Guataxí, and remained there the rest of the month.
Sept. 2-6.	At Gusunáy, on opposite side of river Chanchán, and higher up the valley.
,, 7-25.	At Guataxí and neighbourhood.
,, 26.	From Guataxí by way of Alausí to Ticsán.
,, 27.	To farm of Miraflores.
,, 28.	To Riobamba.
Oct. 8.	From last date to this at Riobamba.
,, 9.	To Mocha.
,, 10.	To Ambato.

AMBATO

,, 11-30.	At Ambato: packing collections from Guataxí, etc.
,, 31.	To farm of Juivis near Baños, and remained there until November 12, making in that time an ascent of Tunguragua by the Alisal.
Nov. 12.	An excursion down to Agoyan.
13-30.	At Ambato: examining mosses, etc. etc.
Dec. 1-31.	All this month at Ambato, occupied as in November.

[The portions of letters selected to form this

chapter comprise such episodes as a war, an earthquake, and an insurrection; but the most important portion of it consists of a very detailed account of a two and a half months' excursion to the Bark forests of Alausí in the western slopes of the Andes, in a letter to Sir William Hooker. This was printed in the Journal of the Linnean Society, but as it is full of interesting matter I include it here, only omitting such passages as refer to his future proceedings in another district, the full account of which will occupy the next chapter.]

To Mr. George Bentham

AMBATO, *March* 3, 1859.

. . . . . .

We are still at war with Peru, and the blockade of Guayaquil continues, the Pacific steamers being allowed to land only the mails and passengers. The indiscriminate pressing of men and horses for the Ecuadorean army, and the scarcity and dearness of the necessaries of life (potatoes, for example, have been at ten times the price they bore when Seemann visited this country), have much impeded and restricted the field of my operations. In the beginning of summer (end of July 1858) I went to Quito, and my first intention was to visit some unexplored localities in that neighbourhood, and thus occupy myself until the next rainy season; but I suffered so much in that rarefied atmosphere that I soon sought a more genial clime, and as I hoped an excellent field of operations, in the forest of Pallatanga, which is near half-way from Riobamba towards the narrow plain bordering the Pacific, and at a height of 5000 to 7000 feet. You will have learnt from a letter I wrote to Mr. Saunders that I found the vegetation there (in October 1858) consisted of so very few species that I judged it expedient to return to the Sierra, where I found the people in great alarm—more at the devastating progress of their own armies than at the threatened invasion of the Peruvians—and ready to desert the towns should hostilities actually commence. So the risk of losing all my goods has kept me from leaving them far, during the remainder of the dry season. But for this, I should, after leaving Pallatanga, have plunged into some other forest, for I find that the woody slopes on both sides of the Andes must be in future my principal field for collecting, the really Alpine plants having been already gathered to a great extent, and having most of them a very wide range.

I am now packing my flowers and ferns, which (especially the former) comprise many interesting things, gathered under disadvantageous circumstances. The difficulties of travelling anywhere out of the central plain of the Ecuadorean Andes is immense. Roads there are none—what go by that name are deep slippery gullies and narrow ledges along steep declivities, where far more lives are annually lost than in navigating the rivers of the plain. . . .

To Mr. George Bentham

AMBATO, *April* 13, 1859.

. . . The collection now sent is not of the class I could have liked, but the unsettled state of the country has prevented me leaving the higher grounds. The facilities of getting about and of procuring provisions have also limited the explorations of all previous travellers almost entirely to the central plain—"callejon" (lane) they call it here—of the Quitonian Andes, and to the adjacent snowy summits; but I am certain that the forests on the eastern and western slopes are still almost entirely unexplored, from a height of 3000 to 8000 feet. I see scarcely any real trees described among Hartweg's plants. These forests contain also the finest ferns. That they are still almost intact is not to be wondered at, when their exploration involves the risk of life, health, and everything; especially those on the eastern side. I hope by little and little to go over them and send you their gleanings.

. . . . . .

To Mr. John Teasdale

AMBATO, *April* 14, 1859.

. . . The introduction of the Christian religion among the South American Indians I have visited has been, for the most part, a decided injury to them. Formerly they had either no religion at all or they were nearly pure theists; now they are decided idolaters, as many Catholic priests have candidly admitted to me. Among the vices they have contracted in their "civilised" state, not the least frightful is the readiness to sell or hire their wives and daughters to the lustful white man. At from 50 to 100 miles from where I am writing, on the eastern slope of the Andes, there are still powerful independent tribes who refuse to receive the missionary, but who would kill any of their women on whom the white man should merely look to lust after her. . . . The term "savages," so glibly bestowed by writers on the Indian races, would be more correctly applied to those Christian nations who play at the game of war, and who, instead of deciding their differences on the principle of "doing to others as you would they should do to you," kill, burn, and waste as many and as much as they can.

. . . Yet the introduction of a pure and simple Christianity might much benefit the Indian; and we must not too harshly judge him for transgressions against our own moral code. The Indian's notion of "crime," for example, is not the same as ours. He feels the disgrace of being found out in a lie or a theft, but if he escapes detection he exults in his adroitness. He is naturally apathetic and dislikes exertion; but he makes his wife work like a slave. On the Rio Negro I have seen the poor women grating mandiocca by moonlight until midnight; and they must be stirring before daybreak to give their husband his morning drink; while he, extended in his hammock, is warming his nether extremities near a fire which must not be allowed to go out. When I had seen this, I felt no pity for the Indian when the white man took him by force to row his boats and do other work for him.

. . . On March 22 of this year a fearful earthquake shook the whole of the Quitonian Andes. The damage done in Quito itself is estimated at four millions of dollars, and some adjacent villages are quite destroyed; but as the shock came by day, only a few people were killed who were in the churches—the buildings that suffered most. On that day Dr. Taylor of Riobamba and his son were my guests, and (along with my lad) we were riding down the valley to eat peaches at a neighbouring farm. Singularly enough, neither we nor our horses felt the shock, although it was a very long one; but all on a sudden we saw people running out of their houses, and clouds of dust rising up among the hills. A little way farther on several tons of earth had been shaken down across our path, and we passed the débris with difficulty, and not without risk that more might fall and crush us. Below the farm, the cliff bounding the valley had slid down for a length of 200 yards, and the people of the farm had been half choked and blinded with the dust raised by the fall. In the town of Ambato itself no damage was done beyond the cracking of a few very old and of some very new walls. On the following day, about 2 P.M., I was startled by hearing the family of my neighbour (and landlord) run shrieking into the yard, crying out "Temblor! temblor!" I ran out myself just in time to see the walls of an unfinished house, which an ambitious shopkeeper had been rearing close by to the imprudent height of three stories, crumble to the ground. The adobes had not got "set," and the earthquake had cracked several of them; hence the downfall of the whole. Fortunately, nobody was injured by the fall.

I have been entrusted by the India Government with the charge of obtaining seeds and young plants of the different sorts of Cinchona (Peruvian Bark) found in the Quitonian Andes for transporting to our Eastern possessions, where it is proposed to form plantations of these precious trees on a large scale. This task will occupy me (if my life be spared) the greater part of next year.

The expedition to the woods above spoken of (in August and September of 1859) was to make myself acquainted with the different sorts of Barks, and to ascertain what facilities existed for procuring their seeds, etc., or, more properly speaking, what difficulties had to be overcome, and I assure you they are not slight ones. I established myself in a sugar hacienda about half-way between Riobamba and Cuenca, and five days' journey from Ambato, and from thence penetrated two days' journey farther into the forest towards the west, or nearly to the roots of the Andes on the Pacific side. The owner of the hacienda, Don Pepé Leon, a descendant of a noble Spanish family, and his wife, Señora Manuelita (a handsome and very clever Ambatiña—most of the handsome women are of Ambato), were very agreeable people, and I spent a pleasant time with them.

Notes of a Visit to the Cinchona Forests in the Valley of Alausí, on the Western Slope of the Quitonian Andes

To Sir William Hooker

Ambato, *Oct.* 20, 1859.

My last letter informed you that I was contemplating an expedition to the forests producing the Cinchona tree on the western slopes of the Quitonian Andes. I was for some time doubtful as to what part I should visit. It was but two or three days' journey to the forests of Jilimbi and Guanujo at the western foot of Chimborazo, but to reach them the Paramo de Puenevata (the northern shoulder of Chimborazo) has to be passed near the snow-limit, and in the months of July and August it snows there almost incessantly, while the winds blow with a violence unparalleled even in this windy region, frequently hurling away both horse and rider, who are either seen no more or their mangled remains are found at the foot of some precipice. Besides, only one sort of Cinchona was known to exist in those forests, whereas by going a few days' journey farther to the southward, to the forests below Alausí, in the valley of the river Chanchán, I might expect to find three sorts, and the road thither nowhere ascends above 12,000 feet. So the latter plan was finally adopted, and on the 22nd of July I sallied forth from the pleasant town of Ambato (8500 feet) along the narrow "callejon" (lane) which separates the eastern from the western branch of the Cordillera. My company comprised five horses and mules, one mounted by myself, another by my servant, and the remaining three laden with my baggage, consisting of drying-paper, clothing and bedding, and a copious supply of tea, coffee, and sugar—articles rarely to be met with in a country where there are no inns, and where the inhabitants with few exceptions use no other beverage than aguardiénte and sour chicha. An arriero took charge of the beasts of burden.

Our first day's stage to Riobamba was a long one, 12½ Columbian leagues (about 40 English miles). The first five leagues, reaching to the village of Mocha, are along a very gradual ascent, varied by a few shallow quebradas (ravines). The soil is what in Yorkshire we used to call "a leight blaw-away sand," which, when the sun and wind are up, scorches and blinds the traveller, though it produces scanty crops of maize, barley, peas, and lupines (eaten here under the name of "chocchos"). The indigenous vegetation is limited to a few insignificant weeds, chiefly Composites, nestling under the hedges of Yucca and Agave. The flowers of the two latter plants—so great a rarity in England—are here to be seen all the year round, and their tall tree-like peduncles are the poles used throughout the Cordillera for all common purposes, such as fences, rafters, and even walls of houses, etc. Long files of asses laden with them enter the towns of Ambato and Riobamba every market-day.

Beyond Mocha we leave the sandy country, and after passing two streams which descend from Mount Carguairazo on our right, we begin to ascend to the Paramo de Sanancajas, the grassy meseta (plateau) which extends along the eastern base of Chimborazo, at a height of from 11,000 to 12,000 feet. Near its commencement the road leading from Quito to Guayaquil branches off to the right, while that to Riobamba and Cuenca continues straight on. The weather had been rainy for many previous days, and we had had drizzling rain all the way to Mocha, so that we were not without apprehension of suffering from the cold on the paramo. Fortunately, just as we reached it, the sun shone forth, the clouds cleared away, and the glaciers of Chimborazo stood out against the blue sky like cut marble; but the ground was still so sloppy that what I had formerly passed over in two hours now took me three. What is called the "road" consists of I know not how many deep ruts, crossing and anastomosing in a very bewildering way, and so muddy and slippery that my horse preferred stumbling along among the hassocks of paja blanca (white grass)—a species of Stipa with feather-like silvery panicles tinged with rose—which forms the mass of the vegetation on the paramo. This grass affords excellent thatch; it is also extensively used in packing, and along all the higher grounds it is almost the only material for fuel. Between the hassocks, especially where there are slight declivities, there is an interesting sub-alpine vegetation—a dense grassy turf is enamelled with flowers, white, yellow, red, and purple, which seem to spring direct from the ground. Three daisy-like Werneriæ, all stemless and solitary, of which *W. nubigena* with its large white stars is the most conspicuous, grow along with a stemless Valeriana, a small Castilleja, a Lupinus, a Cerastium, two species of Gentiana, and two of Azorella. The cæspitose Werneriæ are true alpines, and grow at 2000 feet above the species just referred to. There are many little lakes, frequently bordered by the swelling, glaucous, sphagnum-like tufts of a Plantago, over which creep the silvery threads of a minute Gnaphalium and an equally minute white-flowered Gentiana. In such situations grow also a small Ranunculus, bearing generally a single sessile flower and a pedunculate head of follicles, a Stachys, and several other herbs of humble growth. Heath-like tufts of *Hedyotis ericoides*, often accompanied by a suffruticose Valeriana of similar habit, and sometimes by a Calceolaria, here and there diversify the landscape;

while the hassocks shelter in their bosom purple Lycopodia and other plants.

Having passed Sanancajas, we descend to the sandy plain of Riobamba, whose general character is the same as that of Ambato, save that cactus-hedges often replace those of aloes.

In Riobamba I remained three days with my hospitable countryman Dr. James Taylor, and then proceeded on my way, going the first day only as far as Miraflores, a farm six leagues away from Riobamba, and near the village of Guamóte. On the way we had to climb over a small space of paramo, where we got the benefit of a storm of hail and sleet. The vegetation was scanty, and I gathered only a minute Umbellifer which was new to me. Miraflores is what is called a cold farm, consisting chiefly of pasture and barley fields. A short ascent from it brought us upon the Paramo de Tiocajas, which is full six leagues across. Anything more desolate than this paramo I have nowhere seen. It is one great desert of movable sand, in which the distant patches of Cacti, Hedyotis, and a succulent Composita only render its nakedness more apparent. Where there is a little moisture, solitary plants of a silky-leaved Plantago struggle for existence. The altitude is about the same as that of Sanancajas, and it may be imagined how cheerless was a slow ride of nearly 20 miles over such a waste, rendered all the more gloomy by a leaden sky overhead, and a piercing wind which came laden with mist and fine sand. I was obliged to go nearly at the pace of my loaded beasts, the unsettled state of the country, and the number of deserters from the "constitutional" army roaming about, rendering it unsafe to leave my goods a moment. Yet even such an "Ager Syrticus" has its points of interest, for on this place is seen the dividing of the waters of the Atlantic and Pacific oceans. We passed many small streams, some rising on the paramo and some in the Western Cordillera, but all running eastward to join the Great River, with whose waters and forests I was long so familiar; when, however, we approached the southern side of the paramo, we came on the Rio de Pumacháca (River of the Bridge of Tigers), a considerable stream rising in the Eastern Cordillera and running westward towards the Pacific; it is, in fact, one of the sources of the river Yaguáchi, which enters the Gulf of Guayaquil. From the Pumacháca northward, until very near Quito, all the streams of the central plain between the two branches of the Cordillera flow eastward, and unite in the gorge of Baños to form the river Pastasa, which speedily reaches the Amazonian plain, and thence the Atlantic; but the streams around Quito itself unite to form the river of Esmeraldas, and seek the Pacific. Near the Pumacháca there was rather more vegetation; patches of Cyperaceæ were dotted with the white flowers of a minute Lobelia, which I have seen in many similar situations, and groups of Cactus were draped over by an Atropa, remarkable for its aromatic leaves. It is singular that in so deadly a genus all the species I have seen in the Quitonian Andes have edible though very acid fruit, and that the shoots are cropped by asses and llamas.

As we descended from the southern side of the paramo, the Hedyotis began to be mixed with a small labiate shrub of very similar foliage, and bearing numerous spikes of lilac or violet flowers; and farther down the latter grew so abundantly that it covered the whole hill-side with a mass of aromatic flowers, which was an agreeable change from the sterile paramo. The road ran parallel to the Pumacháca, but at a vast height above it. It was well on in the afternoon when we reached the village of Ticsán, still in the cool region, and, as we calculated on finding more comfortable quarters in Alausí, which was two leagues ahead, we resolved to try to reach it, which we accomplished just after nightfall, having in the day made ten leagues. With some trouble we succeeded in getting a little food for ourselves; but food for our beasts was of more importance, and we could get none. At 4 o'clock the following morning I roused my people and sent them out to the neighbouring farms in quest of alfalfa (lucerne). They returned bringing a mule-load, which, though an insufficient quantity, was better than none, and we delayed our journey until 8 o'clock in order that the poor animals might eat, for we had this day only five leagues before us.

Our road now turned to the right, while that to Cuenca continues southward and crosses the elevated ridge of Azuáy. We still followed the course of the Pumacháca, which gradually turns westward, and bursts through the Cordillera in a gorge so deep and narrow that with difficulty has a narrow path been cut along the declivity on the southern side. The whole five leagues from Alausí to Chúnchi consists of steep ascents and descents, and of perilous crossings of precipitous slopes, not to be passed without a shudder; for the track is in many places so narrow that two persons mounted could not pass each other without endangering the life of one of them. Fortunately, our beasts were sure-footed and the road was dry; in fact, from Ticsán, where we fairly began to descend the western slope of the Cordillera, we found we had got into the height of summer, having left mid-winter behind us at Ambato and Riobamba. The hill-sides were well covered with grass, but all completely withered up by nearly two months of dry weather; so that, except near the streams, where there was a margin of scrub or low forest, the eye rested on nothing green.

Alausí stands at about the same height as Ambato, but is subject to still more violent winds, so that even the crops of maize are rarely to be seen standing erect. As a town it bears no comparison with Ambato either for size or neatness, and, like all the other pueblos of the canton (of which it is the *chef-lieu*), seems to have been for several years in a state of decadence: the houses begin to fall and are merely propped up, not repaired or rebuilt; and yet there are all around valuable farms of wheat and maize.

Throughout the Quitonian Andes a bit of solid rock is rarely seen, save where black, jagged masses of trachyte stand out in the higher peaks, which are all either active or dormant volcanoes; and on a superficial view most of the hills seem to be made up of débris, either, as around Ambato, of calcined and triturated granite and schists, or, as in descending from Alausí, of stones and rude blocks confusedly heaped together. But in one place we saw above us a low cliff of vertical strata, much cracked and bent, as if by some force applied to their ends. The brown hill-sides began to be diversified by an arborescent Cactus, with polygonal stems and white dahlia-like flowers, which, Briareus-like, threw wide into the air its hundred rude arms. Lower down, at about 6000 feet, I saw specimens full 30 feet high and 18 inches in diameter. Along with it grew frequently a Cæsalpinia and a Tecoma, both of which are abundantly planted near Ambato and Guano, the former for the sake of its bark, used in tanning, and the latter because it bears a profusion of ornamental yellow flowers, and is supposed to possess wonderful medicinal virtues.

About two leagues below Alausí the road descends to the margin of the river, where it meets the Chanchán, a larger stream coming from the Eastern Cordillera, near the volcano Sangáy; the two united take the name of the latter, and preserve it until issuing into the plain, where, joined by the Chimbo from Chimborazo, they form the river Yaguáchi, which empties itself into the gulf just above the city of Guayaquil. Crossing the Chanchán by a rude bridge near its junction with the Pumacháca, we entered on a beach clad with a grove of Acacias—low spreading trees with very odoriferous yellow flowers and binate spines sometimes 3 inches long. Near this place, which was still some 8000 feet above the sea, we came on the first sugar-cane farm. The road again leaves the river, and we had finally to climb a long cuesta to reach the village of Chúnchi, which is full 1500 feet above the river.

Chúnchi is the last village on the slope of the Cordillera, and I had calculated on making it my head-quarters, though the forest is still a day's journey farther down. I brought recommendations from Ambato, and the people seemed willing to assist me; but the houses were so miserable, so full of dirt and vermin, and so utterly destitute of furniture (for I could procure neither bedstead, chair, nor table), that I saw I should work on my plants with infinitely less comfort than I used to do in a palm-hut in the warm forest. Another and greater difficulty was the procuring of food for my beasts, for all the pastures were dried up, and a man who sold me alfalfa for two days then told me he could spare no more. About a league from Chúnchi, and 1000 feet lower down, there is a cane-farm called Guataxí, whose owner, Señor José Leon, I had known in Riobamba. Almost in despair, I rode down to consult with him, and he at once invited me to take up my quarters in the hacienda, where he has a good house, with neatly-papered rooms and decent furniture. The cane-grounds extend along the banks of a stream, which before falling into the Chanchán forms a considerable lake, on whose shores there was still a little herbage; besides that a few squares near the house were planted with alfalfa.

On the third day after establishing myself at Guataxí, having procured a guide, I proceeded to Lucmas, a short day's journey lower down the river, where there are a few small chacras tenanted by Indians and Zambos. There I was told I should be near the *Cascarilla roja* (Red Bark), and I was recommended to a person called Bermeo, who had worked a good deal at getting out cascarilla and sarsaparilla. I at once secured his services, and, as he turned out an honest, active fellow, I took him with me in all my subsequent excursions in the district. From him I learnt that the *Cascarilla roja* did not commence until another day's journey downwards, and that to have a chance of seeing it in any quantity (which, he admitted, was at best only problematical), it would be necessary to penetrate at least three days into the forest. As my object for the present was merely to make myself acquainted with the plant, and with the soil and climate in which it grows, I decided on going no farther than until I should meet with it; for the procuring and transporting of provisions necessary for a long stay in the forest is both difficult and expensive.

I remained a day at Lucmas to look around. It is at an altitude of between 5000 and 6000 feet, and produces luxuriant sugar-cane. The small banana called Guinéo flourishes (as indeed it does at Guataxí), but the plantain is near its upper limit, and the fruit is small and scanty. There are tolerably lofty forest trees in the valleys and on the hills, while the steep sides of the latter are often covered with grass, more or less intermingled with scrub, and often with Bromeliaceæ. In descending towards Lucmas, I saw on the bushy hill-sides a great deal of the small tree called Palo del Rosario, a curious, and I believe undescribed, Sapindacea, which I had already gathered at Baños in the Eastern Cordillera. Its most remarkable feature is, that while the layer of wood next the bark is quite white, all the internal layers are purple-brown with a black outer edge—a colour not unlike that of old walnuts; so that articles fabricated of this wood are curiously mottled.

Unfortunately, the trunk never exceeds a few inches in diameter, so that only small articles can be made of it. I have secured a specimen of the wood, and of spoons made from it, for the Kew Museum.

One of the most frequent trees at Lucmas, and the most valuable for its hard wood (though the young branches are brittle), is an Escalloniacea called Ignia. It grows to a good size; the leaves are narrow-lanceolate and very long—the lower ones always red—and the reddish flowers are borne in long pendulous racemes; so that the tree has a very pretty aspect. It abounds along the western slope of the Cordillera, and grows at from 5000 to 9000 feet. It is accompanied by an Amyrideous tree called Alubilla, which the people hold in great dread, as they believe that to touch it or pass beneath its shade is enough to cause the body to swell all over. I had already, at Baños, gathered flowers and fruit of it, and stained my hands with the milk, to the great horror of those who saw me, but without experiencing any ill effects; and I believe that the swelling attributed to it is owing more to sudden changes of temperature, or to alternate scorchings and wettings, for I have seen such an effect follow where there was no Alubilla. Be this as it may, the young man I took as guide felt one of his eyes begin to swell the day we left Lucmas for Guataxí, and in a few hours he was swollen from head to foot. In two or three days he was quite well again, but there are cases of the swelling lasting a month. As might be supposed, the blame was laid on the Alubilla.

Lucmas takes its name from the abundance of

a species of Lucuma, producing an edible fruit; that name is applied to many species of Lucuma and Achras, all natives of warm or hot countries. Another evidence of the approach to a hot climate was in the existence of a species of Echites, twining among the bushes, and in an epiphytal Marcgraviacea, quite similar in its long scarlet spikes to *Norantea guianensis*, though the bracts are small patellæ, not elongated sacs, as in that species. A very odoriferous Citrosma, with large thin leaves, three together, is known by the name of Guayúsa, and is often taken in infusion, like the Guayúsa of Canelos, which, however, is a species of Ilex.

There were a good many herbs, of species not seen elsewhere. One Composita, with virgate stems 12 feet high, large alternate lobed leaves, and from each axil a small leafy ramulus bearing at its apex a corymb of white radiate flowers, was very ornamental. Orchideæ were tolerably abundant, but prettier even than these were two Bromeliaceæ; the one seemed at first sight merely a mass of long scarlet flowers growing out of the moss on old trees and stones, for the leaf-sheaths are imbricated into a little bulb, and the blade is reduced to a spine; the other (apparently an Æchmea) has broadish soft leaves and large violet flowers looking at a distance more like those of an Iris or an Amaryllidea.

On the 4th of August my company started for the forest, our destination being the Rio de Pumacocha, a large stream rising in Azuáy and falling into the Chanchán at about 4000 feet altitude, on the farther side of which much Red Bark has been got in former years. We started on horseback, and a mule carried our necessaries. My counsel was to leave the horses, but Bermeo felt sure I should not be able to perform the distance on foot; we had gone, however, a very short way when we found it necessary to cut our way through the forest, for the track had got overgrown in two years that no one had passed along it; nor was it possible without wasting a good deal of time to open a passage overhead so that a man might pass mounted; I therefore preferred going on foot most of the way. We reached the banks of the Puma-cocha at an early hour of the afternoon, but the ford which Bermeo had passed in former years had been destroyed by the falling of a cliff, and in its place ve found a deep whirlpool; so with the drift-wood along the banks we set to work to make a bridge where the river was narrowed between two rocks, and when completed carried across it our baggage, saddles, etc. Then, after a long search, we found a place where we could swim the horses over, and by rolling down a good deal of earth and stones we made a way for them to ascend on the other side. Once across, we selected a site for our hut among Vegetable-ivory palms, and thatched the hut with fronds of the same. Close by were the remains of a platanal, showing that the spot had formerly been inhabited, and fortunately still bearing a sufficient number of plantains to cook along with our salt meat during the two days we calculated on remaining there. Our horses were taken to the top of a neighbouring hill, where there was a bed of one of those large succulent Panicums called Gamalote, which afford a very nutritious food for cattle, and were there made fast for the night. Here we slept tranquilly, save that we were occasionally aroused by the snuffing of bears around us; and before daylight Bermeo and his companion were on foot, and making their way through the forest in quest of Cinchona trees. They returned at 7 o'clock, having found only a single tree standing, and from that one the bark had been stripped near the root, so that it was dead and leafless. We breakfasted, and then I accompanied them into the forest. We followed the track they had already opened, and then plunged deeper in, meeting every few minutes with prostrate naked trunks of the Cinchona, but with none standing. Bermeo several times climbed trees on the hill-sides, whence he could look over a large expanse of forest, but could nowhere get sight of the large red leaves of the Cinchona. At length we began to tire, and we decided on returning towards our hut, making a detour along a declivity which we had not yet explored. We went on still a long time with the same fortune, and were beginning to despair of seeing a living plant, when we came on a prostrate tree, from the root of which a slender shoot, 20 feet high, was growing. My satisfaction may well be conceived, and my first thought was to verify a report that had been made to me by every one who had collected Cascarilla, namely, that the trees had milky juice, which to me was strange and incredible in the Rubiaceæ. Bermeo made a slit in the bark with the point of his cutlass, and I at once saw what was the real fact. The juice is actually colourless, but the instant it is exposed to the air it turns white, and in a few minutes red. The more rapidly this change is effected, and the deeper is the ultimate tinge assumed, the more precious is the bark presumed to be. It is rare to find shoots springing from an old root, because the roots themselves are generally stripped of their bark, which, along with the bark from the lower part of the trunk, is known by the name of "Cascarilla costrona" (from "costra," a scab), and is of more value than that from any other part of the tree.

The *Cascarilla roja* seems to grow best on stony declivities, where there is, however, a good depth of humus, and at an altitude of from 3000 to 5000 feet above the sea. The temperature is very much that of a summer day in London, though towards evening each day cold mists blow down the valley from Azuáy; and for five months in the year—from January to May—there is almost unceasing rain.

If the *Cascarilla roja* has been almost extirpated at Puma-cocha, there is still left abundance of Salsaparilla, and of a very productive kind, for Bermeo assured me he had once taken 75 lbs. weight of the roots from a single plant; whereas in Brazil the greatest yield I have heard quoted was a little over 30 lbs. The Puma-cocha species has a round stem and few prickles, while that most esteemed on the Rio Negro has a triangular stem thickly beset with prickles.

Let me now say a word about the other plants accompanying the Cascarilla, and first of the Ivory palm, which is known throughout the Ecuador by the name of Cádi. . . . It has a stout erect trunk of 15 or 20 feet; the fronds are 30 feet long.

. . . The nuts are much the same as in the other species, only rather larger; they are extensively used in the Sierra for making heads of dolls, saints, and walking-sticks. The Cádi produces a very excellent "cabbage," but the Indian and other inhabitants are fonder of a large maggot called Majón which is bred in its trunk. I have seen the Indians of the Rio Negro and of Canelos roast and eat the larva of a beetle extracted from the trunk of the Pupunha palm (*Guilielma speciosa*).

. . . . . .

In general, the arborescent vegetation seemed scanty in species and uninteresting. One of the most striking trees was an Erythrina with a slender tortuous (almost twining) trunk, from which sprang long spikes of scarlet flowers, and few branches bearing each a coma of ternate leaves, whereof the leaflets were sometimes 18 inches across. There were also a few Figs, and on the steep declivities there were patches of low forest, consisting chiefly of Clusiæ, Thibaudiæ, and Melastomaceæ. Two small Trichomanes crept along the branches of shrubs, but terrestrial ferns were all but absent.

On returning that evening to our hut, I consulted with Bermeo about our ulterior movements. He told me that if I would go another day's journey into the forest, he could with certainty show me more trees of the *Cascarilla roja*, which he had seen not many months previously; and as on account of the Revolution no one had this year entered the forests to collect Cascarilla, it was probable they were still untouched. But for this our stock of provisions would scarcely suffice, and I saw no probability of adding anything interesting to the general collection; besides, I had to visit other forests in quest of other sorts of Cascarilla, and I saw the season was already passing for the flowers and seeds of most trees. We therefore on the following day retraced our steps up the valley, and after another day spent at Lucmas in drying my paper and adding what I could to my collection, I returned to Guataxí.

I was unable to move far from the farm for above a fortnight afterwards, on account of the passage of the Government troops from Quito to Cuenca. . . .

During this interval I was obliged to content myself with the flora of Guataxí. The cane-farm is about 7000 feet above the sea; the maximum temperature each day was generally about 73°, though it once reached 77°, and the minimum temperature varied from 55° to 60°. A plateau, about a thousand feet higher, belongs to the farm, and produces good crops of grain and potatoes. The hills adjacent to the farm, except where under cultivation and artificially irrigated, are covered with grass, amongst which the withered remains of a good many annuals were visible. Almost the only annual still flourishing was, singularly enough, a species of Monnina, with violet flowers; and, as most of the species of this genus are trees, I took it for a Polygala until I saw the fruit. The "Yerba Taylor" (*Herpestes chamædryoides*, H. B. K.), which has great fame as a remedy for snake-bites, was frequent, but mostly scorched up. Amongst the perennial herbs (most of which were new to me) may be mentioned an Epilobium, a Stachys, a Phaseolus, a Desmodium, two Crotalariæ, a shaggy

Hieracium, a very pretty Leria with large blue flowers, growing on shady banks, and a branched Composita with silky-white leaves and handsome purple flowers, besides several Solaneæ, Labiatæ, Ehretiaceæ, and two Acanthaceæ, which last order seems entirely absent from the cold region; also a suffruticose Lantana with yellow flowers, which I had not seen elsewhere. In moist places a little Cuphea was very abundant. The shrubberies consisted chiefly of Compositæ, whereof one resembled a Spiræa in aspect and in the odour of its numerous small white flowers; but there was also a new Büttneria, and the common Clematis of the warmer parts of the Cordillera climbed about everywhere.

In cultivated ground, especially in the maize and cane fields, two delicate broad-leaved Paspala called Achín spring up in great abundance. Every day I saw the servants of the farm get bundles of them for the cows, pigs, etc., which ate them with greater avidity than even the alfalfa, so that, though weeds, they were nearly as valuable to the owner as the crops amongst which they grew.

Among the trees, which grew chiefly along the banks of the river, were two species of Lycium not previously seen, an Inga, a Mimosa, and a Bignoniacea with broad opposite leaves and cymes of large purple flowers. The last, known by the name of Hualla, is frequent in the Western Cordillera at from 6000 to 9000 feet, and is one of the best timber trees. It is not improbably the little-known *Delostoma integrifolium*, Don; but it is not a Delostoma, for, besides an essential difference in the calyx, the septum is contrary to the valves, as in

Tecoma, not parallel to them, as in Delostoma and Bignonia.

So soon as the last soldier had passed, I put in execution my project of visiting the forests producing the *Cascarilla serrana* or Hill Bark, which is found at 8500 to 9000 feet on both sides of the river Chanchán. I went first to the forest of Llalla, at the foot of Azuáy, and only a little more than two hours' journey from Guataxí. Here there is a cattle-farm and a few Indian chacras, in one of which I established myself. I found a rather interesting vegetation, and this consoled me for my wretched quarters in a hut dark and smoky, and so low that I could not stand erect. We had happened on a windy time, and as the walls and roof were full of chinks, the violent wind which got up at midnight starved us beneath all our blankets and ponchos. After sunrise there was a brief lull, and then it came on again to blow from the same quarter (west, with a slight touch of northing), and so continued through the day. We had no rain during the five days of our stay, although the storms on the farther side of Azuáy often overlap as far as Llalla, so that from Guataxí we could see it raining in this hill-forest, when not a drop fell in the lower grounds; and even when it does not rain the forest is generally enveloped in mist. This constant supply of moisture renders the vegetation more vigorous than in the dry grounds below, and is the cause why the trees are so thickly clad with mosses that it is difficult to push one's way through them. Two mosses, whose long slender stems hang down like a beard from the branches, bore here abundance of fruit, which for two years I had sought in vain in other localities. But I was most pleased to find a moss with large laciniato-ciliate leaves—so novel a feature in this tribe that I took it for a Plagiochila, until I found the capsules nestling amongst the terminal leaves.

To return, however, to our Cascarillas, of which there are two sorts in Llalla, the one called "Cúchi-cára" or Pig-skin, because dried pieces of the bark resemble morsels of pig's skin boiled and then grilled (which is a favourite dish in Ecuador). The same bark is sometimes called Cháucha, a term implying thickness without much consistence; as, for example, in this bark, which shrinks much in drying, and in a sort of large watery potato, called Chauchas. The other bark is called "Pata de gallinazo" or Turkey-buzzard's foot; it does not peel off freely like the other, and when dried generally occurs in small split fragments, but as it is rather deeper-coloured it is more esteemed than the Cuchicara. The same or similar kinds are known in other districts as Cascarilla naranjada. The demand for either kind has of late years been very slight, so that there has not been such destruction of these barks as of the red, and on a stony hill-side not far from the hut I found above twenty large trees of the Cuchicara, from 40 to 50 feet high. All had fruited freely this year, but the capsules were already empty, with the exception of one small corymb. In the forest of Yalancáy, on the opposite side of the river and near the road leading from Alausí to Guayaquil, I afterwards found a tree with recent fruit and even a few flowers. The latter are deep brick-red, and the capsules are usually elongate-oblong, but vary to roundish oblong. Trees of the Pata de gallinazo were scarce, and I did not see any in flower or fruit. Both sorts have the leaves broadly oval, with or without a slight apiculus, and pubescent beneath; but in the Cuchicara the petiole and midrib are red, which is not the case with those of the Pata de gallinazo, nor do the leaves of the latter turn so red with age. . . .

Of the trees growing along with the Cascarillas in Llalla the Motilón was the most frequent and the largest, attaining sometimes 60 feet in height. This is the second species I have gathered under this name; the fruit is an edible drupe, but I hesitate to refer the genus to Amygdaleæ until I see the flower. With the Motilón grew, however, a true Cerasus, with very large leaves; it had flowers and young fruit. Other trees in the same forest were the Hualla, the Ignia, a Berberis, a Rhamnus, a Nonatelia, two Myrtaceæ, and especially an arborescent Loranthus, with dense spikes of fragrant yellow flowers—the leaves on some ramuli alternate, on others opposite, and on others three together. . . . The shrubs included a Barnadesia, two Salviæ, a sarmentose Fuchsia, and most abundant and ornamental, an aphyllous Fuchsia, epiphytal and climbing high up the trees, which it adorned with its large vermilion flowers.

Patches of verdant pasture were scattered in the forests, and in these I gathered a stoloniferous Ranunculus new to me, a small Juncus, a curious Rubiacea allied to Richardsonia, two Ionidia, the one with red the other with scarlet flowers, and some other herbs. In the woods there was also a stinging herb with large white flowers of the order Loaseæ.

The Orchideæ must not be forgotten: they were very numerous and in fine state, especially two large-flowered Odontoglossa, whose liana-like peduncles depended almost to the ground. There were also some Oncidia and Epidendra, and many curious things whose affinities I did not recognise, and which I have not yet examined.

From Llalla I dispatched my men to the adjacent paramos on that side of Azuáy, with instructions to bring me everything they found in flower. They returned bringing a good many alpines, including some pretty Senecios not elsewhere seen, a red-flowered cæspitose Werneria, a small Crucifera, an Alstrœmeria, a Gnaphalium, but especially a beautiful Gentiana, allied to *G. cernua*, and instead of having only one or two pendulous flowers, as in that species, bearing a profusion of erect pyriform red flowers.

. . . .

[In a letter to Sir William Hooker from Ambato (Oct. 10, 1859), the following remark on the vegetation of the two slopes of the Andes is of much interest:—

"As regards the general vegetation, the Amazon side of the Andes is incomparably richer than the Pacific side. In the former a perpetual spring reigns—sun and rain divide each day, rain predominating in what is called winter and sun in summer; but in the latter the ground gets burnt up with seven months of dry weather, and soaked with five months of continual rain. You will therefore be prepared to hear that in my late expedition to the Pacific side I have found scarcely any ferns, and still fewer shells and beetles."

This statement was, however, somewhat modified in the following year, when he found that the Cinchona forests of Limon, about 70 miles to the north-west, had a rich and interesting flora, with an abundance of ferns and orchids. The superior richness of the eastern slopes as a whole seems, however, to be an undoubted fact.]

To Mr. John Teasdale

AMBATO, *Nov.* 15, 1859.

Before I left Ambato for Guataxí (July 22), the first Act of the Revolution was played out on the flanks of Chimborazo, at a site called Tumbúco, where a battle was fought between the Government troops (consisting chiefly of blacks and Zambos from the low country around Guayaquil) and the insurgents, who were "serrános," or people of the hill-country, some whites, some Indians, but the most part of mixed race. The latter were defeated, and the victorious army marched on Ambato. It was something to see the flight of the inhabitants of Ambato, and the files of mules laden with all their movable goods, even to glass windows, when the news of the battle of Tumbúco arrived. I had nowhere to flee to, so I laid in a stock of live pigs and fowls, and of potatoes, stuck out the Union Jack, and prepared for a siege. Well, the turbulent blacks came on us by slow marches, but they respected my house and cattle; and indeed the whole town was let off with a requisition of provisions and horses. Yet the danger was not imaginary, for Riobamba was sacked some time afterwards (it is only a week ago to-day) by the troops stationed there. Not a shop or a warehouse was spared, and eight or nine private houses shared the same fate.

. . . Your sanitary and social reformers seem much occupied with devising suitable habitations for the poor and industrious classes. They would be much shocked could they see the promiscuous way in which people sleep here, even in the wealthiest houses. The other day I remonstrated with my landlord—one of the best men in the place —for allowing a number of people of both sexes to sleep together in the same room—some in beds, some on the floor. "I assure you," replied he, "we throw open both doors and windows at daybreak!" He had no idea, poor man, of any possible vitiation of the moral atmosphere. I thought of the fair (but frail) Pauline Buonaparte, who, when an English lady asked her, "How *could* you sit so naked to that sculptor?" made answer, "My dear madam, you forget I had a fire in the room!"

In January last I spent three weeks with the Cura of Puela—a small village at the western foot of Tunguragua. The parsonage-house consisted of but two rooms, the one a small dormitory occupied by the Padre, and where he had barely room to turn himself; the other a much larger room, where the rest of his family worked and ate during the day, and slept at night. I append a diagram of this main apartment, wherein 1, 2, 3 represents a raised stage made of wild canes (called a barbacóa), extending across one end of the room. No. 1 was my bed, made neat and comfortable with my own bedding (which I always carried about with me, and was half a mule-load). In No. 2 slept two young fellows—the Padre's servants—on sheep-skins; and in No. 3 slept his two maid-servants, at right angles to the men, and with their feet towards them. No. 4 is a bench whereon reposed my lad. No. 5 is a curtained four-post bed,

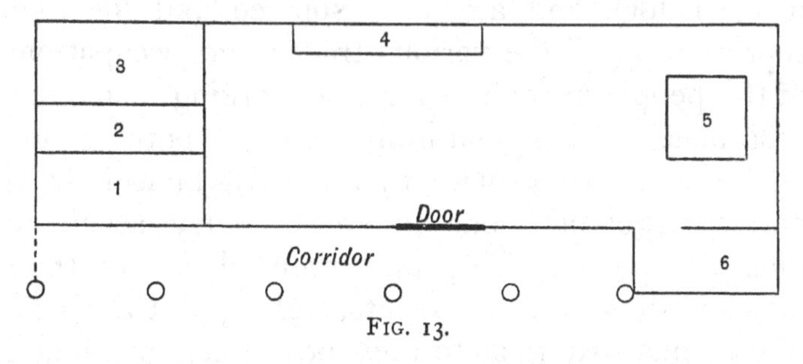

FIG. 13.

occupied by the Padre's maiden sister, of the matronly age of twenty-one years complete; and No. 6 a small recess, jutting on the external corridor, where a young fellow—the Padre's nephew—extended his lazy length on a barbacóa; but even this place was open to the main room, having a doorway but no door. I afterwards transferred my bed to No. 6, on the Padre's suggestion that it was snugger and more retired!

THE INDIANS OF THE ECUADOREAN ANDES

[I cannot find in Spruce's MSS. or notes any account of the natives of the highlands of Ecuador, although he must have seen a good deal of them as muleteers or porters during his very numerous excursions, or as dealers in the various products of the country. He sent home, however, to his friend Mr. Teasdale a set of forty-four coloured drawings of "Costumes and Customs of Qúito," which are now in possession of his son John Teasdale, Esq., Solicitor, of York, and which he has kindly allowed me an opportunity of inspecting. These were executed by a native Indian (though some writers doubt if there are any absolutely pure Indians left in Ecuador), and are very spirited and life-like, representing all the various trades and occupations of the people in their respective working or holiday costumes, and very naturally coloured, both colours and brushes being made by the artist himself from native vegetables or minerals. They serve to illustrate not only the people themselves, but their tastes in dress and ornaments, and support the view of previous writers as to their possession of mental faculties comparable with those of their conquerors and masters.

Yet they appear to be by no means prepossessing, as exhibited in the accompanying portraits of four Quito Indians, reproduced from photographic prints in Dr. Theodor Wolf's *Geografia y Geologia del Ecuador*. These recall in their coarse massive features and stolid expressions many of the natives of the North American plains and mountains, such as the Cheyennes and some others, and suggest an original identity of the mountain as opposed to the forest tribes of both continents.

The following description of the Ecuadoreans in the *Universal Geography* of Élisée Reclus emphasises the several characteristics of these people. "Except during times of frenzy and ecstasy, the Ecuadoreans

FIG. 14.—INDIANS OF THE PROVINCE OF QUITO.
(From Wolf's *Geografia y Geologia del Ecuador.*)

are a sad and sullen people. The features, especially of the women, seem haggard with care and sullen misery. Yet, despite their sordid surroundings, the Quitonians appear to possess the sentiment of form and colour in the highest degree. Notwithstanding the rigid formulas and conventionalities to which the priests have enslaved them, many of the Mestizoes and even of the full-blood Indians succeed in executing really remarkable religious paintings as well as sculptures of Christs and Madonnas, works greatly admired in Peru and other South American countries, to which they are regularly exported. But the natives have lost one artistic industry—inlaid work in costly woods. It has been noticed also that neither his extreme poverty nor the dull existence to which he is condemned has prevented the Ecuadorean from distinguishing himself by the elegant cut and harmoniously-blended colours of his native costumes."

We seem to have here the surviving remnants of a people with high capabilities, who have been so crushed down by centuries of slavery and repression, combined with the struggle against the forces of nature in some of their most terrible aspects, as to have become degraded both physically and mentally, while still exhibiting unmistakable traces of the higher civilisation and more sympathetic government they enjoyed under the Incas.]

CHAPTER XXI

THE CINCHONA FORESTS OF EL LIMON, ON THE WESTERN SLOPE OF CHIMBORAZO

(1860)

[THIS year was wholly occupied in the arduous work of obtaining for the Indian Government seeds and young plants of the best of the medicinal barks, produced by the *Cinchona succirubra*, which was becoming exceedingly scarce owing to the reckless destruction of the trees producing it. The bulk of the present chapter is occupied with a reprint of the more interesting portions of the very full report of his labours made by Spruce. This is of much general and historical as well as of botanical interest, and as it can now only be procured in a very cumbrous form in a large and costly "Blue book" comprising the whole official record of the introduction of the various Bark trees into India, its inclusion in any account of Spruce's botanical work is imperative.

I give first the "List of Botanical Excursions" for the year, which summarise the whole story; and also some short extracts from the few letters he was able to write, which supply some more personal and descriptive features to the narrative.]

LIST OF EXCURSIONS DURING THE YEAR

1860.	
Jan.–Feb.–March.	Chiefly at Ambato, making preparations for entering the forest of Red Bark at the western foot of Chimborazo, to fulfil my commission from the Indian Government to procure seed and plants of Red bark.
,, 24.	From Ambato to San Andres.
,, 25-April 9.	To Riobamba, where I remained till April 10, pursuing my preparations in conjunction with Dr. James Taylor.

Ambato, etc.

April 10.	From Riobamba to Mocha. (Struck deaf in left ear on this journey.)
,, 11.	To Ambato.
,, 18-20.	Excursion to Cusatagua near Pillaro. (See map of Llanganati Mountains to north-east of Ambato.)
,, 24.	To Baños to bathe in hot springs for my deafness.
,, 28.	Returned to Ambato.
,, 29.	The Breakdown. Woke up this morning paralysed in my back and legs. From that day forth I was never able to sit straight up, or to walk about without great pain and discomfort, soon passing to mortal exhaustion.
May–June 1.	All this time under medical treatment at Ambato, but with very little improvement of health.

[Final Departure from Ambato, which had been his Head-quarters for 2½ Years]

June 12.	Started for the Bark Forests of Chimborazo: by way of Mocha to the Tambo of Chuyuipogyo.
,, 13.	Across Chimborazo to Guaranda.
,, 14-16.	At Guaranda.
,, 17.	On the way down the western declivity of Chimborazo. This night slept in the forest.

Red Bark Forests of Chimborazo

,, 18.	To El Limon: a group of small cane-farms in the region where the Red Bark still flourishes. The streams that flow down the numerous valleys combine into the little river Chasuán, which, in the plain, enters the river Ventanas—an affluent of the Guayaquil river.

1860.

June 18-July-Aug.-Sept. 12. From this date to the 12th of September at El Limon, superintending the work of getting plants and seeds of *Cinchona succirubra*. The seeds were all gathered under my eye, and were dried, sorted, and packed by myself. Partly on horseback, and partly dragging myself about on foot by the aid of a long staff, I explored pretty thoroughly the neighbourhood of our hut, and gathered (especially) numerous fine ferns and mosses.

" " Left El Limon and crossed over into the valley of Las Tablas.

" 13. Over another ridge into the valley of San Antonio, to the farm of Tabacál, where I remained till the 28th gathering seeds of Red Bark.

GUAYAQUIL

" 28. Started for Guayaquil, and travelled down the valley to Pozuelo.

" 29. From Pozuelo to Bodegas (on the river Guayaquil), where I remained till October 6.

Oct. 6. Down the river (by steamer) to Guayaquil. From this date till November 24 at Guayaquil; then started up the country to meet Mr. Cross coming down with the Bark plants.

Nov. 24. By steamer to Bodegas.

" 26. In canoe up river Ventanas to Caturáma.

" 27. To Aguacatál, a cacao-farm above the village of Ventanas. Here I remained until December 24, putting together fifteen Ward's cases, preparing a raft to take them down to Guayaquil, and when Mr. Cross arrived from the forest with the plants, superintending the work of transferring them to the cases, embarking the latter on the raft, etc.

Dec. 24. Set out on our raft and this day reached a point a few miles above Caracol.

" 25. Passed Caracol and Bodegas, and anchored in the river Guayaquil.

" 26. Down the river.

" 27. Reached Guayaquil at noon.

" 31. Had the cases embarked on board the Pacific steamer for Panama and England.

To Sir William Hooker

AMBATO, *March* 12, 1860.

. . . .

I have succeeded in hiring the forests producing the *Cascarilla roja* after about ten times as much correspondence as would have been necessary in any civilised country, and I am now getting together a staff of workmen (no easy task in these revolutionary times) with which to enter the forest as soon as the rains abate. I am also in treaty with the owners of the woods near Loja which produce the *Cinchona condaminea*; but as this species seems to flower and fruit exactly at the same time as the *Cascarilla roja*, and the localities of the two species are fifteen days' journey apart (under the most favourable circumstances), it is plainly impossible that I can see with my own eyes the seeds of both species gathered, which is the only way to be sure of having the right sort. . . .

REPORT ON THE EXPEDITION TO PROCURE SEEDS AND PLANTS OF THE *CINCHONA SUCCIRUBRA* OR RED BARK TREE

Towards the end of the year 1859, I was entrusted by Her Majesty's Secretary of State for India with a commission to procure seeds and plants of the Red Bark tree, and I proceeded to take the necessary steps for entering on its performance.

. . . .

Within the ascertained limits of the true Red Bark it exists (or rather existed up to a recent period) in all the valleys of the Andes which debouch into the Guayaquilian plain. Many years ago it was obtained in large quantities in the valley of Alausí, below an Indian hamlet called Linje, on the northern side of the Chanchán (nearly opposite to Puma-cocha, which is on the southern side of the same stream), but it has long been exhausted there.

. . . .

The Bark grounds, which still continue to be worked, form part of five contiguous farms, called respectively El Morado, Matiaví, Sínchig, Talágua, and Salinas, whereof the two former belong to the church of Guaranda, and the three latter (which extend upwards to the paramos of Chimborazo and downwards to the plain of Guayaquil) are the property of General and ex-President Juan José Flores, who, after a banishment of fifteen years, has lately returned to take the chief part in the recovery of Guayaquil from a faction who would have given it up to Peru. Only the high lands of those farms, where there is natural pasturage and ground suitable for the cultivation of potatoes and cereals, have been turned to any account by the proprietors. The middle part is dense, unbroken forest, and in the lower part, which produces the Red Bark, a good many poor people of mixed race from the sierra, and a few liberated slaves from the plain, have formed little cane-farms, without asking leave of the owners or paying any rent. The farms belonging to General Flores have been for some years leased to a Señor Cordovez, who resides at Ambato; and Dr. Francisco Neyra, notary public of Guaranda, rents the farms of the church, but only so far as respects the bark they produce. With these two gentlemen I had, therefore, to treat for permission to take from the bark woods the seeds and plants I wanted. At first they were unwilling to grant me it at any price, but, after a good deal of parley, I succeeded in making a treaty with them, whereby, on the payment of 400 dollars, I was allowed to take as many seeds and plants as I liked, so long as I did not touch the bark. They also bound themselves to aid me in procuring the necessary workmen and beasts of burden. Through the intervention of Dr. Neyra, who has throughout done all he could to favour the enterprise, I engaged with his cascarilleros (who all inhabit the village of Guanujo, adjacent to Guaranda) that whilst they were procuring bark for him, they should also seek seeds and plants for me.

From Dr. Neyra I ascertained that a site called Limon would be the most suitable for the centre of my operations. . . . At Limon existed formerly the finest manchon of Red Bark ever seen. It was all cut down many years ago, but I was informed that shoots from the old roots had already grown to be stout little trees, large enough to bear flowers and fruit, and that the squatters (who are many of them cascarilleros of Guanujo), since they got to know the value of the bark, had carefully preserved such trees as were standing in their chacras or clearings. Messrs. Cordovez and Neyra have made their depot for the bark about four leagues lower down the valley, where a stream called Camaron, running down the next transversal valley to the northward, joins the Chasuán.

The intestine war still continued to rage, and the country was divided into two factions, whereof one held Quito and the whole of the Sierra, and the other Guayaquil and the low country. Both maintained as large an army as they could raise in support of their cause, and pressed into their ranks all the suitable men they could lay hold on. Only those of pure Indian extraction were exempt from forced military service; but, when the troops were marching about, they continually seized on Indians to carry their baggage and to drive laden beasts. . . .

My preparations for entering the forest being completed, I was awaiting the coming of the dry season, when a severe attack of rheumatism so far disabled me, that I determined to delegate my commission to Dr. James Taylor of Riobamba. Animated, however, by his assurance that in the warm forest I might expect to recover the use of my limbs, I finally resolved to proceed thither in his company. . . .

We started from Ambato for the forest on the 11th of June. Our road was the same as I had travelled the preceding year, until reaching the paramo of Sanancajas beyond the village of Mocha, where it turns to the right towards the southern shoulder of Chimborazo. In consequence of my having needed two long rests on the way, night came on and found us still on the paramo. Thin clouds had enveloped Chimborazo most of the day, but towards evening they gradually cleared away, and after sunset the majestic dome was entirely uncovered, though a slender meniscus of cloud, assuming exactly the form of the cope of the mountain, and still illumined by the rays of the sun (which had set for us), hung for some time like a "glory" over the monarch of the Andes. When this at length melted away, the light reflected from the snow by a clear star-lit sky enabled our beasts to pick their way. It was 8 o'clock when we reached the tambo of Chuquipogyo, a solitary house at between 12,000 and 13,000 feet of altitude. The rude accommodation and the inhospitable climate offered no inducement to a prolonged stay at Chuquipogyo, but as I was so much exhausted as neither to be able to sleep nor on the following morning to mount my horse, there was no alternative but to remain there all the day and night of the 12th. At 7 A.M. of the 13th we resumed our march. The day was fortunately fine, and we had only now and then a few drops of small rain and sleet, instead of the snowstorms with which the traveller has too frequently to contend in the pass of Chimborazo. The vegetation consisted chiefly of hassocks of a Stipa and a Festuca, so that the general aspect was that of a grey barren waste; but at short intervals we crossed deep gullies whose sides were lined with mosses, and sprinkled with calceolarias, lupines, and other pretty plants. Towards noon we came out on the Arenal (the moraine of the glacier), near the limit of all vegetation. In a hollow a little below it was a marsh with a rivulet—one of the sources of the

Pastasa—in which I saw, not without surprise, a bed of the large-leaved Rumex, which is frequent in similar situations, at from 8000 to 9000 feet. The Arenal consists of sand and fine gravel of a pale yellow colour. In one place the road, for a considerable distance, resembles a broad, smooth gravel-walk in England, so that the only bit of really good road in Ecuador has been made by nature's hand on the crest of the Andes. The vegetation is limited to scattered tufts, or rather hillocks, of a Valeriana, a Viola, an Achyrophorus, a Werneria, a Plantago, a Geranium, a Draba, a pretty silky-leaved Astragalus, and the elegant *Sida Pichinchensis*, all of which (save the Astragalus) have rigid leaves in the characteristic rosettes of super-alpine vegetation, and send enormously thick roots deep down into the loose soil, although even these do not secure them from being frequently torn up by the violent winds and storms that sweep over them. My attention was so much taken up with these interesting plants, and with the immense mass of snow on our right, and in tracing the downward course of ancient lava-streams, which are as visible on Chimborazo as on Cotopaxi and Tunguragua, that I scarcely felt the wind, which swept us along like a gale at sea, and occasionally lifted small fragments of gravel and hurled them at us. It is scarcely necessary to state that the wind is here always easterly through the day, getting up strong generally about 10 A.M., and rarely continuing to blow with equal force through the night and following morning. Now and then it veers for a moment, and gives the traveller a side blow, which, were he not wary, might unhorse him.

We had left winter behind us on the eastern side of the Cordillera, and on our first day's journey, as we looked down the deep valley of the Pastasa, we saw a mantle of dense cloud and rain spread over the forest of Canelos. Even the eastern side of Sanancajas was wet and muddy, but after passing Chuquipogyo the road became nearly dry, and, on the western side of the Cordillera, it was even inconveniently dusty. In the direction of the Pacific not a cloud was visible, though the great distance and the hazy horizon prevented our actually seeing the ocean. So abrupt is the transition from the rainy season, which prevails on the eastern side of the Cordillera simultaneously with the dry season on the western.

The Arenal must be near a league across. As we descended from it the whole mountain side became covered with flowers, and nowhere have I seen alpine vegetation in such perfect state. *Gentiana cernua*, with its large pendulous red flowers, formed large patches, and was accompanied by three other species of the same genus, with purple and blue flowers, by Drabas, and other alpines. Still descending, the true alpines began to be mixed with half shrubby Fuchsiæ, Calceolariæ, Eupatoria, etc. Even before reaching the zone where these genera grow in the greatest luxuriance, and at less than 2000 feet below the Arenal, we came on the first tree, a Polylepis (allied to our common burnet), forming groves here and there along the declivity. The bark of this tree resembles that of the birch in colour, and in its peeling off in flakes; but if one could suppose an arborescent Acæna, it would give a better idea of the pinnate silvery foliage. On the opposite side—not of Chimborazo, which is bare of trees, but of its sister mountain Carguairazo—a Buddleia approaches nearest the snow-line. In descending the same side of the Cordillera, towards Pallatanga, ten leagues south of Chimborazo, a Podocarpus and a Berberis ascend higher than any other tree, while a Polylepis (distinct from that of Chimborazo) ceases 1000 feet below them. On Chimborazo, on the contrary, the same Podocarpus fails a long way below the first-mentioned Polylepis. An accurate discrimination of the species is therefore needed, before we can compare their climatal distribution.

Still descending, various other trees began to appear, such as Buddleiæ, Myrciæ, and especially an Araliacea, called from its large palmate leaves (which are hoary beneath) Puma-máqui or tiger's paw. Here and there the track rounded the heads of quebradas, deep and dark, and full of low trees, which were laden with mosses.

At about half-way down we came out on a narrow grassy ridge, called the Ensillada (Saddleback), where several long low straw huts had been recently erected for the accommodation of the soldiery when marching that way. As we neared the encampment, four raw-looking youths armed with lances rushed out and confronted us, demanding our passports. We had none to show, but our antagonists did not look very formidable, and a shot from one of our revolvers would probably have put them to flight, had I not been furnished with a weapon which I have found far safer and more efficacious in such contingencies, namely, a bottle of strong aguardiénte, a taste of which dispelled all opposition to our progress, and also served to induce the guardians of the pass to boil us water for making coffee.

Below the Ensillada we came on steeply-inclined strata of schists, across and down which we went on stumbling for at least a couple of hours; for, as the track runs over their projecting and jagged edges, which no pains have been taken to smooth down, we passed them not without inconvenience and danger. At this stage of our journey we became enveloped in cloud, which filled all the valley of Guaranda, so that we could thenceforth only discern objects near at hand.

We reached Guaranda just after nightfall, having travelled eleven weary leagues from Chuquipogyo. Guaranda is a rather neat little town, with good tiled houses built of adobes, and stands on ground which slopes down to the right bank of the Chimbo, at an elevation of about 9000 feet. As it is on the main road leading from Guayaquil to the interior, it presents in time of peace a very lively aspect in the dry season, when it is constantly full of travellers and beasts of burden; but when we reached it there were not the least signs of traffic, and only soldiers were to be seen in the streets. The temperature is slightly warmer than that of Quito, and the adjacent hills are grassy, where not under cultivation. From the little I could see of the indigenous vegetation, it appeared interesting. A large Thalictrum was abundant, as was also a sarmentose Labiate, with spikes of secund scarlet flowers, and a Tagetes, called, aptly enough, Allpa-anis (earth aniseed), from its scent and its lowly habit.

I was detained several days at Guaranda, partly in purchasing provisions for the forest, including an ox to be taken alive to our rendezvous, and partly in the vain attempt to procure licence for our cascarilleros (who had lately all been enrolled either in the line or the militia) to proceed to the forest; but I had to content myself with the assurance that, until the country was delivered from its present straits, not a single citizen could be spared for any other service. Only one of the cascarilleros, whose rancho we were to occupy, actually accompanied us to Limon, whether with leave or without I never knew, and he was there too much occupied in distilling cane-brandy, and in drinking no small portion of it himself, to be of the slightest use to us in seeking plants and seeds. Through Dr. Neyra's intervention, I secured the services of four Indians of Guanujo, and they proved of the greatest use to us, especially after we began to rear the Bark plants.

As far as Guaranda, two of our boxes had been carried by each beast of burden, but thenceforward, on account of the straitness of the path, they had to be carried singly. On the steep, narrow, and slippery tracks which traverse the western slope of the Quitonian Andes, the beasts of burden are chiefly bulls, called cabrestillos, whose cloven hoofs enable them to descend with greater security than even mules. Our provision of potatoes, peas, and barley-meal, etc., had to be carried in sacks so small that two of them placed on the back of each cabrestillo did not project beyond the animal's sides.

We set forth from Guaranda on the 17th of June, the direction of our route being first northerly, as far as the adjacent village of Guanujo, and then north-west to the pass of Llullundengo, on a ridge of Chimborazo, nearly in front of the Ensillada (from which the deep, wide valley of Guaranda separates it), and at a height of about 12,000 feet. Having surmounted this, we entered on the most precipitous and dangerous descent I have ever passed. The track leads straight down a narrow ridge, varied at wide intervals by level steppes, rarely exceeding a hundred yards across. The soil, from the summit down to the very plain, is a yellowish or reddish loam, wherein the sandy element prevails in some parts and in others the clayey, and it is of immense thickness, as we could see in the deep gullies worn in the mountain side by the rains and in the landslips. Angular masses of rock are sparingly embedded in it and scattered on the surface, but rounded pebbles are rare.

The vegetation in the pass consisted of Vaccinia (especially *V. Mortiña*, Benth.), Gaultheriæ, Melastomaceæ, Compositæ, etc., disposed in compact shrubberies, with intervening grassy glades. But we had scarcely turned the ridge before the forest became dense and continuous, at first low and bushy, but increasing in height at every step. At about 9500 feet we came on the first Cascarilla Serrana or Hill Bark, and it accompanied us downwards to, perhaps, 8000 feet. It is called indifferently Cuchicara and Pata de Gallinazo, which I believe to be terms merely indicative of the relative facility with which the bark may be stripped off in different individuals, either of the same or of various species.

At 3 P.M. we reached the Rio de Tablas, a considerable stream of clear water, foaming over large stones; its roar had been audible for the last hour of our steep descent. We crossed it, and on a deserted clearing of some two acres drew up for the night, uniting all our rubber ponchos to make a fall-to roof, to shield us from the night dews. The animals were turned loose to graze on the scanty grass in the clearing and on the leaves of a Chusquea on the edge of the forest.

I have nowhere seen Melastomaceæ so abundant as in the forest surrounding our encampment. One species grows to a stout tree 40 feet high, and bears large *pendulous* panicles (a novel feature to me in this order) of blood-red flowers, with large turgid yellow anthers. A lower spreading tree, apparently a Pleroma, bore numerous large violet flowers. Other smaller sarmentose species had also large rose or violet flowers. Altogether, I have never seen so gay a forest vegetation, except on the river Uaupés.

We were still in a rather cool region, but the night was dry and the wind very slight, so that we had not to complain of cold. After an early breakfast the next morning, we followed our way, which became still narrower and rougher as we proceeded. We had to climb the high ridge separating the valley of the Rio de Tablas from that of the Chasuán, and then to descend to the latter river, but there were many subsidiary ridges, with intervening hollows, or sometimes nearly level crossings (called travesias). The track in the precipitous ascents and descents is mostly a gully worn in the soft loamy soil by the transit of men and beasts, to the depth in some places of 10 feet, and so strait that the traveller, to save his legs from being crushed, must needs throw them on his horse's neck. Here and there a large stone sticks out, forming a high step, in descending which there is risk of both horse and rider turning a summerset. In the travesias there is a considerable depth of black tenacious greasy mould, worn by the equable step of beasts of burden into transverse ridges (called camellones, from their resemblance to the humps on a camel's back), with alternating furrows from 1 to 3 feet deep. This mould is formed in great part of the decayed leaves of the Suru, a bamboo of the genus Chusquea,[1] which forms almost impenetrable thickets, and whose arched stems and intricate branches, overhanging our way, much impeded our progress. In such places there was still a good deal of water and mud, for the rainy season was only just over in the forest.

At 6000 feet we lost the Wax palm (*Ceroxylon andicola*, H. et B.), which had accompanied us, though growing very sparsely, from about the upper limit of the Hill Bark. It descends to the same altitude on the eastern side of the Cordillera. Lower down, palms began to be tolerably abundant, but of few species. . . .

At a very little below 4000 feet we came out on the first chacras at Limon, where I almost immediately noted, and with no small satisfaction, a group of three Red Bark trees, each consisting of from two to four stems of 30 feet high, springing from old stools, and bearing a small quantity of fruit. We had still about two miles of gentle descent to the trapiche (cane-mill) destined for our habitation, and we reached it early in the afternoon, in the midst of a dense fog.

The trapiche stood on a narrow ridge running eastward and westward, sloping gradually on the northern side to the Chasuán, distant half a mile, and very abruptly, or 200 feet perpendicular in about 300 yards, to a tributary rivulet on the southern side. It was merely a long, low shed, and a sketch of its internal arrangements may serve for that of all the other trapiches, of which there were about a dozen at Limon. About two-thirds of its length was occupied by the rude machinery and adjuncts of the cane-mill. The remaining third had an upper story with a flooring of bamboo planks, half of it open at the sides, and the other half with a bamboo wall about 6 feet high, not reaching the roof in any part of it. This was our dormitory, and it was reached by a ladder—merely a tree trunk, with rude notches for steps. On the ground floor was the kitchen, with a wall of rough planks of raft-wood, placed by no means in juxta-

[1] The Chusqueæ are bamboos peculiar to the hills, with solid stems, rarely exceeding 30 feet in height, and not preserving an erect position for more than a few feet from the ground.

position, but not so wide apart that a dog or a pig could have got through the interstices. The whole fabric was, therefore, abundantly ventilated, and only too frequently filled with fog, as we found to our cost, in coughs and aching limbs, and in mouldy garments, saddles, etc.

Having reposed a day at Limon, Dr. Taylor went on with my horses two days' journey to Ventanas, hoping to find Mr. Cross there and to bring him up. During his absence I had to look after killing the ox and drying the beef, and to repair our dwelling, which was sadly fallen to decay, especially as to the roof. I therefore set the Indians to drag bamboos and palm-leaves out of the forest, with which we patched up the hut as well as we could. I visited also all the Bark trees known to exist within a short distance, and was well content to see on many of them a good crop of capsules, which had already nearly reached their full size on the finest trees; on other trees, however, there were only very young capsules, and even a good many flowers, so that I might have obtained at least thirty good flowering specimens; but, wishing to gather as many seeds as possible, I dried only a couple of specimens, which I had afterwards cause to regret, for not one of the late-flowering panicles produced ripe capsules. I learnt from the inhabitants that the trees had been covered with blossom in the latter part of April and beginning of May.

When Dr. Taylor had been ten days at Ventanas, a brief note from Mr. Mocatta was left at Guaranda by the Spanish minister (on his way from Guayaquil to Quito), informing me that Mr. Cross had been taken suddenly ill, when about to start for Ventanas. I therefore sent to recall Dr. Taylor, and, after his return to Limon, our operations were confined to visiting the Bark trees daily, which extended through a zone of about four miles in breadth, and to collecting and studying the accompanying vegetation. As we had a fair share of sun towards the end of June, I was in hopes the fruit would speedily ripen; but nearly all through the month of July the weather was cool, with a good deal of mist and fog, so that the capsules scarcely increased in size, many fell off, and some were attacked by a maggot and curled up. On the tree which bore most capsules they began to turn mouldy, the mould being not fungi but rudimentary lichens. I began to fear we should get no ripe seeds, and as the seeds had been especially recommended to me in my instructions from England, it may be imagined how severe was my feeling of disappointment. I had another motive for fearing the same result. The people of Limon had got a notion that I should buy the seeds of them, and one morning, when I made my round among the trees, I found that two of them had been stripped of every panicle, undoubtedly by some one who calculated on selling me the seeds. This was very provoking, for the seeds were far from ripe, and all the rest might be destroyed in the same way, so I immediately went round to the inhabitants and informed them that the seeds would be of no value to me unless I gathered them myself; and I offered a gratuity to the owners of the chacras where there were trees in fruit to allow no one to approach the trees except myself and Dr. Taylor. This had the desired effect, and I do not think a single capsule was molested afterwards.

Whilst Dr. Taylor was at Ventanas, the troops of the Provisional Government of Quito began to march down from the Sierra to attack the forces which held the low country, and they selected the route by Limon and Ventanas, along which an army had never been known to pass. For six weeks we were kept in continual alarm by the passing of troops, and it needed all our vigilance to prevent our horses and goods being stolen; indeed, one of my horses was carried off, though I afterwards recovered it. It was now clear that, unless there had been two of us, both independent of the political feuds of the country, the enterprise must have fallen through. All our provisions had to be procured from Guaranda, and, as they soon deteriorated in a moist, warm climate, whenever our stock got low Dr. Taylor had to take my horses and an Indian and go all the long distance to Guaranda to fetch more. . . . About half a day's journey down the valley there were a good many plantains on a deserted farm, and at twice the distance a negro had a fine plantation of them, from which I two or three times got up a mule-load; but the hungry soldiery soon made an end of them, and then even that resource was cut off.

. . . .

The view from Limon takes in a vast extent of country, both upwards and downwards, and the whole is unbroken forest, save towards the source of the Chasuán, where a lofty ridge rises above the region of arborescent vegetation and is crowned by a small breadth of grassy paramo. Nowhere are there any bare precipices, and a very steep declivity forming an angle of 60° with the horizon, appearing far away up the Chasuán, is as densely wooded as any other part. The opening at Limon, it will be understood, is purely artificial.

The crystalline waters of the Chasuán and its tributaries, in that part of their course where the Red Bark grows, run over a black or dull blue, shining, and very compact trachyte, which would seem to be the foundation of the Quitonian Andes, for it appears almost everywhere in the lower valleys, on both the eastern and western declivities. In the river Pastasa it occurs at from 3000 up to 7000 feet. Generally it is exposed to view only in the bed of the streams, or on their banks, where it often rises into rugged and fantastic cliffs. Over the trachyte at Limon there is to be seen in the bottom of the valleys a fine-grained ferruginous sandstone of a deep brown colour, in thick strata, and usually in large detached masses, lying either horizontally or variously tilted up. I suppose, therefore, that, so far from having been deposited over the trachyte, it is merely the remains of a large bed of rock which once extended conformably over the whole region, and has been shivered and dislocated by the upheaval of the trachyte itself. It seems the same sort of rock as exists about the base of Tunguragua, and forms the lofty cliffs on the southern side of that volcano, where the cataract of Guandisagua comes down at three bounds from the edge of the snow to the warm valley of Capíl, in which grow oranges and the sugar-cane. I have never been able to find any trace of fossils in this rock. . . . Nowhere in the Quitonian Andes have I seen the stratified rocks—limestones, friable sandstones, and fossiliferous shales—all, I believe, belonging to the lias formation, which constitute the eastern declivity of the Andes of Peru, or, at least, of the Province of Maynas. No Bark tree was seen growing on rock of any kind. The soil at Limon is the same deep loamy alluvial deposit, with very few stones intermixed, as we had seen from Llullundengo downwards, nor does a bit of rock crop out in the whole of the descent. . . . The northern and eastern sides of the trees had borne most flowers, and, except on one tree of more open growth than the rest, scarcely a capsule ripened on their southern and western sides. These phenomena are explained by the fact that, in the summer season, the trees receive most sun from the east and north, for the mornings are generally clear and sunny, whilst the afternoons are almost invariably foggy, and the sun's declination is northerly. Another notable circumstance is that the trees standing in open ground—pasture, cane-field, etc.—are far healthier and more luxuriant than those growing in the forest, where they are hemmed in and partially shaded by other trees, and that, while many of the former had flowered freely, the latter were, without exception, sterile. This plainly shows that, although the Red Bark may need shade whilst young and tender, it really requires (like most trees) plenty of air, light, and room wherein to develop its proportions.

. . . .

The cascarilleros have found out that the bark is worth money, but neither they nor the greater part of the inhabitants of Ecuador have any correct idea of the use that is made of it in foreign countries; the prevalent opinion being that a permanent coffee- or chocolate-coloured dye (still a desideratum in Ecuador) is extracted from it. I explained to the people of Limon how it yielded the precious quinine which was of such vast use in medicine; but I afterwards heard them saying one to another, "It is all very fine for him to stuff us with such a tale; of course *he* won't tell us how the dye is made, or we should use it ourselves for our ponchos and bayetas, and not let foreigners take away so much of it." There is to this day the same repugnance to using the bark as a febrifuge as Humboldt remarked sixty years ago, and as exists also in New Granada, where Cedron and various other substances are preferred to Quina. I think I can explain this repugnance. The inhabitants of South America, although few of them have heard of Dr. Cullen, have a theory which refers all diseases to the influence of either *heat* or *cold*, and (by what seems to them a simple process of reasoning) their remedies to agents of the opposite complexion; thus, if an ailment have been brought on by "calor," it must be cured with "frescos"; but if by "frio," with "calidos." Confounding cause and effect, they suppose all fever to proceed from "calor." Now they consider the cascarilla a terribly strong "calido," and justly; so, by their theory (which is the reverse of Hahnemann's), its use could only aggravate the symptoms of fever. . . .

Even at Guayaquil there is such a general disinclination to the use of quinine that, when the physicians there have occasion to prescribe it, they indicate it by the conventional term "alcaloide véjetal," which all the apothecaries understand to mean "sulphate of quinine," while the patient is kept in happy ignorance that he is taking that deadly substance.

The lowest site of the Red Bark at Limon is at an elevation of 2450 feet above the sea, where the Chasuán receives the rivulet already mentioned as running below our hut. It is precisely the point where the track from Ventanas leaves the Chasuán (along whose margin it had run thus far, with a gentle ascent from the plain) and begins to ascend the steep cuesta separating the Chasuán from its tributary, the ascent being 350 feet in the first 500 yards; so that where the real ascent of the Andes begins there also begins the Red Bark. At San Antonio, however, I saw a tree at a height of no more than 2300 feet; and, if I might believe my informants, trees of immense size have been cut down at points whose height I estimate at barely over 2000 feet. Following the track leading to Guaranda, the last Bark trees growing by the roadside are at a height of 3680 feet; but leaving the track, and following the hill-side on the left bank of the Limon, there are Bark trees scattered about for a distance of a league, and up to a height of near 4500 feet. On the opposite ridge, or that separating the Limon from the Chasuán, there are also several trees ascending to a still greater elevation, or nearly to 5000 feet; but I did not take the barometer to these latter, which were all sterile, in consequence of growing in lofty forest.

The cascarilleros do not usually go in quest of Bark trees before August, there being generally less fog in that and the following month than at any other period of the year. . . . The trees being cut down and the roots dug out, the bark is stripped off much in the same way as oak bark in England, but no other tool than the machéte is used. . . . For drying the bark a stage 3 feet high is erected, called a tendál. Care must be taken that the flame from the fire beneath the tendál does not reach the bark, and if rain be apprehended the whole has to be roofed over. When the bark is perfectly dry, they have only to convey it to the depot at Camaron and receive their twenty dollars per quintál, which is the price usually paid them by Messrs. Cordovez and Neyra; or rather, they have generally received the value in advance, according to the custom of the country.

. . . .

In the valleys of the Chasuán and Limon I saw about 200 trees of Red Bark standing. Out of the whole number, only two or three were saplings which had not been disturbed; all the rest grew from old stools, whose circumference averaged from 4 to 5 feet. I was unable to find a single young plant under the trees, although many of the latter bore signs of having flowered in previous years. This was explained by the flowering trees growing uniformly in open places, either in cane-fields which had been frequently weeded or in pastures where cattle had grazed and trodden about. The young plants, which I had been assured I should find abundantly, proved to be either stolons or seedlings (very few of the latter) of the worthless *Cinchona magnifolia*, which grows plentifully at Limon, and must have fruited during the rainy season, as the capsules were all burst open when I arrived there.

Cinchona succirubra is a very handsome tree, and, in looking out over the forest, I could never see any other tree at all comparable to it for beauty. Across the narrow glen below our hut, and at nearly the same altitude, there was a large old stool from which sprang several shoots, only one of which rose to a tree, while the rest formed a bush at its base. This tree was 50 feet high, branched from about one-third of its height, and the coma formed a symmetrical though elongated paraboloid. It had never flowered, but was so densely leafy that not a branch could be seen; and the large, broadly oval, deep green and shining leaves, mixed with decaying ones of a blood-red colour, gave the tree a most striking appearance. *C. magnifolia*, called here Cascarillo macho (male bark), grows rapidly to be a large tree. I saw one which must have been over 80 feet high, and I cut down a young tree which measured 60 feet. Saplings of 15 to 20 feet have a very noble appearance, from the large heart-shaped leaves, little short of a yard long; but in full-grown trees the ramification is so sparse and irregular, and the leaves are so much mutilated by caterpillars, that all beauty is lost. This species sends out stolons from the root, which sometimes form a matted bed, looking like a growth of seedlings. I have not observed the same peculiarity in any other Cinchona.

. . . .

I proceed now to give some account of the other indigenous inhabitants of the Red Bark woods, animal and vegetable.

The Andine Bear, chiefly inhabiting the middle wooded region, descends to the lower limit of the Red Bark. On the eastern side of the Andes it rarely goes as low as 3000 feet.

The Jaguar (*Felis onça*), chiefly inhabiting the plain, does not yet seem to have climbed as high as Limon, but at Tarapoto, in the Andes of Maynas, it was abundant up to more than 3000 feet elevation. The Puma or Leon (*Felis concolor*) exists not only in the plain but throughout the wooded slopes of the Andes; it is only too abundant in the roots of the Cordillera, and I have seen its footsteps on recent snow at a height of 13,000 feet on the high mountains to the eastward of Riobamba. "Puma" is the Indian generic name for every sort of tiger, but the Spanish colonists limit it to the red tiger, and call the spotted jaguar "tigre." Bears never troubled our hut, but we had two nocturnal visits from the puma. On one of these occasions the puma seized and was carrying off a little dog, but a very large and fine black dog sprang on the puma and forced him to let go his hold. . . . The screams of an animal seized by a tiger are about the most doleful sounds one ever hears in the forest, and after being once heard their cause can never be mistaken.

The Wild Pig (Peccary) frequently ascends to Limon, where there are also two or three smaller pachyderms.

Two sorts of Monkeys are common, one of them almost as noisy as the howling monkey, but of a different genus. I do not know of any monkey which ascends to the temperate region of the Andes.

A pretty red-headed Parrot, so small that it might be taken for a paroquet, arrived in immense flocks about the end of July and took up its summer residence in the Red Bark woods. The same species abounds in the valley of Alausí, where it makes sad havoc of the maize crops, and ascends by day to 8500 feet, but always descends to Puma-cocha to roost. Along with the parrots came Toucans of two species.

Snakes are very frequent, and some of them venomous. Limon seems to be the highest point to which the Equis ascends, a large and deadly snake which is a great pest in the plains of Guayaquil; it takes its name from being marked with crosses (like the letter "x") all along the back.

Butterflies I have rarely seen in greater number, and they include at least four species of those large blue butterflies (probably species of Morpho) which, on the eastern side, are seldom seen above the hot region. Cockroaches, too, ascend higher than I have elsewhere observed. We had four or five species in our hut, none of them large, and one very minute species which often damaged my fresh specimens of plants by mutilating the flowers. It is so abundant at Camaron, 1000 feet lower down, that it fills the pease and barley meal and renders them uneatable. Ants are far more frequent than in the temperate region, but less so than in the plains. House flies are as great a nuisance as at Ambato, and though fleas are not quite so numerous as in the cool sandy highlands, there were yet plenty of them (as the Spaniards say) "para el gasto."

. . . .

As above indicated, Limon was once entirely clad with forests, in which respect it contrasts strongly with the valley of Alausí, where the slopes on both sides are covered with grass, even down to the hot region, and only the lateral valleys and the plateaux are wooded. I cannot doubt that the difference arises from the former being situated in the roots of a snowy mountain, while there is no perpetual snow within a long distance of the latter. I have observed the same difference, referable to the same cause, along the eastern side of the Andes. After passing the valley of San Antonio, to the southward, there is this intermixture of woods and pajonales all the way to the frontier of Peru. As would naturally be expected, the vegetation at Limon is far more luxuriant, and the abundance of ferns, especially in the narrower valleys, is in striking contrast to their scarcity at Puma-cocha. Tree-ferns, of five species, are everywhere scattered in the forest, and add a feature of beauty to the scenery quite wanting in the valley of Alausí.

I estimate the average height of the virgin forest at Limon at 90 feet; but, as everywhere else in the tropics, there are here and there trees which stand out far above the mass of the forest. The monarch of the forest at Limon is an Artocarpea, which, from the leaves and from flowers picked up beneath the trees, I have little hesitation in referring to Coussapoa. The following are the dimensions of a tree of this species which I found prostrate in a recent clearing. Length 120 feet, not including the terminal branches, which had been lopped off, still 20 inches in circumference, and which would have made it at least 20 feet more. Circumference at 10 feet from the ground 12 feet 4 inches; from that point narrow buttresses were sent off to the ground on all sides. At 25 feet the trunk was forked, and the ramification was thenceforth dichotomous, at a narrow angle.

No other tree reaches the dimensions of the Artocarpea. A Lauracea, called Quebra-hacha (Break-axe), rises to 110 to 120 feet; its exceedingly hard wood is the usual material for the cylinders of the trapiches. My collection contains unfortunately very few of the larger trees. On the western slope of the

Quitonian Andes there is a great burst of blossom at the commencement of the dry season, that is, towards the end of May; and another of less extent after the rains of the autumnal equinox; so that, as my visit fell between those two epochs, many of the trees were in the same unsatisfactory state as the Hill Bark already mentioned, and others had not yet begun to flower. Besides, I should hardly, under any circumstances, have been at the trouble of cutting down a large tree for the sake of only two or three specimens; and, after we began to prepare the Bark plants, the Indians could hardly be spared for any other service.

In proceeding to give a classified list of the plants collected and observed, I shall generally limit myself to indicating their natural order. In order that my attention might not be called away from the main object of the enterprise, I collected very few (often unique) specimens of each plant. . . . The general character of the vegetation may, however, be sketched very intelligibly with very little reference to species.

[The following account of the vegetation of the Red Bark forests has been reduced by the omission of all passages not directly bearing on the subject, or dealing only with botanical details. It is, however, so full of information on points of geographical distribution and of examples of unusual plant-structure, and also contains so many short descriptions of strange or beautiful flowers still unknown to our horticulturists, as to make it both interesting and instructive to all who study or appreciate the beauty and variety of the vegetation of tropical regions. It is therefore, with these exceptions, printed entire.]

SKETCH OF THE VEGETATION OF THE RED BARK FORESTS OF CHIMBORAZO (alt. 2000 to 5000 feet)

Gramineæ, 4.[1]—A good many species of this order were observed, but, as is mostly the case in the dry season, nearly all partially dried up and out of flower; besides that, even in the

[1] The number affixed to most of the orders indicates how many species of that order I gathered in a perfect state.

recesses of the forest, they were sought out and cropped by the starving animals. After the bamboo above spoken of, the Arrow-cane (*Gynerium saccharoides*) is the most notable grass, and forms considerable beds, especially near streams. This species is abundant enough on low shores and islands of the Amazon, but it has nowhere spread far from the river-bank, nor (so far as I can ascertain) is it found wild on any of the tributaries of that river, but those which rise in the Andes. . . . Even on the Amazon it looks dwindled, and rarely exceeds 18 or 20 feet high; but on reaching the roots of the Andes of Maynas, one begins to see this noble grass in its true proportions. . . . It attains its maximum of development on stony springy declivities, at an elevation of about 1500 feet above the sea, where a forest of Arrow-cane, with its tall slender stems of 30 to 40 feet, each supporting a fan-shaped coma of distichous leaves, and a long-stalked thyrse of rose and silver flowers waving in the wind, is truly a grand sight. The longest stem I ever measured was one I met a man carrying on his shoulder at Tarapoto. From that stem had been cut away the leaves and peduncle, and the base of the stem, which is generally beset by stout-arched exserted roots (serving as buttresses), to a height of 1 to 3 feet; yet the residue was 37 feet long, so that the entire length must have been at least 45 feet.

The other grasses accompanying the Red Bark comprise several of those rampant forest Panica which thread among adjacent branches to a height of 15 feet or more. The long internodes serve as tubes for tobacco pipes and for other similar uses. There are also two broad-leaved Gamalotes of the same genus. Of grasses frequent in the hot plains I noted only *Dactyloctenium Ægyptiacum* and *Paspalum conjugatum*.

Cyperaceæ, 1.—This order is scarce, both in individuals and species. The half-dozen species observed belong chiefly to Scleria and Isolepis.

Araceæ, 4.—As abundant and varied as in the forests of the plains. An arborescent species, called Casimin by the inhabitants, grows everywhere, even on hills where there is little moisture. The stems reach 10 feet, and are sometimes thicker than the thigh, though so soft that a very slight stroke of a cutlass suffices to sever them. The small spathes are fascicled in the axils of the leaves, but of all that I opened the contents were so injured by earwigs and other insects that it was impossible to ascertain the structure of the flowers. . . . Species of Anthurium and Philodendron are frequent, and their deeply-cloven or perforated leaves often assume grotesque forms. One very beautiful climbing Aroidea, with shaggy petioles and leaves streaked with deep violet above, purple beneath, I could never find in flower.

Cyclanthaceæ.—Three scandent species of Carludovica, all with bifid leaves.

Palmaceæ.—Frequent enough, but of few species. The Cadi or Ivory palm is everywhere dispersed, and is precisely the same species as I saw at Puma-cocha. I gathered and analysed the male inflorescence, but the stripping off the fronds for thatch is unfavourable to the development of the fruit, which I never saw in a perfect state. A very prickly Bactris, 20 feet high, with five or six stems from a root, grows here and there; and in shady places three or four Geonomæ are frequent. The Euterpe grows chiefly at the upper limit of the Red Bark. A noble Attalea (called Cumbi and Palma real) extends up the valley of San Antonio to the lower limit of the Bark region. It has a slight beard to the petiole.

Bromeliaceæ.—Many species are perched about on the trees, but none of striking aspect. The presence or absence of this family affords no indication of climate on the equator, for trees of Buddleia and Polylepis, at the upper limit of arborescent vegetation, are as thickly hung with a Bromeliacea as any trees on the Amazon.

Amaryllideæ, 2.—Both herbaceous twiners, the one a Bomarea, with pendulous umbels of showy flowers, calyx red, corolla white, with violet spots; an order, so far as my experience goes, entirely absent from equinoctial plains, but tolerably abundant in the temperate and cool regions of the Andes.

Musaceæ.—Heliconia, two species.

Zingiberaceæ.—Cossus, three species. This is about the highest point at which I have seen any Cossus or Heliconia, twò genera frequent in the plains.

Marantaceæ.—Two or three species of Maranta were observed.

Orchidaceæ, 28.—Tolerably abundant, but comprising few handsome species. Most epiphytal Orchids love light, and in the dense lofty forest they are rarely seen, and often inaccessible, for they grow on the upper branches of large trees, and descend to the lower branches only on the margin of wide streams, where the whole of one side of the trees is exposed to the light. At Limon, however, in ancient clearings, now become pastures, where a few trees of the primitive forest have been left, and where others have here and there sprung up, despite the treading about of cattle, the branches are laden with Orchids and Vacciniums; and although none of the former be of remarkable beauty, yet they are in so great variety, and there is such a charm in seeing them on the rugged mossy trees in their native woods, that to me they were always objects of interest. The finest Orchid, as to its flowers, is an Odontoglossum, with large chocolate-coloured flowers, margined with yellow. As respects foliage, a fairy Stelis (*S. calodyction*, MSS.), with roundish pale green leaves, beautifully reticulated with the purple veins, far excels every other plant seen in the Cinchona woods. I found but a single tuft, almost buried in moss on the trunk of a tree. An Orchid (genus unknown), with thick coriaceous leaves, curiously spotted with white—a rare feature in epiphytal Orchids—was discovered by Mr. Cross. Very remarkable was an Oncidium, with numerous peduncles, 10 feet long, twining round one another and on adjacent plants. Besides the Orchids growing on trees, a good many species, allied to Spiranthes, grow on the earth and on decayed logs.

The 28 Orchids gathered in flower are, perhaps, scarcely a third of the whole number observed. On the slopes of the Andes some Orchid or other is in flower all through the year, and almost every species has its distinct epoch for flowering.

. . . .

Commelynaceæ.—Three species of Commelyna seen, chiefly near streams and in cultivated places.

Pontederiaceæ, 1.—A small creeping plant, with white or very pale lilac flowers, probably a Pontederia, in moist springy situations by the Chasuán.

Dioscoreacea.—Only the male plant seen of a Dioscorea.

Smilaceæ, 2.—Species of Smilax, both with roundish stems and a few prickles.

Gnetaceæ, 1.—A Gnetum (*G. trinerve*, MSS.), apparently parasitic, and remarkable for its three-ribbed leaves. It is the first species of this genus I have seen in the hills, though Gneta are common enough in the plains, and especially on the Rio Negro, where the kernel of the fruit is eaten roasted.

Myricaceæ, 1.—A wax-bearing Myrica, which descends to 2000 feet on open beaches of the Rio San Antonia, but was not observed by the Chasuán. The same or a very similar species grows on wide gravelly beaches of the Pastasa, Morona, and other rivers which descend the eastern slope of the Andes, and a good deal of wax is obtained from its fruit, principally by the Jibaro Indians, who sell it to traders from Quito, Ambato, etc., under the name of "Cera de laurél" or laurel wax.

Urticaceæ, 2.—Two or three fruticose Pileæ were observed, but the only plant gathered was a tree 25 feet high (growing by the Rio San Antonio), which seems a species of Sponia, a genus placed by some authors in Ulmaceæ.

Moraceæ.—Here and there grows a parasitical Ficus, but the species seemed much fewer than I have observed in other similar localities.

Artocarpeæ.—None gathered, although, as above remarked, the tallest tree of the forest belongs to this order. Two Cecropiæ are not infrequent, and another tree, with a tall white trunk and large hoary pedatifid leaves, looking quite like a species of the same genus, extends up the slopes of the mountains to 8000 feet, and has its lower limit above that of the Cinchona; but as I have never seen its flowers, and as the Cecropiæ are apparently confined to the hot and warm regions, I suppose it may be generically distinct.

Euphorbiaceæ, 3.—The species gathered comprise an Acalypha, a Phyllanthus, and a small tree of unknown genus. . . .

Callitrichaceæ.—A Callitriche, in pools by the Rio San Antonio.

Monimiaceæ.—Three species of Citrosma are frequent.

Menispermaceæ.—A woody twiner of this order was noted, probably an Abuta, but without flower or fruit.

Cucurbitaceæ, 8.—Plants of this family are abundant, and, besides the eight species gathered, some others were seen in a barren state. I gathered two Anguriæ, with trifoliolate leaves, and the characteristic scarlet flowers of the genus. One plant, apparently of this order, puzzled me much, for the woody stems, partly twining and partly climbing by means of radicles, and no thicker than packthread, bore a bunch of slender flowers (calyx scarlet, corolla yellow) near the base; but though I pulled down some stems of enormous length, I could see no traces of leaves on them. At length I succeeded in getting down an entire stem, 40 feet long (by no means one of the longest), which had a couple of trifoliate leaves near the apex. . . .

Begoniaceæ, 4.—Two climbing and two terrestrial species. Of the latter, one is a large coarse plant 10 feet high, with leaves resembling those of *Heracleum giganteum.* I have gathered the same, or a very similar species, on the eastern side of the Cordillera. One of the climbing species is very ornamental, from its long pinnate shoots bearing a profusion of roseate flowers and generally purplish leaves. This genus, entirely absent from the Amazonian plain, though it has one representative in that of Guayaquil, abounds on the woody slope of the Andes, especially in the warm and temperate regions.

Papayaceæ.—Two species of Carica were seen, both slender simple arbuscles of 5 to 6 feet, the one by the Chasuán, the other by the San Antonio. The leaves of the former are boiled and eaten by the inhabitants under the name of "col del monte" (wood cabbage). . . .

Flacourtiaceæ, 1.—A small tree, probably a species of Bonara.

Samydeæ.—A Casearia, which seems to be *C. Sylvestris*, grows in some abundance, but the fruits were open and empty. This is the highest point at which I have seen a species of Casearia, a genus abundant in the plains, especially in woods of secondary growth.

Passifloreæ, 2.—Both species of Passiflora; the one a woody twiner (frequently found on the Red Bark tree), with entire leaves, smallish green flowers, and globose berries the size of a cherry;

the other a beautiful arbuscle, seen only in the valley of San Antonio, where it grows from the very plain up to 2600 feet. The slender stems, of from 8 to 14 feet, are usually simple and arched, and the large white flowers grow in small pendulous corymbs from the axils of very large, elongate, glaucous leaves. . . .

Cruciferæ, 2.—Apparently species of Sisymbrium, the one growing near streams, the other in open situations; both in very small quantity. . . .

Capparidaceæ.—The only species observed was a Cleome, a genus which extends from the plain to a great height on the wooded hills.

Sterculiaceæ.—A raft-wood tree, Ochroma, is pretty abundant. Another tree of the same order (not seen in flower) appears to be a Chorisia.

Büttneriaceæ, 1.—A rampant Melochia. *Muntingia Calaburu*, a tree found in the plains on both sides of the Cordillera, grows abundantly by the Rio San Antonio, up to 2500 feet.

Malvaceæ, 2.—Four or five common weeds, whereof *Sida glomerata*, Cav., is the most plentiful, comprise all that was seen of this order.

Tiliaceæ, 1.—A very handsome tree, with a slender straight trunk, reaching 60 feet, very long branches, large, ligulate, serrated, distichous leaves, and terminal panicles (sometimes 4 feet long) of yellow flowers, scented like those of *Tilia Europæa*; it is abundant and ornamental about the middle region of the Red Bark. Besides this tree, another of the same order (apparently a Heliocarpus), growing to about 30 feet, is also frequent. . . .

Polygaleæ, 2.—A Monnina and *Polygala paniculata*, L. . . . The Polygala of the Bark woods is the common and almost the only species of the equatorial Andes, on whose western slopes it descends to the plain, and does not seem to ascend higher than 7000 feet on either side, nor is it abundant at any elevation. When I recollect the abundance of *Polygala vulgaris* on cold English moors, I am struck with this paucity of Polygalæ in the Andes, and still more when I compare it with their abundance and variety on hot savannas of the Orinoco, and in hollows of granite rocks by the Atabapo.

Sapindaceæ, 1.—A woody climbing Serjania, a fine plant. A Paullinia with trigonous stems is frequent, and is the common substitute for rope, where much strength is not required. I saw no flowers of it, and only empty capsules. There is also a Cardiospermum, which I have seen on both sides of the Cordillera up to 7500 feet, and this is the greatest elevation at which I have noted any Sapindacea, an order which abounds in the hot plains.

Malpighiaceæ, 1.—A twiner, with fruit too young to enable me to speak positively of the genus. Plants of this order, which constitute so large a proportion of the vegetation of the plains, diminish rapidly in number and variety as we ascend the hills, and beyond the warm region of the Andes the scandent species entirely disappear; but a Bunchosia (probably *B. Armeniaca*)—a tree about the size of our pear trees—ascends high into the temperate region. On the hills which slope down to the left margin of the Pastasa this tree grows up to 8000 feet, and in some places forms large continuous patches, unmixed with any other tree. The edible, though rather insipid drupes, as large as a peach, are exposed for sale in large quantities in Ambato and the adjacent towns, under the name of "ciruelo de fraile" or friar's plum. . . .

Ternströmiaceæ, 1.— . . . Two species of the anomalous genus Saurauja form trees of about 30 feet, and are conspicuous from their abundance, from their large lanceolate serrated leaves, and axillary panicles of white flowers resembling those of *Fragaria vesca*. One of the two, with ferrugino-tomentose leaves, seems quite the same as I have gathered on Tunguragua up to 7000 feet (Pl. Exs. 5089). A Freziera descends on the banks of the Rio San Antonio to 2300 feet. . . .

Clusiaceæ, 3.—One of them, a Clusia, abundant and ornamental from its numerous rose-coloured flowers, but the plants nearly all males. . . . Two or three other Clusiæ were seen, not in flower or fruit.

Marcgraviaceæ, 1.—A Norantea, the same as that gathered in the Bark woods of Puma-cocha. *Marcgravia umbellata* is very abundant, and climbs to the tops of the loftiest trees.

Anonaceæ, 2.—The one a Guatteria, rather scarce at about 3000 feet, the greatest elevation at which I have ever observed the genus; the other a small Anona, also scarce; it bears an edible fruit, called "cabeza de negro," the size of an orange but longer than broad. This order has its chief site in the hot plains.

Ericeæ, Subordo *Vacciniaceæ*, 6.—Four Vaccinia, one Thibaudia, and one Macleania, all epiphytal shrubs. One of the Vaccinia, with fleshy rose- or blood-coloured leaves, densely (almost teretely) imbricated on the branches, and with slender red flowers in their axils, looked very pretty on the old trees; but the Thibaudia was still more ornamental, from the profusion of its large tubular flowers—calyx and corolla at first yellow, turning red after the bursting of the anthers, and persisting a long time; they unfortunately turn black in drying, so that my specimens give no idea of their beauty. In Thibaudia we have a remarkable example of a genus which ascends from the very plain (where, however, it is very scarce) nearly to the extreme limit of lignescent vegetation. Ericeæ, on the contrary, according to my observations, do not descend lower than 6000 feet, on the equator.

Amyrideæ.—Two small trees, of the genus Icica, were seen in flower; and some of the tallest trees with pinnate leaves,

I have no doubt, from their resinous juice, belong to the same order.

Meliaceæ, 1.—A species of Trichilia, called Muruvillo, whose bark is held as a febrifuge, barely enters the Bark region at San Antonio, but does not extend up to it at Limon. A tallish tree, with pinnate leaves and very large serrated leaflets, which was putting forth large terminal panicles when I left the woods, probably belongs to this order.

Zygophylleæ, 1.—A fine tree of 40 feet, with large opposite pinnate leaves; it is closely allied to Guaiacum, though scarcely referable to that genus.

Podostemaceæ.—The withered remains of at least three species were observed on granite rocks in the river San Antonio, and they are the first of the family I have seen in the Andes.

Oxalidaceæ.—At San Antonio grow two species of Oxalis, both of which I have previously gathered, the one on the eastern side of the Andes near Baños, and the other at Pallatanga on the western side.

Caryophyllaceæ.—A solitary species of each of the genera Stellaria and Drymaria grows very sparingly. In ascending the eastern side of the Andes, I first came on a Stellaria at between 2000 and 3000 feet. This order, frequent enough in the upper regions of the Andes, seems to exist in the plains at their base only in the genera Polycarpæa, Drymaria, and Mollugo, all three very scarce on the Atlantic side, but the last-named very abundant on the Pacific side.

Portulaceæ.—A Portulaca grows in sandy places inundated by the Rio San Antonio.

Polygoneæ.—A Triplaris, apparently identical with that observed at Puma-cocha, and possibly distinct from *T. Surinamensis*, extends a little way into the territory of the Red Bark, and in descending from thence becomes more abundant all the way down to the plain, where it is called by the Guayaquilians Arbol de frios or Ague tree. Its presence, indeed, is a pretty sure indication of a humid site.

Amarantaceæ, 1.—A woody twiner. There are besides two or three weedy plants of this order, probably species of Telanthera.

Chenopodeæ.—Two common weeds; one of them being the ubiquitous *Chenopodium ambrosioides*, which grows with almost equal luxuriance in the elevated central valley of the Andes and in the plains of the Amazon and Guayaquil.

Piperaceæ, 5.—Species of this order are very numerous. I saw perhaps as many as twenty, belonging chiefly to the genera Artanthe and Peperomia. A very fine pepper, resembling *Artanthe eximia*, Miq., but a still handsomer plant, grows towards the lower limit of the Bark region. The stem is 20 feet high, slender and perfectly straight, and beset with short, distant, nearly horizontal ramuli, from which hang almost vertically the large, Pothos-like, coriaceous, shining, deep blue-green leaves. A multicaul Artanthe, 15 to 25 feet high, springs up abundantly in the pastures, where trees of it grow at such regular distances, and are so conspicuous by their yellow-green foliage, that one would suppose them planted. Their ashes afford an excellent lye for soap. On stones by the Rio San Antonio grows a stout Peperomia, 1 to 2 feet high, subramose, and putting forth axillary fascicles of slender white spadices, which exhale a strong odour of aniseed. When in the midst of a dense patch of it, the scent is almost stifling, though pleasant enough at a short distance. Peppers are equally plentiful in the plains and throughout the wooded slopes of the Andes.

Lauraceæ, 3.—All small trees, not exceeding 40 feet; but a great many more were observed, including some of the loftiest trees of the forest.

Leguminosæ (Subordo *Papilionaceæ*), 3.—Several others were observed, but either in poor state or inaccessible. Some of the lofty trees with pinnate foliage, which were not seen in flower or fruit, probably belong to this order. The commonest Papilionaceæ is a Mucuna, with herbaceous twining stems, without tendrils, and large yellow flowers. It is the first Mucuna I have seen in the hills, but it is equally abundant by the river Guayaquil. Five species of Erythrina were seen, two at Limon (one of them being the same as that gathered at Puma-cocha) and the remaining three by the Rio San Antonio. There are also two Phaseoli, one Dioclea, and another Phaseolea with slender spikes of small pale yellow flowers and hard scarlet seeds, of which I have not yet determined the genus. An Indigofera, with small pink flowers, was gathered at San Antonio, and the same is frequent in the plain of Guayaquil.

Leguminosæ (Subordo *Cæsalpinieæ*), 1.—This fine tribe, so abundant in the Amazonian plain, becomes scarce the moment we enter the hills, and is very poorly represented in the Bark woods. My specimens were gathered from the only tree I saw of an obscure-looking Cassia. There is, however, one very fine Cæsalpinieous tree, extending up the hills to 4000 feet, but much more abundant at 2000 feet. The trunk grows to from 20 to 60 feet, and the branches each bear a coma of very long, elegant, pinnate, pendulous leaves, like those of a Brownea. . . .

Leguminosæ (Subordo *Mimoseæ*), 4.—Three Ingæ and one Calliandra. Other two Ingæ were seen, without flowers. *Mimosa asperata*, perhaps the commonest of all plants on the muddy shores of the Amazon and the river Guayaquil, struggles up the Rio San Antonio to the lower limit of the growth of the Red Bark, but never seems to flower at that elevation.

Rosaceæ, 1.—A Rubus, with numerous small flowers, apparently distinct from *R. Urticæfolius*, Poir., which I gathered in Maynas at the same elevation (3000 feet), and these are the lowest points at which I have observed any Rosaceæ near the equator; although plants of this order, especially of the tribe Sanguisorbeæ, constitute a considerable proportion of the vegetation of the open highlands.

Hydrangeaceæ, 1.—A Cornidia. The same, or a very similar species, of this truly Andine genus grows by the Pastasa, on the eastern side of the Cordillera, at about 4000 feet, and other three species were gathered on Mount Campana, in Maynas, at 3000 feet. I have never seen any Cornidia either above or below the warm region.

Cunoniaceæ.—A pinnate-leaved Weinmannia, sometimes reaching 80 feet high, is very frequent, and extends down the banks of the Chasuán to perhaps 2200 feet. A humbler species descends nearly as low on the Andes of Maynas. On the wooded declivity of the volcano Tunguragua, Weinmanniæ constitute a considerable proportion of the vegetation, and extend upwards to at least 11,000 feet.

Lythraceæ.—A Cuphea, a small, weak, much-branched undershrub, with purple flowers, grows gregariously in the pastures, generally accompanied by *Sida glomerata* and a Stachytarpheta. By the Rio San Antonio grow other two Cupheæ, one of which grows also in the valley of Alausí. This genus, abundant in the plains on both sides of the Cordillera, spreads up the hills to 7000 feet, or through the region of the Red Bark, but scarcely up to that of the Hill Barks. *Adenaria purpurata* grows by the Rio San Antonio up to 2500 feet, and descends on its banks into the plain, the same as it does by streams on the eastern side of the Cordillera.

Onagraceæ, 1.—Three species of Jussiæa grow by the Rio San Antonio. In the warm and hot regions this genus takes the place of Œnothera, which is frequent in the hills, but rarely descends below 6000 feet. In other parts of South America, as for instance along the coast of Chili, Jussiææ are found inhabiting a cool climate. A single plant of a large-flowered Fuchsia was gathered at about 2700 feet. A similar species occurs very rarely on the eastern side of the Cordillera, at a little higher elevation. These are the only instances I know of Fuchsias descending so low, their favourite climate being found in the temperate and cool regions of the Andes, say from 6000 to 11,000 feet.

Melastomaceæ, 9.—The first plant which took my attention at Limon, after the Cinchona, was a beautiful epiphytal Blakea, growing from 12 to 18 feet high, with broad coriaceous leaves and large rose-coloured flowers, from which features, and from its often sitting high up the trees, it has almost the aspect of a Clusia. At the base of each flower is a turgid involucre, of four large, orbicular, widely and closely imbricated leaves, within which is secreted a limpid fluid. When the corolla falls away, the involucral leaves close firmly over the calyx, and do not open out, nor does the contained fluid dry up, until the globose roseate berry, the size of a pea, is quite ripe. Another singular character is the syngenesious anthers, with a minute pore at the apex of each cell, through which not a grain of pollen ever escapes, as I satisfied myself by repeated observation; fertilisation being effected through the agency of minute beetles, which abound in the flowers, and eat away the inner edge of the anther cells, probably part of the pollen also. . . .

The remaining Melastomaceæ offer nothing noticeable, except the scarcity of Miconia, the South American genus most abundant in species and individuals, and occurring from the plain to the limits of true forest on the hills. I gathered but one species, which I refer doubtfully to Miconia.

Myrtaceæ, 1.—Two or three Myrciæ, which are rather scarce. A fine Eugenia, called "Arrayàn" (but different from the Arrayàn of Quito), with very hard, durable wood, and exfoliating bark, grows to a tree of 60 feet or more. Two Psidia are frequent; the one (on the beaches by the Rio San Antonio) seems the common Guayaba of the temperate region; the other is a timber tree called Guayaba del Monte, which, although of very slow growth, ultimately reaches the dimensions of the Arrayàn, and yields equally valuable timber.

Barringtoniaceæ.—A Grias, with the characteristic coma of large elongato-lanceolate leaves, seems to reach its upper limit at about 3500 feet. . . .

Loasaceæ, 1.—A weak branching herb with small white flowers, probably an Ancyrostemon. There grows also in the cane-fields a virulently stinging Loasa, which is too common a weed on the eastern side, at about 5000 feet. This order, quite absent from the Amazonian plain, accompanies woody vegetation from about 1200 feet up to 11,000 feet at the least, and many of the species are climbers.

Umbelliferæ, 4.—Whereof three are Hydrocotyles, one of them departing from the habit usual to the South American species, in putting forth erect stems of 3 to 12 inches from a trailing rhizome. There is also a fourth Hydrocotyle (*H. pusilla*, A. Rich.), distinguished by its minute leaves and scarlet fruit, which I gathered at the same elevation on the Andes of Maynas. I have nowhere seen such abundance of Hydrocotyles in the forest as at Limon, where they constitute a notable proportion of the ground vegetation. In moist, open situations, on the higher grounds, they are common enough. . . .

Araliaceæ.—Two species of the fine genus Panax are not uncommon.

Rubiaceæ, 19.—I think I gathered every plant of this order I saw in tolerable state, but a good many more were observed, on the whole about 30. Of plants peculiar to the warm and temperate valleys of the Andes, never descending to the plain, at least in this latitude, the following may be mentioned: *Cinchona succirubra* and *magnifolia*, two Hameliæ (one with larger flowers than I have seen in any other species), a Gonzalea, and *Rubia Relboun*. Of genera abundant in the plains and rarely climbing the hills are Randia, Uncaria, Nonatelia, Faramea, and Cephaëlis. *Uncaria Guianensis*, a twiner with formidable aculeiform stipules, has a very remarkable distribution. I have thrice met with it on the Atlantic side of the Andes, viz. first, at Pará near the mouth of the Amazon; secondly, towards the head of the Orinoco; and thirdly, on the hill of Lamas, in the Andes of Maynas. In each of these three localities, so widely separated, it occupies a very limited area. I again met with it about the lower frontier of the Bark region, and on the rivers entering the Gulf of Guayaquil it is so abundant as to form a serious obstruction to navigation, especially in the upper part of their course, where the current is rapid and canoes ascending the stream must necessarily keep close inshore. . . . Of plants allied to Cinchona, the most remarkable is a fine epiphyte, resembling Buena and Hillia in the large white salver-shaped odoriferous flowers. . . . There is also a handsome tree, growing from 4000 feet upwards, perhaps allied to Lüdenbergia, but with a curious bilamellate crest on the apex of each segment of the corolla. I have previously gathered a congener at Tarapoto, and another on Tunguragua. Two very fine and closely-allied species of the tribe Gardeniæ I can refer to no described genus. One of them has leaves of immense size, near a yard long, and they are aggregated at the apex of a usually simple stem, so as to give it the appearance of a palm. The moment I saw it, I recollected having observed the same or a very similar tree near Santarem, where I could never find flowers, nor did I meet with it elsewhere on the Amazon.

Loranthaceæ, 1.—A Loranthus, with numerous small, yellow, sweet-scented flowers, growing abundantly, especially on Inga trees. There are many other species, but no large-flowered ones.

Aristolochiaceæ, 1.—Two Aristolochiæ were seen, but in a barren state. A third species, scarcely referable to Aristolochia, was gathered with young flowers. None of the three were seen climbing on the Red Bark tree.

Lobeliaceæ, 3.—One Centropogon and two Siphocampyli. One or two other species of the latter genus were seen. The only Lobeliacea I have seen in the plain is *Centropogon Surinamensis*, which I gathered at the foot of the granitic mountain Iméi, at the source of the river Pacimoni.

Valerianaceæ, 1.—A slender twining Valeriana. This genus, absent from the plains, begins to be met with in the hills at about

3000 feet, and extends thence to the very snow-line, going through more phases in external appearance than I know in any other genus.

Compositæ, 3.—So long as I herborised only in the plains, I could never understand how Humboldt had assigned so large a proportion of equinoctial vegetation to Compositæ, for, from the mouth of the Amazon to the cataracts of the Orinoco and the foot of the Andes, with the exception of a few scandent Vernoniæ and Mikaniæ, and of a few herbs on inundated beaches of the rivers, the species of Compositæ that exist are weeds, common to many parts of tropical America, nor did I meet with more than one arborescent Composita (*Vernonia polycephala*, DC.) in the whole of that immense area. But in ascending the Andes, from 1200 feet upwards, Compositæ increase in number and variety at every step, and include many arborescent species. About midway of the wooded region, and especially in places where the trees form scattered groves rather than continuous woods, Compositæ are more abundant than any other family, both as trees and woody twiners, and in the latter form extend nearly to the limit of arborescent vegetation, especially as species of the fine genus Mutisia; while on the frigid paramos no frutescent plants ascend higher than the Chuquiraguas and Loricarias, and as alpine herbs, the Achyrophori, Werneriæ, etc., reach the very snow-line. In the Red Bark woods Compositæ are plentiful, and I should estimate the number of species at near 50. The trees of this order are chiefly Vernoniæ, and they abound most in deserted clearings. During my stay, a plot was again brought under cultivation which had remained desert for twelve years, during which period it had become so densely and equably clad with a Vernonia, whose slender white stems had reached a height of 40 feet, that at a distance it looked like a plantation. Many of the woody twiners are Compositæ, chiefly Senecionidæ, and as herbaceous or suffruticose twiners there are several Mikaniæ. The young shoots of a species of Mikania bear very large cordate leaves, usually white over the veins and purple or violet on the whole under-surface. . . . Among shrubby Compositæ I noted some Eupatoria and two Baccharides, but no Barnadesia; nor among herbs any Gnaphalium, although on the eastern side of the Cordillera the two latter genera descend nearly to 3000 feet. *Tessaria legitima*, DC., is abundant by the Rio San Antonio. I have come on this tree in the roots of the Cordillera on both sides, by all the streams which have open gravelly or sandy beaches laid under water by occasional or periodical floods.

. . . .

Apocyneæ, 2.—One Peschiera and one Echites. This order rarely ascends up out of the hot region in the Andes, and in the temperate region I have seen only a single species.

Asclepiadeæ, 4.—All milky twiners. This order, like the preceding, has its principal seat in the hot region, but is by no means confined to it, for two or three slender Cynoctona are frequent in the cooler parts of the Andes, trailing over the hedges of Cactus and Agave.

Solanaceæ, 5.—In this order, also, my collection contains a very small proportion of the species existing in the Red Bark woods. Shrubby Solana are almost endless, and two species rise to trees. Two or three species of Cestrum also occur as slender trees.

Cordiaceæ, 1.—A Cordia, a stout sarmentose species, which threads about among the trees up to a considerable height, though it never actually twines.

Convolvulaceæ.—This order seems confined to a couple of Ipomææ, both occurring very rarely.

Myrsineæ, 2 (or perhaps 3).—The most remarkable of all the plants I gathered is a Myrsinea, though, as it grows only at from 5000 to 7000 feet, it barely touches the frontier of the Red Bark region. It is an arbuscle of 8 to 10 feet, bearing a coma of large, long, deep green coriaceous leaves, so that without flower it has quite the aspect of a Grias; but above the leaves there is a mass, the size of the human head, of densely packed panicles and minute flowers, all of the same deep red colour. I have not previously seen any Myrsinea at all resembling it in habit; but I have examined it sufficiently to state with confidence that it belongs to this order, although probably to an undescribed genus.

Labiatæ, 1.—Besides the solitary species gathered, there exist two species of Hyptis, one of them apparently *H. Suaveolens*; but this order is always scantily represented in the forest. In cane-fields at San Antonio I saw a Stachys with small white flowers.

Verbenaceæ, 2.—One of them a prickly suffruticose Lantana, threading among the bushes up to 18 feet in height; the other a woody twiner, with pretty waxy flowers, flesh-coloured externally, but the limb purple within; it is probably a Citharexylon, allied to *C. scandens*, Benth. (gathered on the Uaupés), though the habit is totally different from the arborescent Citharexyla which grow in the cooler parts of the Andes. A Duranta was noted at San Antonio. A Stachytarpheta, which I take to be *S. Jamaicensis*, and is known in Peru and Ecuador as "Verbena," seems to follow the steps of man in the Cordillera from near the plain up to 10,000 feet. At Limon it exists sparingly as a weed. Another species of the same genus, with very slender spikes and small lilac flowers, abounds in open places.

Gesneraceæ, 17.—The abundance of this family is one of the distinctive features of the Red Bark woods. One group, comprising several species, has a woody rhizome, creeping up the trees, and few long sarmentose leafy branches. The leaves of each pair are very unequal, and the smaller one sometimes obsolete; the larger one is long, lance-shaped, and, while the rest of the leaf is green, the apex and sometimes part of the margin are stained of a deep red, so as to resemble a lance dipped in blood, whence the native name "punta de lanza." The axillary flowers are comparatively inconspicuous, and they are partially concealed by large red or blood-stained bracts; they seem to vary considerably in structure in the different species, but I have scarcely examined them, and cannot, therefore, refer these plants with certainty to their proper genus. Another group, whereof two species were seen and gathered, has the long tubular corolla subtended by pinnati-partiti sepals, which are so densely beset with stout jointed hairs as to resemble the calyx of a moss rose, a peculiarity which I do not find noted in any described species of this order. One of the two is a small under-shrub, with the calyx and the corolla yellow; the other a slender herbaceous twiner with a scarlet calyx and a dull violet corolla. An Achimenes, with pretty scarlet flowers, abounds along the declivities.

Bignoniaceæ, 2.—The one a Bignonia, with round stems; the other an Amphilophium, with 6-angled stems; both twiners. Another Bignonia was seen, not in flower. I saw no tree of this order, though Tecomæ exist both in the plain and in the cool hill forests. I have never seen any climbing Bignoniaceæ at a greater elevation than about 3500 feet, but they form a large proportion of the scandent vegetation of the hot plains.

Acanthaceæ, 9.—This order is tolerably abundant, and two under-shrubs growing about the lower boundary of the Bark region bear spikes of large handsome scarlet flowers, in appearance like those of a Justicia, but different in character. A Mendozia, with woody twining stems and umbels of small white verbena-like flowers, grows everywhere.

Scrophulariaceæ, 4.—All humble herbs, two of them species of Herpestes, and all rather scarce.

Of Ferns and their allies I gathered the following:—

	Species.
Equisetum	1
Lycopodium	2
Selaginella	6
Polybotrya	1
Rhipidopteris	1
Elaphoglossum	5
Lomaria	2
Blechnum	1
Xiphopteris	1
Gymnopteris	1

	Species.
Tenitis	3
Adiantum	6
Hypolepis	1
Pteris (including Litobrochia)	5
Meniscium	1
Asplenium (including Callipteris, Diplazium, and Oxygonium)	21
Hemidyctium	1
Didymochlæna	1
Polypodium	4
Phegopteris	5
Goniopteris	2
Dictyopteris	1
Goniophlebium	2
Campyloneuron	5
Niphobolus	1
Pleopeltis	3
Anapeltis	3
Dipteris	1
Aspidium	5
Nephrodium	6
Lastrœa	11
Nephrolepis	3
Davallia	2
Cyathea	1
Hemitelia	1
Alsophila	4
Gleichenia	1
Trichomanes	4
Hymenophyllum	5
Lygodium	1
Total	131

From these should be deducted 10 or 12 species gathered beyond the limits of the Red Bark, which will leave (say) 120 species. Within those limits the following Ferns were seen, but not gathered, either because they are common throughout tropical America or from the specimens being imperfect: *Azolla Magellanica*; Equisetum sp.; *Pteris aquilina*, var. *caudata; Gymnogramme calomelanos*, and another species of that genus (Pl. Exs. 4153) which grows everywhere in the roots of the Cordillera on gravelly beaches; *Cyclopeltis semicordata*, a common fern in the hot and warm regions, wherever there are rocks; a loosely pilose Pteris, in very ragged condition, gathered previously at Tarapoto (Pl. Exs. 4667); a Dicksonia, of which I saw only young plants and old frondless trunks; several species of Elaphoglossum, of which the fertile fronds were shrivelled up, having been in perfection in the wet season, and two or three Hymenophylla in the same state; so that if we make allowance for the few species which must have eluded my search, we may safely assume that I left at least 20 ferns ungathered, and the whole number may be taken at 140, that is, of ferns existing in a space not more than four miles long by three-quarters of a mile broad, or of *three square miles*. Perhaps few parts of the world possess so many species of ferns growing naturally in so small an area.

The five species of tree-ferns gathered in fruit all grow in tolerable abundance, and one of them, an Alsophila, with a trunk 40 feet high, large, stout, pale green fronds, and exactly opposite pinnæ, is perhaps the handsomest tree-fern I ever saw. The Cyathea has almost constantly, below its own fronds, a supplementary crown of numerous deep green, widely arched, sterile fronds of a Lomaria, among which spring vertically the slender, pectinate, fertile fronds; while the trunk is enveloped in a continuous sheath of the soft, pale, but clear green foliage of *Bartramia viridissima*, C. Müll.; the whole forming one of those lovely pictures which only those who seek out Nature in her remotest recesses are privileged to see.

Musci

This Bartramia was in good fruit, but the great part of the mosses had fruited during the rainy season, and the number of species was by no means so great as one would have supposed, to see the dense festoons of moss depending from old trees. They are in main part composed of two or three species, which modern botanists would refer to Trachypus, of as many Meteoria, and of a Frullania. *Rhacopilum tomentosum* is frequent, as it is all through the roots of the Cordillera, on both sides; and another Rhacopilum (*R. polythrincium*, MSS.) grows in some abundance. Orthotricha, common enough in the region of the Hill Barks, scarcely descend below 6000 feet, and at Limon their place is supplied by Macromitrium and Schlotheimia, both very sparingly represented. Hookeriæ, so abundant and ornamental on the eastern slope of the Cordillera, in the same latitude and altitude, barely exist at Limon.

Hepaticæ

Hepaticæ are rather more varied than mosses, and the genus Plagiochila, especially, is well represented. Notwithstanding the vast variety of Plagiochilæ I have gathered on the Amazon and on the eastern side of the Andes of Peru and Quito, I still found new forms at Limon. The favourite site of this genus is in the warm and temperate region of the Andes. Lower down the number of species diminishes rapidly, and higher up, towards the limit of the forest, the huge masses of robust Sendtneræ, Lepidoziæ, and in some places of Frullaniæ, leave little room for the delicate Plagiochilæ. Lejeuniæ, on the contrary, are hot country plants.

Lichens

Of Lichens, the foliaceous species are remarkably scarce. Epiphyllous lichens, whose abundance and variety is so notable a feature of the vegetation of the Amazon, seem to attain their upper limit in the Red Bark woods. The trunks of the trees are generally too well covered with mosses to leave much room for the development of crustaceous lichens. Still, a good many species exist, chiefly Graphideæ, and I did not notice any lichen on the Red Bark which does not grow indifferently on other sorts of trees. . . .

Reserving the important subject of climate to be last discussed, I resume my narrative of operations.

In the month of July a report reached us that an Englishman, bringing with him a number of boxes, had arrived at Ventanas. On the strength of this I immediately sent Dr. Taylor thither with horses, and he had the great satisfaction of finding the Englishman to be Mr. Cross. Ventanas, however, was so full of soldiery, and was so likely to be soon the theatre of a conflict (for the opposing army lay encamped only a few leagues lower down the river), that Dr. Taylor very wisely had the materials for the Wardian cases removed about three hours' journey up the river, to a farm called Aguacatál, where they were not likely to be molested.

Mr. Cross had had all sorts of obstacles thrown in his way by the forces that held the river, and with the greatest difficulty had found men to row his canoes, so that the distance from Guayaquil to

Ventanas (which appears so short on the map) had taken him thirteen days to travel. He finally reached Limon on the 27th of July, looking pale and thin from his recent illness and from the sleepless nights passed on the river, but anxious to set to work immediately. We had no young plants for him, nor any expectation of obtaining them, but I was satisfied that cuttings would succeed, although it would necessarily be a tedious process to root them well. The owner of the chacra of Oso-cahuitu showed me some sprigs, cut from an old stool of Red Bark, which he had stuck into the ground by a watercourse four months previously, and they had all rooted well. Mr. Cross also agreed with me that the success of the process was certain, and that the question was merely one of time, which only experience could solve. After reposing the following day (Sunday), we had a piece of ground fenced in, and Mr. Cross made a pit, and prepared the soil to receive the cuttings, of which he put in above a thousand on the 1st of August and following days. He afterwards put in a great many more, subjecting them to various modes of treatment; and he went round to all the old stools, and put in as many layers from them as possible; but only those who have attempted to do anything in the forest, possessing scarcely any of the necessary appliances, and obliged to supply them as far as possible from the forest itself, can have any idea of the difficulties to be surmounted. Glass was the only thing for which we could find no substitute, and to get up to Limon the glasses of the Wardian cases was not to be thought of, over roads so narrow and rough, where even the surest-footed beast goes on continually stumbling. So we made our frames of palm-fronds, our buckets of bamboos, and invented similar contrivances for other needful articles. The closed communication with Guayaquil was felt to be a sore obstacle, as we might have sent thither for canvas and other things required for the plants, and also for a little wine and porter for the invalids.

The mornings were always cool and sometimes dull, but at 7 o'clock or so the sun would often come out blazing hot. In the afternoons, when the fog seemed to have set in for the day, it would sometimes clear away for a brief space, and admit the scorching rays of the sun. On these occasions, and on the days of sustained heat, the only means of keeping the plants from withering was to give them abundance of water; and then there was the risk, on the other hand, of their damping off. Water was supplied to the trapiche, for the service of the still and for culinary purposes, by a small acéquia (canal) carried along the hill-side from the head of a rivulet about a mile off. We had by this means generally sufficient water for our plantation, but as the acéquia was ill made and protected by no fence, the cattle, roaming about, generally trod and dammed it up at least once every day, when the Indians had to seek out and repair the damaged spots. But when the supply of water failed just at the moment of one of those outbursts of sun, there was no alternative but for all hands to run with buckets down to the deep glen, where there was a considerable stream, although the steep ascent from it was very toilsome. In a few weeks the cuttings began to root, and then they were attacked by caterpillars, which also had to be combated. In short, it is impossible to detail here all the obstacles encountered, and which only Mr. Cross's unremitting watchfulness enabled him to surmount. As his labours have been crowned by success, he may perhaps give a separate account of them, which will necessarily be fuller and more accurate than any I could furnish.

The passage of troops still went on for some days after Mr. Cross's arrival at Limon. A good deal of rain had fallen in the upper woody region and the roads were horrible. The poor beasts of burden, ill-treated and with their heavy loads ill-adjusted, had their backs worn into sores, and many of them sank under their burdens. Wherever a beast gave in, there it was turned adrift. In the warm forest, maggots soon filled their sores and ate into their very entrails ; so, after wandering about for a time, most pitiable objects, they at length nearly all died. Between Guaranda and Ventanas not fewer than 300 dead horses and mules strewed the track and the adjacent forest, and above 20 carcasses were laid within nose-shot of our hut. I set the Indians to roll them into ditches and hollows, and cover them with branches and earth, but the horrid smell turned their stomachs and they never half performed the task. During the day, whilst we were going about, we did not feel so much inconvenience, but when the night breeze filled our hut with the vile odour we found it impossible to sleep. Now I smoked awhile, and then I lay down, covering my face with a handkerchief wetted with camphorated spirit, but all in vain. When I considered the fate of those poor animals, and still more that of their unfortunate owners, from whom they had been taken by force, and who, in losing perhaps their only mule, had no means left of conveying to market the produce of their industry, and thereby supporting their families, it will not be wondered at that I cursed in my heart all revolutions. Grave indeed must be the motive of complaint which a people can have against its rulers to justify it in taking up arms to obtain redress.[1]

Towards the end of July the weather improved, and in a few sunny days the fruit of the Bark trees made visible advances towards maturity. On the 13th of August I noticed that the finest capsules were beginning to burst at the base, and on the following day I had all taken off that seemed ripe, gathering them in this way: an Indian climbed the tree, and breaking the panicles gently off, let them fall on sheets spread on the ground to receive them, so that the few loose seeds shaken out by the fall

[1] I may here relate an incident bearing on the same subject. Whilst Dr. Taylor was bringing up Mr. Cross from Ventanas, a body of some 800 men, whose commander I had known at Ambato, arrived from Guaranda. As usual, they bivouacked at Limon, and when I turned out on the following morning, I saw my four Indians prisoners in the hands of the soldiery, and one of them, with his hands tied behind him and a rope round his body, about to be dragged off towards Ventanas. Among the beasts of burden which accompanied the troops, this poor fellow had recognised his own mule —his *only* mule—as dear to *him* as Sancho's ass was to Sancho, and, with the aid of his companions, had contrived to abstract it during the night and hide it away in the forest. In the morning the mule was missed, and my Indians were immediately denounced as the delinquents, for they had been seen handling the mule the previous evening. I confess my indignation was at that moment at the boiling-point, and I wished for a hundred "Rifle Volunteers" to put the whole disorderly rabble to rout. However, I had given up half my dormitory to the colonel, and had treated him with as much hospitality as lay in my power, so that I had some right to expect he would not deny any request of mine; and accordingly, after a short parley with him, he ordered the Indians to be released. Thus I kept my Indians, and the Indian kept his mule, which was all we wanted.

were not lost. The capsules were afterwards spread out to dry on the same sheets, and the drying occupied from two to ten days. The first seeds were gathered at Limon on the 14th, and the last on the 29th of August. Early in September they were all dry.

Mr. Cross sowed, on the 16th of August, eight of the seeds I had gathered; one of them began to germinate on the fourth day, and at the end of a fortnight four seeds had pushed their radicles. On the 6th of September one had the seed-leaves completely developed, and by the 9th of the same month, or on the twenty-fifth day after sowing, the last of the eight seeds pushed its radicle. One of the seedlings was afterwards lost by an accident, but the remaining seven formed healthy little plants, and when embarked at Guayaquil, along with the rooted cuttings and layers, bid as fair as any of the latter to reach India alive. He had previously sown, at Guayaquil, eight Cinchona seeds gathered by me in 1859, and which had remained nine months in my herbarium; even of these, four germinated, and the remaining four might possibly have grown also, had they not been carried off by mice. It is therefore clearly proved that well-ripened and properly dried seeds do not lose their vitality for a much longer period than their excessive delicacy would lead one to suspect.

Having learnt that there were a few seed-bearing trees at Tabacal, a farm in the San Antonio valley, near the deserted village of San Antonio, I determined to go there while Mr. Cross and Dr. Taylor were attending to the work at Limon. The distance is not perhaps more than 15 miles in a straight line, but there is no road unless by way of Guaranda, which would take four days. I therefore followed a route already taken by Dr. Taylor, namely, along the path to Guaranda as far as the first ridge, and thence down to some cane-farms on the Rio de Tablas. From this point Dr. Taylor with an Indian had opened a track. I remained at Tabacal from the 14th to the 28th of September, collecting seeds as the capsules ripened and drying them carefully before packing.

I had now gathered about 2500 well-grown capsules (without enumerating many smaller ones), namely, 2000 from ten trees at Limon, and 500 from five trees at San Antonio. Good capsules contain 40 seeds each—in some I have counted 42 —so that I calculated I had (in round numbers) at least 100,000 well-ripened and well-dried seeds. Some small turgid (almost globose) capsules contained only from two to four seeds, as large and ripe as any in the largest capsules, while other capsules of the ordinary length, but slender, proved to contain only abortive seeds and were accordingly rejected in the drying. Had the month of July been as sunny as it is said usually to be, many more capsules would doubtless have ripened; as it was, only about one flower in ten produced ripe seeds.

I had scarcely finished drying my seeds at Tabacal, when I received the welcome intelligence that the army of General Flores had obtained possession of Guayaquil, and that the communication between the coast and the interior was reopened. I therefore resolved to proceed to Guayaquil and dispatch from thence a portion of my seeds by the first opportunity.

I started from Tabacal on September 28. The road thence to Guayaquil follows the right bank of the river, as far as to where the latter is confined to a deep chasm, and then crosses to the left bank. The descent is really very gradual, but seems more steep than it is, because the river tosses and foams among the huge stones which impede its course. As we descended, it was interesting to mark the gradual transition to the vegetation of the hot region. Leguminous trees, so scarce in the hills, began to be frequent. A bombaceous tree here and there adorned the forest with its numerous purple flowers. *Cinchona magnifolia* was budding for flower; it accompanied me to within 1000 feet of the plain. Enormous figs, with a long cone of exserted roots, straddled over the decayed remains, or often only over the site, of the tree which had served to support them in their infancy, and which they had strangled to death after establishing for themselves a separate existence.

At about 1500 feet elevation, I met with a Myristica, which grows about Tarapoto at the same altitude. A little lower down I saw the first Neea, and near it a Vismia, not one of those weedy species diffused throughout tropical America, but a handsome tree, resembling *V. uvulifera* (from the Casiquiari). These three genera seem rarely to ascend above the hot region.

Five leagues below Tabacal the road again passes, by a broad pebbly ford, to the right bank at Pozuelos, where we drew up for the night, thoroughly wetted by a soaking shower which had accompanied us for the last hour and a half. Pozuelos is a miserable little bamboo village, but notable for its extensive orangeries, which produce the finest fruit in Ecuador. Here the valley opens out wide, and by an almost imperceptible descent mingles gradually with the plain. The river became muddy, still, and tolerably deep. The vegetation is now unmistakably tropical, and there is as noble forest around Pozuelos as I have anywhere seen. Palms are far less varied than on the Amazon, but the Attalea above mentioned grows immensely tall and stout. An Astrocaryum, whose clustered trunks are perfect *chevaux de frise*, from the long flat prickles with which they are beset, is very frequent. Mimosæ are abundant, and so are papilionaceous twiners, among which I noted an Ecastaphyllum. The beautiful arborescent Passiflora (Astrophea) grows far larger than at San Antonio, and I could not help now and then stopping my horse under its stems, which here and there bent gracefully over our path, to admire the large pendulous glaucous leaves and the clusters of white flowers; but I sought in vain for ripe berries. In marshy places there are beds of rank ferns, and in pools an Eichhornia and a Pontederia. The common weeds of hot countries begin to appear, such as *Asclepias curassavica* and *Tiaridium Indicum*, the latter of which I had not seen since leaving the Amazon.

. . . . . .

[After much delay at Bodegas, waiting for the small steamer, Guayaquil was reached on October 6, and a portion of the ripe seed sent, as instructed, to Jamaica. The young plants were not ready for transmission till the end of November, when Spruce returned up the river to Aguacatál, the nearest port to Limon, where a negro carpenter put together the Wardian cases, and a raft was purchased to take them down to Guayaquil. The construction of this raft was interesting, and the description of it and of the dangerous voyage down the river will complete the essential portions of this Report.

I will first give, however, a short letter written while Spruce was delayed in the city.]

To Mr. John Teasdale

GUAYAQUIL, *Nov.* 6, 1860.

The town of Guayaquil extends about half a league along the margin of the river, which is here two miles broad. The principal street, called the Malecón (or Mole), runs by the river throughout that distance; but the town is narrow, and at the back stretches a wide, and what is now an arid, plain; but in the rainy season (which will shortly set in) all this plain is water and mud. Beyond the plain a salt creek impedes further progress in that direction. The houses are built of a framework of timber, neatly overlaid with bamboo-cane, and plastered within and without. The rooms are mostly papered and painted, and are often elegantly and even richly furnished—although sparsely, as befits a hot climate. The upper rooms project so far over the lower that they form a broad covered footway, which has a boarded floor, and affords a welcome shade in the heat of the day. A town built of such combustible materials is constantly exposed to conflagrations, and although there are several fire-engines, two of which are manned entirely by foreigners, the fires cause fearful ravages. A few days ago we had within twenty-four hours two fires and a smart earthquake. The latter did slight damage here, but half destroyed the town of Tumbez, which lies farther south and is the first port in the Republic of Peru. In the month of October we had several earthquakes, in one day no fewer than four. So you see that what with commotions below and above ground—earthquakes, revolutions, fires, etc.,—people live here in continual alarm. Guayaquil is, in fact, a town purely commercial, and the people work as if at the bottom of a mine, seeking gold, and in the hope of one day emerging to the light in some place where they may live in peace and comfort.

Since I came to Guayaquil, I have been a day's journey up one of the numerous rivers that empty themselves into the Gulf, to visit a large village called Daule, where I had been recommended to pass the winter. The river Daule is exceedingly pleasant—at least now in the dry season—and almost Chinese in its character. At every turn, groups of Coco palms, Orange trees, Plantains, etc., come in sight, with their accompanying cottage of bamboo-cane—or perhaps a more substantial edifice with a tiled roof, on some sugar plantation. The object of my journey was to inspect a house which is offered me by a gentleman, Dr. Aguirre, who has travelled much in Europe and speaks English, French, and Italian. The house is new—neat and commodious—but I can see that in winter the whole surrounding country will be inundated, which means abundance of mud and stagnant pools at the beginning of the dry season. My present notion is that I had better pass the winter at Piura, which is just within the rainless region on the coast of Peru.

REPORT (*continued*)

. . . . . .

The raft was composed of twelve trunks of raft-wood, 63 to 66 feet long, and about a foot in diameter, ranged longitudinally, so as to occupy a width of 15 feet, and kept in their places by five shorter pieces tied transversely and widely apart, extending nearly to the root end of the trunks, but leaving a considerable space free towards their point, for the convenience of working the raft. The five cross pieces were covered with bamboo planking, so as to form a floor 36 feet long by 10½ feet broad, which was fenced round with rails to a height of 3 feet, and the whole roofed over and thatched with leaves of *Maranta Vijao*. For carrying cacao, the fence has to be lined with bamboo boards, so as to form, with the flooring, a sort of large bin. The rope used in binding together the constituent parts of the raft was the twining stem of a Bignonia, nearly terete, but marked by four raised lines, overlying four deep grooves in the substance of the stem, and alternating with four shallower grooves. When the stem is twisted, to enable it to be tied, it splits lengthwise along those grooves into eight strips, which, however, still pull together, and offer very great resistance to transverse fracture.[1]

[1] I have long known that the strongest of all lianas are Bignonias, and I have many times trusted my life and goods to their strength. In the malos pasos of the Huallaga, canoes are dragged up the most dangerous places by means of from one to four stems of Bignonia, according to the size of the

The cases were all in readiness, and the raft brought down the river and moored in front of the farmhouse, but Mr. Cross did not arrive with the plants until the 13th of December. Some difficulty had been experienced in procuring the requisite number of beasts of burden, and the making of cylindrical baskets to contain the plants had proved a tedious task; besides that, the tying up each plant in wet moss, and the packing them in the baskets, were delicate operations which Mr. Cross could trust to no hands but his own. There had been not a few falls on the way, and some of the baskets had got partially crushed by the wilfulness of the bulls in running through the bush; but the greater part of the plants turned out wonderfully fresh. We had the cases taken down to the raft, and Don Matias lent us a couple of men to carry thither the earth, sand, and dead leaves necessary for making the soil to put in the cases. Mr. Cross put as many plants into the cases as he could possibly find room for, and only rejected a few that were so much injured by their journey from Limon that they were not likely to survive the voyage to India, the whole number put in being 637. As we might expect some rough treatment on the descent to Guayaquil, we did not venture to put on the glasses, but in their stead

canoe and the weight of its cargo. I have never known the lianas to break; and as I have sat in my canoe, anxiously watching its slow upward progress, my only care was that the lianas were securely fastened to the prow, or lest the sudden bursting of a whirlpool beneath the canoe should tear them from the hands of the Indians, as they with difficulty held their way along the rocky shore.

In the Guayaquil district, as on the Amazon, the aerial roots of various Aroideæ and Carludovicæ are the common substitutes for string, but Bignonia stems are always preferred wherever strength is essential.

stretched moistened strips of calico over the cases, which seemed to answer admirably. As Mr. Cross wished the plants to be firmly established in their new residence before removing them from Aguacatál, I determined to delay our departure until the latest possible moment, that is to say, so as to reach Guayaquil and fasten up the cases before the arrival of the steamer on the 28th.

. . . . .

After my arrival at Aguacatál the weather was occasionally showery, but the rains were evidently heavier towards the source of the river, which would suddenly rise several feet, and then rapidly lower again; so that we had to watch our raft night and day, lest on the one hand it should be carried away by the floods or the onslaughts of driftwood, or on the other hand should be left high and dry by the sudden receding of the waters. At 11 P.M. of December 22, heavy rain came on at Aguacatál, and did not cease until 9 A.M. of the following day, when the river had risen much, and continued rising through the day. The next night still heavier rain fell, clearing off at about 8 A.M. of the 24th, which was the day fixed for starting on our voyage. Our raftsmen were three in number as far as to Bodegas, but thence to Guayaquil, where the river is wider, and is therefore not subject to sudden rises and falls, we needed only two. As soon as the rain ceased, we got the glasses of the cases put on board, and when our raftsmen had taken their last trago with their friends, and said their last adios (always a lengthy process), we left Aguacatál; Don Matias, at parting, foretelling us a speedy but perilous voyage.

The oars used in navigating these rafts are merely bamboos, about 20 feet long, half their thickness being cut away for about a yard at the outer end, so as to form a sort of scoop. Two oars were fixed in the prow, and a third oar in the stern, the latter being worked by the old black who had sold me the raft. The river had risen almost to its winter level, and we swept along rapidly. At 2 P.M. we were already eight leagues away from Aguacatál, near a site called Cataráma, below which the river is narrowed in some places to 30 yards, and the navigable channel is further straitened by the trees (chiefly species of Inga) which hang far over the water. Add to this that the river ran like a sluice, and that the turns were frequent and abrupt, and it will be seen how difficult it was to maintain our clumsy craft always in the mid-stream. Although the men tugged hard at their oars, they could not save us from being frequently brushed by the trees; and at length, at a sharp turn, the raft went dead on, and through a mass of branches and twiners that hung over to the middle of the river. The effect was tremendous: the heavy cases were hoisted up and dashed against each other, the roof of our cabin smashed in, and the old pilot was for some moments so completely involved in the branches and the wreck of the roof, that I expected nothing but that he had been carried away; he held on, however, and at last emerged, panting and perspiring, but with no further injury than a smart flogging from the twigs, which indeed none of us entirely escaped. There have been instances on this river of a man being hooked up bodily by the formidable *Uncaria guianensis*, and suspended in mid-air, whilst the raft passed from under him.

Our deck now presented a lamentable sight, but we had little time for ascertaining the amount of damage, as at every turn a similar peril awaited us. We, in fact, twice again ran into the bush, not quite so violently as before, but each time adding to the damage already sustained. We had calculated on reaching Caracol that day, and might still have done so before nightfall, but that there were some bad turns ahead, which, as the men were already much fatigued, we could not expect to pass without very great risk; so at 4½ P.M. we brought to, with some difficulty, at a place where the bank was free from trees, and made fast for the night. We then set to work to clear away the wreck of sticks and leaves which strewed the raft, and to repair the roof, which was completed by moonlight. The cases had received only a few slight cracks, and had none of them turned over, but the leaves of the precious plants were sorely maltreated. . . . As far as Caracol the river continued narrow and winding, and at various points we barely cleared the bushes, but nothing more serious happened to us than the loss of a few loose cloths, which were hooked up by a pendulous mass of the Uncaria. From Caracol downwards the river grew wider, and the banks were less overhung with wood, so that we went on with more security. . . . Soon after nightfall we had got as far as to where the influence of the tide was still felt, and as it was ebbing we profited by it to hold on our way until 2 o'clock of the following morning, when the flood-tide obliged us to lay by. Thenceforward we got on slowly, on account of having to wait between tides, but we reached Guayaquil at noon on the 27th without any further accident, and I immediately went on shore and sought out a carpenter, to assist Mr. Cross in nailing laths over the soil and in fixing on the sashes. By 5 P.M. of the 28th everything was completed. The plants, thanks to Mr. Cross's tender care of them, bore scarcely any traces of the rough treatment they had undergone in their descent from Limon, and in their late voyage from Aguacatál, and the only thing against them was that they were growing too rapidly, owing to the increased temperature to which they had lately been subjected.

. . . . . .

On the 29th, a large goods steamer came in, which goes to and fro between Lima and Guayaquil. She was not to sail again until the 2nd of January, and the plants, if sent by her, would have to remain at Payta until the 13th or 14th, when another steamer should pass from Lima to Panama; but, as there was no alternative, we had them put on board her, and commodiously arranged on the poop-deck. I then took leave of Mr. Cross and the plants, satisfied that so long as they were under his care they were likely to go on prosperously, and having done all I could on my part to conduct the enterprise to a successful issue. During its performance, all engaged in it had run frequent risk of life and limb; but a far greater source of anxiety to me were the contretemps (a few only of which have been indicated in the preceding pages) that every now and then threatened to bring our work to naught. It is difficult for those who live in a country of peace and plenty, but above all of good roads, to appreciate the obstacles that beset all undertakings in countries where none of those blessings exist. . . .

[It only remains to say that Spruce's long-continued labour and extreme care were crowned with success. The young plants reached India in good condition, and the seeds germinated and served as the starting-point of extensive plantations on the Neilgherry Hills in South India, in Ceylon, in Darjeeling, and elsewhere.

The latest reports from the India Office, which I owe to the kindness of Sir Clements Markham, seem to show that none of the districts where the plantations have been made are really suitable, either in climate or soil, to the natural requirements of the trees. This is indicated by two facts. It is stated that although the trees grow well when young, yet they suffer from dryness of the soil in the dry season, so that artificial watering sometimes has to be resorted to. It is also stated that it has been found necessary to resort to the application of large quantities of stable manure and lime to keep the plants healthy.

In Sikhim the rainfall of 125 inches is said to be distributed over less than 150 days, so that the larger part of the year is rainless. In the Neilgherries violent winds are said to be very hurtful and sometimes destroy the larger trees.

Now the great feature of the native Cinchona forests as described by Spruce is the prevalence of rains almost throughout the year, and especially of a constantly moist soil, kept so in dry weather by

Map to illustrate Mr. SPRUCE'S REPORTS on the "RED BARK" REGION OF ECUADOR.

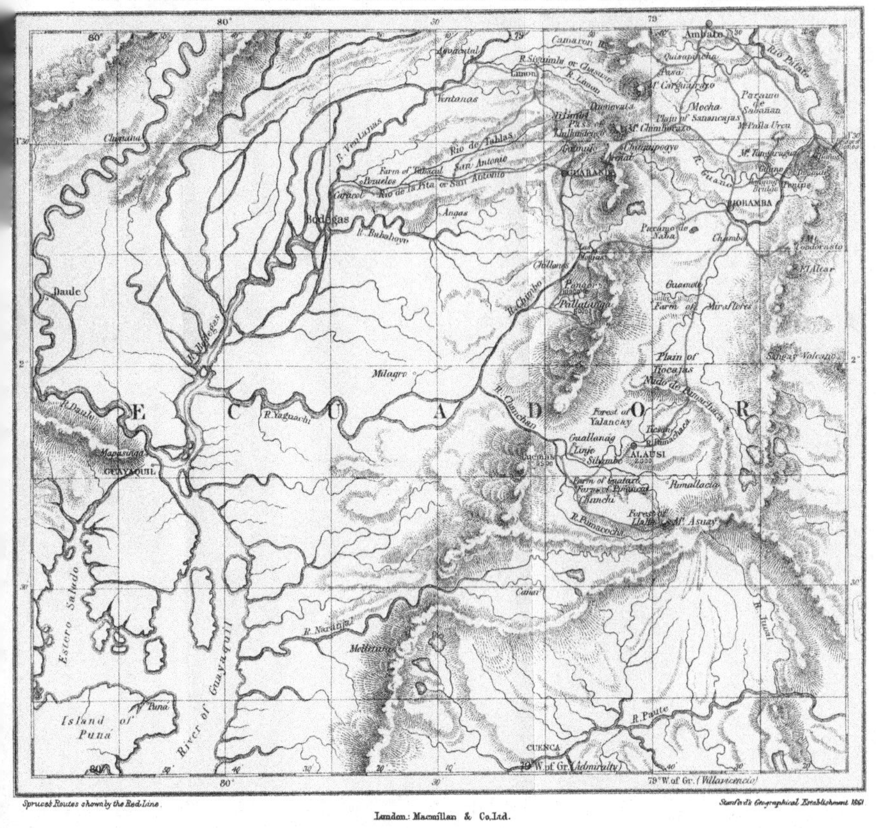

Spruce's Routes shown by the Red Line.

London: Macmillan & Co. Ltd.

Stanford's Geographical Establishment 1861

The material originally positioned here is too large for reproduction in this reissue. A PDF can be downloaded from the web address given on page iv of this book, by clicking on 'Resources Available'.

the surface covering of leaves and leaf-mould. He states that during the dry season, from June to December, there was rain (more or less) on about ten days in each month, and that during the whole six months there were only thirty-one days on which there was neither rain, mist, nor fog. This would appear to be a very different type of climate from that of either Sikhim or the Neilgherries, although the mean temperature may not be very dissimilar. It seems to me probable that the districts most nearly approaching in climate to that of the Cinchona forests would be the mountain slopes above 2000 feet in the Federated Malay States, or in the Sarawak territory in Borneo, both of which have a similar distribution of rainfall throughout the year.

The official Report of June 1907 states "the Cinchona industry in the Neilgherry is rapidly diminishing," and that many of the estates are being abandoned, which can only be due to its being not permanently profitable. Everything therefore seems to point to the fact that the best natural conditions for the growth of these valuable trees has not yet been found.]

CHAPTER XXII

ON THE SHORES OF THE PACIFIC: SPRUCE'S LAST THREE YEARS IN SOUTH AMERICA

[DURING the whole of this period Spruce was struggling hard against the severe illness which prostrated him for the remainder of his life. The list of his Botanical Excursions gives a connected view of his movements in search of health, and the few letters he wrote to his friends give a sufficiently vivid picture of his life and occupations, when he could do little more than rest and make those minute observations on the country and the people which were his chief consolation during the wearisome years of forced inactivity.

One result of these observations was an elaborate paper of 80 pages, on the district of Piura, in which he resided for nearly two years, more especially in relation to the cultivation of cotton there. This paper was published by the Foreign Office, but is now out of print; and as it describes a district very rarely visited by European travellers, I here reproduce those portions of it (about one-third of the whole) which are of general or botanical interest. They also serve to show how carefully Spruce utilised his opportunities for scientific observation, even under the most adverse conditions.]

BOTANICAL EXCURSIONS, 1861 TO 1864

1861.

Jan. 2. On this day the steamer left for Panama with plants of Red Bark on board.

„ 6. Voyage up river Daule.

„ 7- Feb.– March– April– May– June. Reached the village of Daule, where I established myself for the rainy season at a farmhouse called La Bella Union. Remained at Daule until June 11, collecting a little now and then when breaks in the weather allowed me to wander about, which I did to the limits of my strength, viz. within a radius of half a mile.

„ 11. Descended to a small village called Pascuales.

„ 12. From Pascuales to Guayaquil.

„ 25. Embarked this evening at Guayaquil and reached Daule the following morning.

July 30-31– Aug.–Dec. Remained at Daule till the end of July; then, being very sick, I descended to Chonana—the farm of the late General Illingworth, where his son-in-law Dr. Destruge (my physician) was residing along with Mr. William Illingworth. There I remained till the end of the year. In September the house of Gutierrez failed at Guayaquil, whereby I lost 6000 dollars, nearly all I had.

1862.

Jan. 1-31. Remained at Chonana until the middle of this month. I collected very little there, and my chief occupation (when able to work at all) was writing out the Report of my Expedition to procure seeds and plants of the Red Bark. From Chonana I descended to Guayaquil, and near the end of the month proceeded thence by sea to Chanduy, on the arid coast of the Pacific, a little way out of the Gulf of Guayaquil to the north.

Feb.– March– April– May– June– July– August. All these months at Chanduy, on the very borders of the sea, making desperate attempts to take exercise, but on the whole going back rather than forward. Unexpected heavy rains in the month of March brought out an interesting vegetation on the desert. Even lakes were formed there, which soon became peopled with aquatics. With great toil I managed to collect and preserve specimens of everything. I obtained also a few seaweeds and zoophytes on the rocky shore.

„ 18. Started on return voyage to Guayaquil.

„ 19. Passed the Isle of Puná and lay by to await the tide a little below Guayaquil.

1862.	
Aug. 20– Dec. 31.	Reached Guayaquil this morning. From this date to the end of the year at Guayaquil trying to redeem some of my lost property.
1863.	
Jan. 1.	At 11 A.M. embarked on board the steamer for Peru.
„ 3.	Reached Payta at 9 A.M. Hired mules, and at 6 P.M. started to cross the desert by night.
„ 4.	Reached Piura (48 miles) at 10 A.M.
Feb.– March– April– May– June–July– August– September.	From this date I remained at Piura until the 10th of October. When, in consequence of rains in the Andes, the dried-up bed of the river became overflowed (March 14) and ran with a considerable stream to the sea for a few months, a scanty vegetation appeared on its banks, of which I secured specimens.
Oct. 10.	Travelled from Piura back to Payta.
„ 23.	From Payta to Amotape on the river Chira.
„ 24– Nov.– Dec. 22.	From this date until the end of the year on the river Chira; until December 22 at the village of Amotape, afterwards at Monte Abierto, higher up the valley.
1864.	
Jan. 26–	This day returned to Amotape.
February– March– April 20.	Remained there through the following months until April 20, then journeyed to Payta by way of Colan.
„ 30.	Rest of month at Payta.
May 1.	Embarked for England on board the Pacific mail-steamer.
„ 5.	Reached Panama, 6 P.M.
„ 6.	Across the Isthmus.
„ 7.	Sailed.
„ 28.	This morning landed at Southampton, after an absence from England of 15 years all but 10 days.

To Mr. George Bentham

DAULE, NEAR GUAYAQUIL, *March* 9, 1861.

. . . . . .

My mode of working is this. When I bring home freshly-gathered plants, I make notes on them in books prepared for the purpose, and add numbers. If any plant seems strange to me, I keep flowers, etc., in water to await a spare interval when I can analyse them microscopically. So soon as the plants are dried I pack them into other paper and add the labels from my notes. As it often happens that, at each packing, I have not two plants of even the same natural order, the risk of transposition is very small. Indeed, so completely does the reading over of my notes recall the features of the plants, that I feel sure if I were shown the whole of my plants classified in your herbarium, and on blank paper, I could, from consulting my notes, put to them the proper numbers and localities without making perhaps a single mistake. As to positive errors of observation, I am as liable as any other mortal: I would wish to speak with all modesty on that head; and working often in boats, or in dismal huts where a squall would suddenly enter the open doorway and disperse both specimens and labels, there must occasionally have been some transposition of both in gathering them up again. This risk of the blowing away or dropping out of labels was, in fact, what made me give up putting labels to the plants as they were drying.

I have gathered a few plants since I came here, but the rainy season is now reaching its height and all around I have deep mud and water. The village is scarcely 300 yards from the farm-house where I live, yet I cannot go thither on foot, except with india-rubber boots. *Capraria peruviana* (Scroph.) grows about in moist places as *Coutoubæa spicata* (Gentians) does on the Amazon, and looks not unlike it. The arborescent vegetation is scanty but novel. The finest tree is a Cæsalpinia with bipinnate mimosæoid foliage—I cannot reduce it to any described genus. There are several arborescent Capparides—all new to me; but *Cratæva tapioides* is an old acquaintance, and abounds as it did on the Amazon. The Leguminosæ are mostly out of flower now, but I recognise none of them by the foliage, unless one be *Bowdichia pubescens*. Guayaquil is noted for its fruits, and the abundance and variety brought to the port for sale every morning in summer are truly astonishing. Many of them come from a good way up the various rivers, and there are many wild fruits. The latter include two "cherries"; one of them is so like the fruit of *Averrhoa Bilimbi* (Oxalideæ) in appearance that I did not think it could be anything else. The tree abounds at Daule and is now in flower; it is a Combretacea, allied to Terminalia! The other cherry is a Malpighiacea—very different from the Bunchosias or "Friar's plums," and probably a Byrsonima. Two "plums" are surely species of Spondias. A drupe, called Pechiche, the size of a large cherry, but black, and with a mawkish sweet taste, though excellent for preserve, is the fruit of a Vitex. There are also many Sapotaceous fruits not seen elsewhere. I hope to make them all out and to send specimens of the fruits in spirit. I have unfortunately very little strength left for work of any kind, and the squalls that come on suddenly when the sun is hot and penetrate the chinks of these bamboo walls make me feel sometimes as "roomackity" as I did in the Sierra. Piura would have been the place for me—they say the most obstinate rheumatisms can't withstand the climate of Piura. But I do not like the idea of living in the midst of a desert.

I was beginning to work *à mon ordinaire* when I had the misfortune to scald my right foot severely, and had to endure a tedious vesication and afterwards a painful ulceration. Eighteen days of it stretched in a hammock, and unable to tread the ground. I did not mind the pain so much as the lost time.

. . . .

To Mr. John Teasdale

GUAYAQUIL, *June* 22, 1861.

. . . It is singular that the greatest range of temperature occurs here in the summer or dry season, while in the wet season it is more equable but more oppressive. We are now entering on the summer, and it is surprising how rapidly the water and mud dry up off the savannas; for no more rain falls, and we begin to have strong westerly breezes, continuing sometimes through the night. Guayaquil is not unhealthy from June to January, and if they had built the city lower down the Gulf it might have been healthy all the year round. The island of Puná, where Pizarro first landed, is very healthy, but now almost uninhabited. The little towns along the coast to northward are also healthy, and noted for the longevity of their inhabitants.

The vegetable products of cool regions become excessively scarce and dear here in the rainy season, when all import of goods from the highlands of the Andes is suspended, although those mountains lie within sight when the weather is clear. At Daule potatoes were sold at 2¼d. and apples at 5d. apiece; while at the same time potatoes were selling at Ambato, only 80 miles away, at 1s. 3d. the sack.

To Mr. John Teasdale

CHANDÚY, NEAR GUAYAQUIL, *May* 14, 1862.

. . . The rains—or, as we say here, the winter—came on at Chonana in the middle of January, when I descended to Guayaquil, and shortly afterwards went on to Chandúy—a small village on the shores of the Pacific, at 2½ days' journey by sea from Guayaquil, and a little north of the island of Puná. Here it scarcely ever rains, beyond a slight drizzle in the morning, occasionally—the same as at Lima—and throughout the year 1861 there was but one day of heavy rain. This present year, however, we have had a real rainy season that began in February and lasted through most of March. It has been the first rainy season since 1845, and we had actually one night a thunderstorm, a phenomenon that had not previously been witnessed here by even "the oldest inhabitant" (and there are some centenarians). With so dry a climate normally, you may well suppose the vegetation is very scanty; yet there are even a few scattered trees, of humble growth, some of which grow down to the very beach. The species that most abound are a stout branched Cactus (*Cereus peruvianus*), growing to 30 feet, truncheons of whose trunk serve the people for stools; and a beautiful Jacquinia (*J. armillaris*) of the same height. The latter has somewhat the aspect of the Holly, from the dark green, rigid, spiny-pointed leaves; but the flowers, which are very numerous, are of a deep vermilion and very sweet-scented; and they are succeeded by fruits resembling small oranges in colour and shape, although uneatable and narcotic, and used by the inhabitants for stupefying fish.[1] When I arrived here, with the exception of these and a few other shrubby trees, and of a winding green line of mangroves (marking the course of a creek), the whole country had the aspect of a barren sandy waste. Even the range of hills that runs parallel to the coast at a distance of one to two leagues showed only brown and withered shrubs. But when it began to rain a change came o'er the face of nature more sudden and surprising than even that of a bright spring succeeding a severe winter in Europe. "The desert blossomed like the rose." The sandy plains became in a few days clad with verdure: curious and pretty grasses, most of which I had not even seen elsewhere; flowering annuals, including a Polygala—not prettier than the Milkwort of our English heaths, but of nobler growth (1 to 2 feet) and bearing long spikes of roseate flowers; patches of apparently dead brush, scattered

[1] The genus Jacquinia belongs to the Myrsinaceæ, an order allied to the primroses.—ED.

over the savanna, burst suddenly into leaves and flowers. All this was interesting enough, but there was a reverse to the picture. As a shower of rain is such a rare event at Chandúy, the inhabitants think their houses sufficiently protected by a slight roof of the leaves of Arrow-cane (Gynerium sp.), through which the heavy and continued rains of the present year have passed as through a sieve. Figure to yourself, then, my dwelling flooded by night—bed and everything else soaked—so much wet out of doors that I could not take even such exercise as my slender forces permitted, and it will not surprise you to learn that I had a severe attack of jaundice. A little after the equinox the weather grew drier and cooler, and my illness began to leave me, although I have still not quite shaken it off.

The sea-breezes, which blow from the west and south-west, are strong and cool. We have already had the thermometer once down to $66\frac{1}{2}°$, and in June and July we may expect to see it still lower. I walk about as much as I can, and amuse myself with gathering and preserving the flowers, although they are now fast drying up. The beach is rather too steeply inclined to be pleasant to walk on, and shells and seaweeds are rather scarce; but the antics of the burrowing crabs are diverting, and especially their battles with my dog, who disinters them from their holes in the sand. It is singular, however, to have been nearly four months by the seashore and only to have eaten fish three times, nor once to have gone out in a boat. . . .

The industry of the Chandúyenians, who are nearly all pure Indians, is almost limited to the plaiting of Panama hats and to gathering Orchilla (*Roccella tinctoria*), which abounds here on the trees—especially the Cactuses—as it does also in the neighbouring islands of Galapagos. They fish very little, and that merely for their own eating.

. . . The failure of the house of Gutierrez and Co. at Guayaquil was a heavy blow to me. When it suspended payment (October 11, 1861) I had in their hands very nearly a thousand pounds—£700 at interest and the rest in deposit. I have received the balance of interest due to me at that date, but the residue, viz. 5550 dollars (Peruvian or Equatorian), remains to share the fate of the other debts of the firm, and if I ultimately recover a thousand dollars of it I shall think myself well off. The blow was so sudden that I had no time to withdraw my property, especially as I was at two days' distance from Guayaquil (at Chonana with the Illingworths). Even Gutierrez himself did not comprehend how it had happened; but all has come to light now, and it is proved to have been caused entirely by the roguery of the cashier (Gavino Icaza) and of the head book-keeper (Thomas Viner Clarke, an Englishman, I am sorry to say), who, acting in collusion, have robbed Gutierrez to the amount of 360,000 dollars, and possibly more. Not only had they from time to time appropriated large sums of ready money—making the monthly balance (shown to Gutierrez) always tally with the cash in the cash-box, but they had shipped vast quantities of cacao and other produce from the warehouses of Gutierrez (unknown to him) under feigned names, and consigned to houses abroad which had no existence; and Clarke, in whom his patron reposed unbounded confidence, having been sent to Europe last year to purchase goods, returned with a quantity of unsaleable trash and with forged invoices. Clarke took himself off to England immediately after the smash, and was bearing off also 7000 dollars from the cash-box; but Gutierrez missed the money, followed him on board, and took it off his person in the sight of many witnesses. In almost any other country he would immediately have been incarcerated, but they manage matters otherwise here. Icaza walks about Guayaquil holding his head as high as ever; and as he is a scion of one of the *noble* houses of the country, Gutierrez dare not proceed against him by law, which would expose him to the risk of having a knife stuck into him at the turning of some street corner after nightfall. . . .

To Mr. Daniel Hanbury

GUAYAQUIL, *Nov.* 29, 1862.

MY DEAR SIR—Your last letter shows plainly that you consider your correspondent both listless and dilatory. He confesses to both, and can show ample cause. If you knew how entirely disabled I am; how rarely I can sit to a table to do anything, but must write, eat, etc., in my hammock; how I cannot walk except for short distances, nor ride on horseback without being in danger of falling from an arm or a leg suddenly turning stiff, you would surely not be surprised at my want of activity. I had never calculated on losing the use of my limbs, and yet nothing was more likely to happen, if the sort of life I led be considered. When after loss of health came wreck of fortune, simple though my wants be and modest as were my aspirations, I felt for a time completely prostrated. The fact is, I have been too constant to botany; several times in the course of my travels I might have taken to some occupation far more lucrative; and I have met many men who, beginning without a cent, have made more money in two or three years than I in thirteen, and that without being exposed to thunder-storms and pelting rain, sitting in a canoe up to the knees in water, eating of bad and scanty food once a day, getting no sleep at night from the attacks of venomous insects, to say nothing of the certainty of having every now and then to look death in the face, as I have done.

Excuse these personal details, which I have not entered into with any hope or desire of exciting sympathy, but simply to explain that, although still in the midst of objects interesting to the inquirer into the productions and processes of nature, I can pay little heed to them.

[Spruce then describes how he tried to obtain specimens of the flowers, etc., of a particular balsam tree Mr. Hanbury was very anxious to obtain; but after paying the owner of the forests ten dollars to send an Indian to fetch them, he received a mule-load of branches none of which possessed a single flower or fruit, to obtain which one or two more journeys would have to be made at different seasons. He then proceeds:—]

When I came out to the Amazon I resolved never to take a specimen of a gum or resin without gathering specimens of the tree producing it; in which I did very wrong, for I thus lost the opportunity of securing good specimens of many gums, etc., brought by the Indians to the towns for sale;

and when I afterwards fell in with the trees producing them, there was either no gum to be had or merely small fragments, sufficient for identification with larger masses, but not worth sending as specimens to England. The collection of balsams, gums, resins, etc., is a task requiring an Indian's patience. Mostly they must be gathered drop by drop, or incisions must be made and the trees visited after the lapse of months to get the lumps of coagulated juice. I was unfortunate in some things I tried to collect on a large scale. For instance, I took with me down the Rio Negro a demijohn of the Sassafras of that river, and several demijohns of a beautifully white and transparent Oil of Copaiba, procured on the Siapa, intending to send them to England and ascertain their commercial value; but the person who took them down to Pará not only received the freight beforehand, but sold the articles there on his own account instead of delivering them to my correspondent.

[With this letter Spruce sent dried specimens of a gum-producing tree which grew about a mile and a half from the village of Chanduy, and which after several attempts he succeeded in reaching—"though I had to lie down many times by the way." He then concludes thus :]

This is, I think, all your correspondent has to send you this time. You will see he is now good for little else besides talking and writing—even the latter is painful to him and can be done only reclining in the hammock; but if you will have patience with him, he will still try to obtain for you any information within his reach.—Very faithfully yours,

RICHARD SPRUCE.

[Who can wonder that, after the receipt of such a letter as this, Hanbury and Spruce became, for the remainder of their joint lives, the most attached and sympathetic of friends!]

To Mr. John Teasdale

PIURA, PERU, *Jan.* 12, 1863.

I embarked at Guayaquil, on the night of the 1st of January, on the steamer that plies between that port and Lima, my destination being Payta, and thence overland to Piura.

. . . .

At 9 A.M. of the 3rd we reached Payta, and by noon I had got my baggage through the custom-house, and hung up my hammock in the only fonda in the place. But I only remained there a few hours to get together the mules required for the journey to Piura—45 miles across the desert. It is usual to travel here by night, the burning heat of the desert by day causing great (and sometimes mortal) fatigue to man and beast. I was myself conveyed in a litter, being unable to sit on a horse for more than an hour at a time. We started at six in the evening, and at nine on the following morning reached Piura, having rested three hours at a tambo erected at midway of the route, where lucerne and water can be had for the beasts, and coffee, bread, and chicha for their riders, by paying a high price for them. The track is still indicated, in some parts, by long poles stuck in the ground, as it was in the time of the Incas; in other parts by the rare Algarroba trees, which are almost the sole vegetation, where there is any at all. Woe to the traveller who strays from these landmarks: he soon gets bewildered among the médanos or shifting hills of sand, and finds his grave in one of them. To see the sun setting over this desert is like looking into the red-hot mouth of a furnace, and there is usually a lull in the wind at that hour; but he has barely disappeared when a rapid refrigeration sets in, the night-wind sweeps over the desert, and at daybreak the cold is as sensible (of course not so intense) as on the paramos of the Andes.

Piura is one of the driest places in the world, and in "winter," as it is called (December to April), one of the hottest. Yet it is very healthy, catarrhal complaints, caused by the violent winds charged with sand, being the only prevalent ones. The site is a very curious one to have been chosen for a city. There is a river, it is true, but for six months in the year its bed is dry. It is now raining hard in the Andes, where its sources are, and some time next month the water is expected to reach Piura.

March 27, 1863.—. . . We are just now passing through the hottest fit of weather I ever experienced. Fancy a minimum thermometer at 85°, which has usually been the lowest temperature in the twenty-four hours ever since the 1st of March; indeed, up to the present date, it has only three times been as low as 83°. It is true that throughout the same period the thermometer has never risen higher than 89°; but such sustained and nearly uniform heat induces great languor. The hottest part of the year is considered to be almost past, and the months to come will get gradually cooler. Although Piura cannot be said to have a pleasant climate, there can be no doubt of its general healthiness when one sees how many very old people are in it. On the 16th of the present month an old lady died here at the age of *one hundred and eleven*! Her living descendants, including some great-great-grandchildren, are said to be exactly as many as the years of her life.

Piura is considered the sovereignest place on earth for the cure of "rheumatic" (*lege* "syphilitic") affections. Many wonderful cures are reported; but the treatment is rather severe. It is as follows: First, you pay the priest to say "novenas"—that is, masses on nine consecutive days—on your behalf; on each of these days you drink copiously of a warm decoction of sarsaparilla towards midday, and then your friends take you outside the town and bury you up to the neck in the burning sand, shielding your head with a broad straw hat and an umbrella. There you perspire in such a way as to bring out all the mercury you may have taken, and to reduce your swollen joints to their proper dimensions. Now you may see the use of the masses, for if you survive the operation (which is not always) they serve to express your thankfulness; and if you die under it, you will need not only those nine masses, but several additional ones—for which you make due provision in your last will and testament—to secure the repose of your soul.

Piura is perhaps the most superstitious place I have seen in South America, although Quito is far gone that way; but I can tolerate even superstition when it is harmless and picturesque. As I write, at 8 P.M. of the eve of "Nuestra Señora de los Dolores," the bells are ringing to call the devout to a procession in her honour, and gaily-dressed, life-size figures of that sorrowful lady are set up under sparkling canopies at the corners of the principal streets. On this day, ladies who rejoice in the name of "Dolores" (and in Catholic countries they are legion) invite their friends to eat sweetmeats and to drink wine and chicha with them. Even I, old bachelor and foreigner as I am, have received several such invitations, and one Dolores has gone the length of sending me a pair of garters embroidered in blue, red, and white silk! *Vanitas vanitatum!* Yet even this mediæval fooling is better than the unmitigated money-seeking (by fair means or foul), and as reckless and luxurious spending, of Guayaquil.

Sometimes our superstitions are rich in historical souvenirs. When that most valiant of Pizzaro's warriors, Pedro de Candia, leapt on shore at Tumbez, he carried in his hand *a cross*, extemporised from two bits of firewood. The inhabitants let loose on him "a lion" and a tiger, who, instead of attacking him, prostrated themselves before the cross, etc. etc. A piece of that famous cross is preserved on the altar of one of the churches of Piura, and the church itself is dedicated to La Santa Cruz del Milagro (The Holy Cross of the Miracle).

NOTES ON THE VALLEYS OF PIURA AND CHIRA, IN NORTHERN PERU[1]

Topography and Mineralogy

Along the western side of South America, extending from near the Equator on the N. to about Coquimbo in Chile, latitude 30° S., there is a strip of land, included between the Pacific and the

[1] Extracts from the Foreign Office Paper by R. Spruce.

Andes, which has been upraised from the ocean at no very remote period, and is still nearly as destitute of vegetation as the Sahara of Africa. It is, however, watered by a few rivers, some of which rise in the summits of the Andes, and run with a permanent stream into the ocean, diffusing fertility and perennial verdure throughout the valleys they traverse; others, rising in the lower hills which form an outwork of the great chain, carry a considerable volume of water to the sea during the rainy season, but for the rest of the year the lower part of their course is dry. By far the most important of these rivers is the Guayaquil, whose affluents drain the slopes of the loftiest portion of the western equatorial Andes (including the mighty Chimborazo), and on issuing into the plain form a network of navigable streams which at the city of Guayaquil combine into a noble river.

The northern limit of the Peruvian desert is usually placed about Tumbez, at the southern extremity of the Gulf of Guayaquil, in latitude 3° 30′ S., but I now know, from personal inspection, that the coast of the Pacific north of the gulf has the same geological conformation, the same climate, and almost as scanty a vegetation as it has south of it. At what point to northward the struggle between barrenness and fertility begins to be equally balanced, I am unable to say, but I am inclined to place it about Cape Pasado, at the mouth of the river Chones. Guayaquil itself, as seen from the river, with its groves of coco palms and fruit trees, and its picturesque wooded hills, might be supposed a region of forests; but the moment we pass the skirts of the city to westward we find that the country is nearly all savanna, either open and grassy or scattered over with bushes and low groves, and that the woods are confined to the hills and to the borders of salt-creeks. As we descend the river from Guayaquil (*i.e.* to southward), the ground on the right margin, beyond the mangrove fringe, grows more and more open, and at the southernmost point of the mainland, or the northern entrance to the gulf, where stands the village of El Morro, at the foot of a steep rounded hill, the ground is already nearly as bare of vegetation as the coast of Peru. Throughout this distance, and thence northward along the shores of the Pacific to beyond Point St. Elena, there is no stream of fresh water, although there are a few salt-creeks; but in latitude 1° 55′ S. we come on the river Manglar-alto, along whose banks there is vigorous vegetation, as there is also on similar small streams entering at wide intervals to northward; while the intermediate ground is either nearly desert or is a sort of savanna, sparsely set with bushes and cactuses, and bare of herbs except after the rare and exceptional rains.

About Cape San Lorenzo (latitude 1° 5′ S.) the coast is bold and broken, and almost completely clad with low bushy vegetation. In the village of the same name, which nestles in the bay to southward of the cape, at the mouth of a small stream, the houses stand mixed with Coco palms and Plantains, and steep wooded declivities rise at the back. Yet on rounding the point to northward, we come again to a half-open country at the village of Manta and the town of Monte Cristo, a few miles inland; or, as Funnell says of it, "the land hereabout is very barren, producing only a few shrubby trees and some small bushes."[1]

A little farther northward, on the river Chones, there is real forest, from which much timber is obtained for Guayaquil. The Chones falls into the Bay of Caraques, which was an important harbour in the early days of Spanish rule, but has now become useless to navigators through the gradual accumulation of sand at the mouth of the river. The northern extremity of this bay is Cape Pasado, whereof Funnell says: "This Cape Passao is a high round cape, with but few trees on it. It lies in the latitude of 0° 8′ S. . . . . within the cape the land is pretty high and mountainous and very woody."

From Cape Pasado to Cape San Francisco would seem to be the real neutral ground, the heavy rains which prevail every year from April to November along the coast of New Granada and Mexico, up to latitude 23° 30′ N., reaching to southward in some years as far as Cape Pasado, and in others stopping short at Cape San Francisco.

The coast we have been considering stretches out to westward, and recedes from the western ridge of the Andes at least 150 miles; but if we return to Guayaquil and descend the gulf or estuary along its left or eastern bank, we find that at a very few miles inland the ground begins to swell, and rapidly rises to the lofty ridges of the Andes, having the frigid paramo of Azuay to the north. From these mountains descend several streams to the gulf, and the atmosphere is highly charged with humidity, in consequence of which this coast is clad with lofty continuous forest. At Tumbez, the southern entrance of the gulf, where the shore again trends to westward and recedes from the Cordillera, the intervening plain becomes wider, drier, and barer of vegetation as we advance to southward, save where a broad verdant band marks the course of the river Tumbez, whose sources lie in the paramo of Saraguru and other highlands to northward of Loja.

The coast continues to extend to westward until reaching Capes Blanco and Pariña, the westernmost land in South America; then turns southward, and in latitude 4° 55′ S. the river Chira enters the bay of Payta, which, although a mere open roadstead, affords the most secure and commodious anchorage of any port along the whole coast of Peru. Beyond Payta is the mouth of Piura and the town of Sechura, which sometimes gives its name to the whole

---

[1] *A Voyage Round the World*, by W. Funnell, mate to Capt. Dampier.

river. I propose here to treat of the lower part of these two rivers, and especially of the Chira, in some detail.

. . . .

The configuration of the coast-region from Cape Blanco to and beyond the Piura is as follows:—On the western margin rise steep cliffs to a height of from 200 to 300 feet, either directly from the sea or with an intervening beach uncovered at low-water, and usually with a low reef of rocks at about half-tide, whereon even the gentle waves of the Pacific break in a dangerous surf. Having surmounted the cliff, we are on what is called the tablazo, a plateau rising very gently to eastward, in some places slightly undulated, and in others with ridges of considerable height rising out of it, the whole so bare of vegetation that there are places where not a single tree, much less an herb, can be distinguished within the limits of vision. A bold abrupt ridge, called the Silla de Payta, rises immediately to southward of that town to a height (according to Captain Kellett) of 1300 feet; but a far more important range of hills, beginning from near the sea, a little to northward of the mouth of the Chira, runs with a direction of E.N.E. all the way up between the rivers Chira and Tumbez, till it mingles with the Andes towards the sources of the latter river. . . . I suppose these hills to rise, even in their western part (which is all I have seen of them), to from 2000 to 3000 feet; to eastward, as they near the Andes, they must be far higher. Viewed from the south, they appear entirely bare of vegetation, but when they come to be examined their deep ravines are found to contain a few scattered Cactuses, Algarrobos, and other trees; and I am told that on their northern slope there is considerably more permanent vegetation, much as on the hills of Chanduy and St. Elena, which, although of far less extent, have quite the same aspect and structure.

The country to southward of the river Piura is known as the Despoblado (or Desert) of Sechura; but in reality that term might be extended to the whole desert region which stretches northward to the skirts of the forests of the Gulf of Guayaquil, for the narrow strip of vegetation along the courses of the Chira and Piura are mere oases in that vast desert.

. . . .

The deep valley along which the Chira flows to the sea has plainly been excavated by the action of water, and if any depression have originally existed on the tablazo along the same line it must have been very slight, as there is now no appreciable sloping towards it. Its sides are steep cliffs, scarcely at all furrowed transversely on the southern side, but on the northern side in most places very much broken up into ravines and alternating peaked ridges, whose origin may be traced to the effect of the rare but torrential rains descending the rugged slopes of Mancora. The peaks are often truncated cones, so symmetrical that until closely examined they might be supposed the work of art.

. . . At a little way within its mouth the river is only from 80 to 100 yards wide, and this average breadth is preserved, so far as I can learn, for at least 50 miles up. It is of no great depth, for, when at its lowest, a man may wade over it in most places with at least his head out of water; but as the current is pretty strong, and there are some deep holes, it is considered unsafe to ford it on horseback. . . . Very rarely, and with risk and difficulty, are heavy goods conveyed on a raft for a few miles up the stream. There are no bridges across it, but ferries are established at the villages and principal farms. The fluctuations of level throughout the entire year rarely reach 10 feet, but in the años de agua or rainy years there have sometimes been floods to a much greater height.

In ascending the valley of the Chira we come on a series of alternating contractions and lake-like expansions, the latter at one period no doubt really lakes. A little above the village of Amotape, 11 miles from the sea, following the course of the river, but only 7 in a straight line, the valley contracts, so that from the base of the hills on one side to the base of those on the other there is barely half a mile. From this point to above the small village of Tangarará, on the right bank, a distance of 15 English miles along the course of the river, there has been a large lake of a long oval form, the ancient margin retiring from the actual river-bank at one point on the north side nearly 3½ miles, and having an average distance of 2 miles. Deep furrows, like river-courses, extend from the widest part (called Monte Abierto) to the adjacent hills, and in the rainy years rivers again run along them and enlarge their beds. On the south side the space between the river and the base of the cliffs is also of considerable breadth, and has on it the villages of La Huaca and Bibiate in its lower part, and higher up the large farm of Macacará, 10 miles from Amotape.

. . . .

There are similar contractions of the valley, with intermediate lake-like expansions, up to 52 miles from the coast.

On examining the cliffs that bound the valley of the Chira, we find them to consist chiefly of alternating horizontal layers of very various composition, some of them apparently repeated at various depths. The uppermost stratum is in many parts a calcareous sandstone, of minute fragments of shells, grains of quartz, etc., more or less compactly welded together. When of open texture it is the material for the filtering-stones, which are largely manufactured at Payta, and are not only used throughout the province, but are exported to Guayaquil and other ports along the coast. . . .

Below the sandstone (which is repeated lower down) there are alternating layers of pudding-stone and shell-marl, the former consisting of rounded pebbles united into a compact but sometimes fragile mass by an argillaceous cement. The pebbles are nearly always egg-shaped, often the size of an ordinary hen's egg, and might seem water-worn, until being broken across they are found to consist of concentric coatings, varying in their mineral constituents but all more or less ferruginous. . . .

The shell-marl, or shell-rock as it might more properly be called, is one mass of fragmentary crushed fossil molluscs, chiefly bivalves and cirripeds, welded together by a tenacious ochry cement, from which they are often with difficulty separated even by the hammer. Rarely do both molluscs and cement yield to the action of water. . . .

Beneath all these strata, which are so nearly horizontal that there has plainly been no great convulsion since they were deposited—and they are at least 200 feet thick—there is a bed of compact argillaceous shales, which are tilted up at a considerable angle. At Payta, where this deposit is of immense thickness and apparently forms the great mass of the mountain called the Silla, it puts on the appearance of slate, being of a dull dark blue colour, and almost as hard as primary slate; but at Amotape what is evidently the same formation is usually of a greyish colour, and much more easily broken.

. . . .

Returning to the surface—the plateau or tablazo—the most remarkable feature is the quantity of white sea-sand that is accumulated and driven about by the winds in many parts of it. The whole country, however, is by no means covered with sand-hills, as one might suppose from some accounts that have been given of it. The great accumulation is in depressions and hollows towards the northern and eastern sides of the desert, whither it has been borne by the prevalent southerly and south-westerly winds. . . .

In proceeding from Payta northwards towards the valley of the Chira, we find the tablazo strewed with fragments of filtering-stone, clay-stones, etc., but we come on no sand until nearing the valley of the Chira, or even in some places (where the cliff is steep) until descending into the valley itself. We then find the cliff faced with sloping ridges of sand, blown over it by the wind, sometimes reaching into the river itself, whose waters are continually carrying off portions of them towards the sea. It is curious to see old Algarrobo trees with merely their heads out of the sand, but still growing and verdant; while others, entirely suffocated, show no more than a few dead twigs above it. These enormous ridged heaps are found all along the southern side of the valley, but nowhere pass the river to northward, for the sand once blown over the cliff is sheltered by it from the further action of the wind.

Piura lies nearly east from Payta, at a distance of 14 leagues, during the first seven of which the tablazo rises gently and equably, and the road is stony, or in some places dusty, but nowhere sandy. At midway, which is also the highest point of the route, there is a tambo or hospitium, where a supply is kept of water and food for man and beast, chiefly brought from the Chira with great trouble and expense. There the traveller, having started from Payta about sundown, reposes during the midnight hours, and starting again at 2 or 3 A.M., reaches Piura before the sun has risen high enough to heat the desert. From the tambo of Congorá the ground descends for the remaining seven leagues in gentle undulations towards the Piura (whose valley has no steep limiting cliffs like the Chira), and the sandy dunes at once begin, increasing in size and frequency as we descend. These dunes, or médanos as they are called, are notable for their lunate or half-moon shape, sometimes beautifully symmetrical, and having their convex side towards the trade-wind. They are continually shifting and advancing, but in general it is necessary to watch them for weeks to appreciate their motion. If a day's wind of more than usual violence disperse any of them, then soon re-form to north-eastward; a casual protuberance of any kind—a large stone or a mummified mule—being a sufficient nucleus for a new médano. On such days the sand which fills the air has all the appearance of a dense fog, and indeed at Piura the sky is generally more or less obscured from the same cause between 2 and 5 P.M. of every day.

The médanos I have seen near Piura are only from 8 to 12 feet in height, and yet that is quite high enough to render it difficult for the horseman entangled among them to find his way out, for one médano is almost the exact counterpart of another. On the desert of Sechura, however, which is a vast plain apparently depressed below the land immediately bordering the coast, the sand is heaped up to a far greater height, and I have been assured by an arriero that he has found shelter there for the night, on the lee side of a médano, for his company of ten men, thirty to forty mules, and all their baggage.

. . . .

Indigenous Vegetation

Any person, even one accustomed to the study of and search for plants, might travel through the whole extent of the deserts of Piura and Sechura, and (excepting the strip of verdure along the banks of the rivers) would confidently assert them to be entirely destitute of herbaceous vegetation; and yet three kinds of herbs exist there, which, burying themselves deep in the earth, survive through the long periods of drought to which they are subjected. Some of the smaller médanos, especially those under the lee of a low ridge of land, may be seen to be capped with snowy white, contrasting with the yellowish or greyish white which is the ordinary colour of the sand, and yet at a short distance liable to be taken for sand a little whiter than common. The whiteness, however, is that of the innumerable short cylindrical spikes of an Amarantacea, whose stems, originating from beneath the médano, ramify through it, and go on growing so as to maintain their heads always above the mass of sand, whose unceasing accumulation at once supports and threatens to overwhelm them.

The other two herbs of the desert are known to the natives, the one as Yuca del monte or Wild Yuca, the other as Yuca de caballo or Horse Yuca, from their having roots like those of the cultivated yuca (*Manihot Aypi*), or not unlike parsnips, but three times as large. Both roots are edible, and the former is sometimes brought to market at Piura when the common yuca is scarce. The Yuca de caballo is too watery to be cooked, but is sometimes chewed to allay thirst by the muleteers and cowherds, who detect its presence by the slightest remnant of the dried stump of a stem; for both kinds maintain a purely subterranean existence during many successive years, and only produce leafy stems in those rare seasons when sufficient rain falls to penetrate to the roots. A few animals that roam over the desert, such as goats, asses, and horses, obtain a scanty supply of food and drink from these yuca roots, which they scrape out with their hoofs. The fruit of the Yuca de caballo may frequently be seen blowing about the desert, looking more like a pair of very long hooked bird's claws than anything vegetable. It is an elongated capsule with a fleshy pericarp (incorrectly described as a drupe), terminating in a beak several inches long, and when ripe splitting into two valves, which remain united at the base and curl up so as to resemble claws or ram's horns. At Piura it is known by the not very apposite name of espuelas or spurs. In Mexico the fruit of an allied species is called Uña del diablo or Devil's Claws. The Yuca de caballo is a Martynia, of the family of Gesnereæ (or, according to some, of Cyrtandraceæ). I was fortunate enough to see a single plant of it with leaves and flowers in 1863, near the river Piura, on ground which the inundation had barely reached, but had sufficed to cause the root to shoot forth its stems, which spread on the ground, branching dichotomously, to the distance of a yard on all sides. The roundish leaves, clad with viscid down, are lobed much in the same way as those of some gourds, but the large sweet-smelling flowers are like those of a foxglove.

I have never seen either leaves or flowers of the Yuca del monte; but, from the description given me of it, I should suppose it a Convolvulacea, allied to the sweet potatoes (Batatas), and the lanceolate leaves point to the genus Aniseia.

The arborescent vegetation of the desert, although perhaps really more scanty than the herbaceous, is from its nature more conspicuous wherever it exists. There are points from which not a single tree is visible all around the horizon, but they are rare; generally the view takes in a few widely-scattered trees growing in basin-shaped hollows or towards the base of slopes, where at a certain depth there is permanent moisture throughout the wide interval between the años de aguas, at which epochs the supply is renewed. Wells dug in such sites reach water (too brackish for drinking) at various depths, the first deposit often at only a few feet from the surface. The moisture derived from the garuas, scanty as it is, no doubt aids in keeping the desert plants alive; and we have already seen that the air is never so excessively dry as might be supposed, but, on the contrary, sometimes approaches complete saturation. The trees of the desert are the Algarrobo (*Prosopis horrida*), the Vichaya (*Capparis crotonoides*), the Zapote del perro (*Colicodendrum scabridum?*), and an Apocynea with numerous slender branches, bright green lanceolate acuminate leaves, axillary clusters of small white flowers, and fruits, consisting of small twin drupaceous follicles, which are slender, curved, and coated with a thin white flesh. The Capparis and the Apocynea, although they grow to be trees in favourable situations, as in valleys near the sea, are mere shrubs on the desert; and the Prosopis and Colicodendron are low trees of very scraggy growth, their branches all bent one way by the prevailing wind, and the trunk itself often semi-prostrate.

Far away over the desert a tall branched Cactus begins to be met with; the same species abounds on the desert-coast of Ecuador. Farther still, near the roots of the Cordillera, the vegetation becomes gradually more dense and varied, comprising several other kinds of trees, and amongst them most of those about to be mentioned as denizens of the valleys.

When the traveller across the despoblado comes suddenly on one of the valleys, he passes at once from a desert to a garden, whose charms are enhanced by their unexpectedness. Standing on the cliff that overlooks the Chira, about Amotape, he sees at his feet a broad valley filled with perpetual verdure, the great mass of which is composed of the pale green foliage of the Algarrobo; but the course of the river that winds through it is marked (even where the river itself is not seen) by lines or groups of tall Coco palms, here and there diversified by the more rigid Date palm, both growing and fruiting in the greatest luxuriance, their ample fronds never mutilated by caterpillars as they are wont to be in other regions. On the river-bank grow also fine old Willows (*Salix Humboldtiana*), noticeable for their slender branches and long, narrow, yellow-green leaves, contrasting strongly with the dark green of the spreading Guavas (Ingæ sp.), and with the bright green foliage (passing to rose at the tips of the branches) of the Mango (*Mangifera indica*). Mingled with these, or in square openings in the Algarrobo woods, are cultivated patches of sweet potatoes, yucas, maize, and cotton plants, the latter distinguishable by their pale but fresh green colour. It was a magnificent sight to look from this cliff towards the mouth of the Chira when the sun was just setting over it, steeping the hills of Mancora in purple and violet, and gilding the fronds of the palms and the salient edges of the adjacent cliffs, while the deep recesses of the latter and the Algarrobo woods were already shrouded in gloom.

On descending into the valley, the natural forest of Algarrobo is found to occupy a strip of from a few hundred yards to three or four miles in width, extending from the river on each side as far out as there is permanent moisture at a moderate depth. It is divided by fences into plots of various sizes, all private property, except a small breadth of common lands adjacent to each village. I was surprised to hear these plots called not "woods" but "pastures" (potreros), for the trees grow in them as thickly as trees do anywhere, and there is not underneath them an herb of any kind. They are so called because the fruit of the Algarrobo is the main article of food for most of the domesticated animals, and therefore corresponds to the pasturage of other countries. The Algarrobo is a prickly tree, rarely exceeding 40 feet in height, with rugged bark not unlike that of the elm, but more tortuous, and with bipinnate foliage like that of the Acacias, to which it is closely allied. The roots penetrate the soil to only a slight depth, but extend a very long way horizontally. On the desert I have seen an Algarrobo root, no thicker than the finger, stretch away to a length of 40 yards, evidently in quest of moisture. As the trunks never grow straight, and soon become tolerably corpulent, and their roots take too little hold of the friable earth to sustain them against the squally winds, they very generally fall over in age either into a reclining posture or quite prostrate, but immediately begin to turn their heads upwards, send off new roots from every part of the trunk in contact with the soil, and thus get up anew in the world; so that an old potrero or Algarrobo wood has a most irregular and fantastic appearance. Twice in the year the Algarrobo puts forth numerous pendulous racemes of minute yellow-green flowers, which nourish multitudes of small flies and beetles, that in their turn afford food to flocks of birds—most of them songsters, and all of them more pleasantly garrulous than any similar assemblage of little birds I have met with elsewhere in the world. The flowers are followed by pendulous, flattish, yellow pods, 6 to 8 inches long, about a finger's breadth and half as thick, containing several thin flat seeds, immersed in a sweetish mucilaginous compactly spongy but brittle substance, which is the nutritive part. These pods are greedily devoured by horses, cows, and goats, but especially by asses, which are more numerous than any other domestic animals. It is a very concentrated and heating kind of food, and I have seen horses after eating it chew the leaves of the castor-oil plant, or any kind of rubbish, to counteract its stimulating properties. . . .

The Algarrobo secretes an inflammable gum-resin, which exudes from cracks in the bark and coagulates into a blackish mass. Advantage is taken of it to prostrate the trees by fire, when it is required to clear the ground for cultivation. Cutting them down is scarcely ever resorted to, the timber being so hard as soon to render useless the best-tempered axe. The method employed is this: A truncheon of wood, alight at one end, is laid on the ground with that end touching the tree to windward. The trunk soon takes fire, and (especially if the wind be strong) is in a few hours burnt right through nearly horizontally, the part destroyed rarely exceeding from half a foot to a foot in breadth; and being thus prostrated, its still burning end is covered with earth to extinguish the fire. There is no better material for fuel than Algarrobo wood, and its very great hardness and durability would make it a most desirable timber for any kind of construction, were it not that it grows so crooked and is so intractable to work.

Potreros from which animals have been long excluded sometimes grow so thick, from two kinds of lianas which fill up the intervals of the trees, as to be impassable. A species of Rhamnus, called Lipe, armed with formidable decussate spines, and producing minute 4–5-merous flowers, followed by small edible black berries, supports itself against the Algarrobos and climbs high among their branches. When it grows alone and has room to spread, it forms large round bushes, each many yards in diameter, and 12 to 15 feet high. Bushes of Lipe, scattered over the bare ground, look at a distance not unlike the small groves of hollies or other evergreens that stud the sanded or gravelled surface of an English shrubbery. In these bushes hide by day numerous foxes, which come out by night in quest of food. They are as fond of melons as Æsop's fox was of grapes, and do not despise them even when green, so they can get at them. Lizards and a few snakes also seek the shelter of the Lipe. Flocks of small birds roost there by night, and by day pick the berries.

The companion of the Lipe is a rampant Nyctaginea (Cryptocarpus). It climbs to the tops of the Algarrobos, and often hangs therefrom in dense masses. It has heart-shaped stellato-pubescent leaves and panicles of minute green flowers, which persist on the enclosed black utricle. A stout parasitical Loranthus, with small yellowish flowers, often forms large bushes on the Algarrobo, and generally ends by destroying the tree whereon it has established itself.

A far handsomer tree than the Algarrobo sometimes grows along with it, especially where there is rather more moisture than usual; this is the Charán (Cæsalpinia). It is a widely-spreading tree, often branched from the very base, and the shining reddish bark is being constantly renewed. It has exceedingly graceful bipinnate foliage—roseate at the tips of the branches—panicles of yellow flowers, spotted with red, and thick deep-purple pods, which are extensively used in tanning.

The Azota-Cristo or Whip-Christ (*Parkinsonia aculeata*), so called from its excessively long pendulous leaves, from whose thong-like rachis the small leaflets often fall away, is less handsome but still more uncommon-looking than the Charan, and it is also much rarer in this region. It reappears in the Antilles.

A few other trees are occasionally met with, such as a Calliandra, conspicuous for its numerous flowers—green tinged with rose—out of which hang the long, silky, straw-coloured stamens, and for its curled scarlet pods; two Acacias, one of them the widely-dispersed *A. Farnesiana*; a Maytenus, which is especially abundant at the mouth of the Chira, and is common enough along the coast of Ecuador as far north as the Equator; and the Oberál (*Varronia rotundifolia*), a solanaceous tree or shrub, with numerous bright yellow trumpet-shaped flowers and white berries, abounding in a viscid juice, which is used by the dusky beauties of Guayaquil to straighten out their hair and hide its natural crispness. . . .

The trees mentioned above as belonging to the desert grow also in the valley, and far more luxuriantly there, but generally scattered along the outer margin of the Algarrobo belt, especially wherever the soil is much impregnated with salt. The Zapote de perro bears a large berry, not unlike a smallish melon in size, shape, and the alternating green and white streaks. Its taste is disagreeable, and I have not seen it touched by any animal, although it is said to be eaten by dogs (as its name implies), and also by foxes and goats. The Vichaya, a dense growing bush, with oval hoary leaves, has yellow berries the size of a damson, containing a few stony seeds involved in a mawkish sweet pulp. Another Capparis, which scrambles up into the trees, also grows here, but rarely; it is much more frequent near Guayaquil, as is also the Vichaya, which is there called Cuchuchu. In fact, all the trees and shrubs hitherto mentioned (with one or two exceptions) grow also on the desert coast of Ecuador, along with a few others not found in Northern Peru.

In the ravines which run from the tablazo down to the valley, besides a few stunted Algarrobos, there is another small prickly tree, a species of Cantua, with black stems and branches, which becomes clad with fugacious, roundish, Loranthus-like leaves and pretty white flowers only in the rainy years. There also grows a Cactus called Rabo de zorra (fox's brush), from its usually simple stems being densely beset on the numerous angles or striæ with reddish bristle-like prickles.

On the margin of the river, except where the banks are unusually high, there is a narrow strip of land, called the vega, which is overflowed every year about February or March by the flush of water from the Andes, although no rain may have fallen in the plain. The vega is in many parts of the valley the only ground kept under cultivation, and the indigenous vegetation there is of a quite distinct character. Instead of the Algarrobo, we have the Willow and a small Composite tree, *Tessaria legitima*, with leaves very like those of *Salix cinerea*, and soft brittle wood, which is the common fuel at Lima and elsewhere on the coast, where it is called Pajaro bobo. Less abundant than those two trees are *Buddleia americana*, a pretty Cassia, two species of Baccharis, two rampant Mimosæ (one of them *M. asperata*), *Muntingia Calaburu*, and *Cestrum hediondinum* (called Yerba Santa), of which only the two last grow to be trees of moderate size, the rest being weak bushes or shrubs. Over trees and bushes climb a half-shrubby Asclepiadea (Sarcostemma sp.), with very milky stems and umbels of pretty white flowers, a Cissus, a Passiflora, allied to *P. fœtida*, a pretty delicate gourd plant, and a Mikania.

It is usually only on the vega that we find any herbaceous vegetation, except in the rainy years. There the Caña brava, a Gynerium, with a stem 15 feet high and leafy all the way up, and with smaller and less silky panicles than the other species, grows in large patches. The huts of the Indians and Mestizos in the suburbs of Piura have often nothing more than a single row of Caña brava stems stuck into the ground for walls, and others laid horizontally over them for roof, affording, of course, little protection from sun and wind, and none at all from the rain, which happily falls so very rarely.[1] Along with it grow a

[1] It does not enter into the scope of this memoir to describe the towns of North Peru and the customs of their inhabitants, but it might leave a false impression were I not to add that all the better class of houses are as solidly constructed as almost anywhere in South America. At Piura they have thick walls of adobes, and are built round patios or courts, over which awnings are stretched in the heat of the day. Glass windows, verandas, and balconies are almost universal.

few other perennial grasses, chiefly species of Panicum and Paspalum, besides the Grama dulce (*Cynodon dactylon*), originally brought from Europe, but here so completely naturalised that, if allowed to spread, it would exclude almost every other plant. It is valuable as an article of fodder. A few annual grasses, chiefly species of Eragrostis, grow about the outer margin of the vega. Of sedges also (species of Cyperus and Scirpus) there are four or five species.

Other herbaceous or suffruticose plants are a tall Polygonum, the handsome *Typha Truxillensis*, the Yerba blanca (*Teleianthera peruviana*), several species of Chenopodium, including the strong-smelling Paico (*Ch. ambrosioides* and *multifidum*); a Cleome, a Portulaca, *Scoparia dulcis*, a Stemodium, and three or four other Scrophulariaceæ; a Melilotus, a Crotalaria, a pretty Indigofera, with numerous prostrate stems spreading every way from the root, and pink flowers, a Desmodium, a sensitive-leaved Desmanthus, a Sonchus, *Ambrosia peruviana*, and a few other Compositæ; a Datura, two species of Physalis, *Dictyocalyx Miersii*, Hook. *f.* (exceedingly variable in the size and shape of its leaves), and the ubiquitous *Solanum nigrum*; *Verbena littoralis*, two species of Lippia, *Tiaridium indicum*, a Heliophytum, three Euphorbiæ, a small Lythracea allied to Cuphea, and a few others.

In the river itself occasionally grows a Naias, in dense masses, like those of *Anacharis alsinastrum* in English streams and ponds. . . .

Two mosses, both species of Bryum, are occasionally found on the banks of the river Chira, and on the filtering-stones kept in houses, but only in a barren state.

I did not remain long enough in the country to witness the full effect of the rains of 1864 on the desert. The first plant to spring up, in the ravines leading down from the tablazo to the valley, and then on the tablazo itself, were two delicate Euphorbiæ, distinct from those of the vega. A little later on they were followed by a fragile dichotomously branched Scrophulariacea (which is common on the coast to northward of Guayaquil); two viscid Nyctagineæ (species of Oxybaphus) with pretty purple flowers; and two or three grasses (one of them an Aristida), but very sparingly. The Yuca de caballo (Martyniæ sp.) also began to put forth its leaves, but the Yuca del monte had not, up to the 20th of April, shown itself above ground. I had seen far more wonderful effects of the rains of 1862 at Chanduy, where a desert nearly as bare as that of Piura became clad in a month's time with a beautiful carpet of grasses, of many different species, over which were scattered abundance of gay flowering plants. Something similar must have occurred this year to northward of the hills of Mancora, for people who travelled between Amotape and Tumbez in the middle of April reported the whole country clad with verdure, and the grass in the hollows up to the horses' girths.

. . . .

[The following extract from a letter to Mr. Bentham, written a few weeks before finally leaving South America, explains the reasons for his return home, and concludes his correspondence while abroad :—]

To Mr. George Bentham

AMOTAPE, NEAR PAYTA, PERU,
April 13, 1864.

During the last twelve months I have experienced some relief from my pains, and life has not been so barely tolerable a burthen as during the three preceding years; but I see plainly I can never hope to regain my former activity, or indeed be able to undertake any occupation whatever, and I have made up my mind to return to England, my present intention being to embark at Payta for Southampton on the 1st of May. . . .

[The following extract from a letter to Mr. Daniel Hanbury, written from Hurstpierpoint two years after his return to England, gives a curious piece of information as to his friend the late Dr. Jameson of Quito, which is to some extent a vindication of that botanist's character and abilities.

Referring to Dr. Jameson's *Flora of Ecuador*, which Spruce says is extremely imperfect, and mostly a translation from other works, with no original descriptions of plants, and whole genera altogether unnoticed, he has the following remarks which may be of value as showing why a man with (apparently) such fine opportunities, and who was so interested in botany, yet did so little :—]

Jameson told me he had been to Baños only once in his life, although he has been over forty years in Ecuador. He would have liked to go again to gather some of the Orchids I found on Tunguragua, but could not spare either the time or the money. Suppose he were to write to ask you just to step over to the Shetland Islands and get him a form of *Stereocaulon paschale* which grows there—you could do it more easily than he could go to Baños and back. Yet Jameson is one of the most amiable of men, an ardent collector (for other people—much the same as I have been), and a very fair botanist and mineralogist. But what can a poor fellow do who has had a drunken (and worse) wife hanging on him for forty years, who burns his dried plants, whenever she can get hold of them, so that he can keep no herbarium, and who has often had to struggle with absolute want?

[This is the Dr. Jameson after whom was named the beautiful greenhouse shrub *Streptosolen Jamesonii*, as well as many other plants.

The remainder of this volume consists of extracts from letters to Mr. Hanbury, having special reference to matters connected with his residence in the Andes; together with six essays on various subjects relating to his travels, which have either been hitherto unpublished or are almost unknown to English readers. They have been condensed where necessary, but are otherwise as Spruce left them.]

CHAPTER XXIII

ASPECTS OF AMAZONIAN VEGETATION AND ANIMAL MIGRATIONS

(ENGLAND, 1864–1873)

[ON reaching England in May 1864, Spruce remained for some time in London, at Kew and at Hurstpierpoint, with short visits to Mr. Daniel Hanbury and to myself. He thus had frequent opportunities of seeing most of his botanical friends, and his further correspondence with them was of little general interest. There is an exception, however, in the case of Mr. Hanbury, with whom he at once established an intimacy which quickly ripened into a close friendship; and as this gentleman thenceforth acted as Spruce's informal agent in London, supplying him with medicines, books, and any special delicacies he required (always on a strict business footing), while Spruce was always ready to give botanical or other information on Mr. Hanbury's special pharmaceutical researches, letters passed between them weekly, and often daily, for many years, amounting in all to nearly a thousand, all of which were carefully preserved and were presented by Sir Thomas Hanbury (after his brother's death in 1875) to the Pharmaceutical Society. These were kindly lent me, and a few of them are so interesting, and have such a close relation to his work in South America, that I give here some extracts from them, adding a few explanatory words where necessary.

The first is from one written about six months after his return home, and is characteristic of his intense love of nature.]

KEW, *Dec.* 20, 1864.

I am thankful we are so near the shortest day. It is an awful sight to me to see that the sun at noon barely rises as high as the weathercock on Kew Church steeple (seen from the opposite side of the green)—and the poor skeletons of trees! I have not seen trees without leaves for more than fifteen years.

[This was specially interesting to myself because, on my return from the Amazon in October 1852, I was at once struck by two things—the general smallness of the trees, and even more by the low sun at noon, and especially by the fact of its giving hardly any heat, so that it seemed most surprising how any vegetation could continue to grow and thrive under such harsh conditions.

Although Spruce had made Ambato his headquarters for nearly three years, I have found in none of his letters any reference to what accommodation he had there or to the people he lived with, except the one remark (in a letter to his friend Teasdale) that his landlord there was "one of the best men in the place." But as he was often away collecting at Baños, Quito, Riobamba, and other places, as well as in the forests around Tunguragua and the

Cinchona forest to the west of Chimborazo, for days, weeks, or even months at a time, and never makes any mention, on his return, of any injury by damp, insects, etc., to his plants or his books, he was evidently sharing a house with some family (or renting an adjoining house), where he himself and all his belongings were carefully attended to. This mystery is now cleared up by a series of letters to Mr. Daniel Hanbury, enclosing translations of letters he had received from his old landlord in Ambato, Manuel Santãnder, to whom Spruce promises to write (at Mr. Hanbury's request), with a commission to obtain, if possible, dried specimens of the flowers, fruit, and foliage of the "Quito Cinnamon." The result of Santander's repeated attempts for over two years was a small quantity of branches with leaves only, which are now preserved in the Herbarium of the Pharmaceutical Society.

On inquiry, I have learnt that no authentic specimens exist at Kew, and, presumably, there are none in any other European herbaria, so that the tree producing this cinnamon-like bark is still botanically unknown.

Santander's letters show the great and genuine affection which Spruce had inspired in this excellent man and his whole family.]

Spruce's Account of Santander

(*Letter to Mr. Daniel Hanbury*, Feb. 1, 1866)

. . . .

Santander is a remarkable man. In youth he was a soldier, and rose to be a captain. He was in many battles, and during a "revolution" at Guayaquil he was wounded by two musket-balls in the hip and thigh (which still give him periods of torture), was made prisoner, and banished to the coast of Peru without a cent. At Payta he set up a school which gave him a bare living. While a soldier he had taught himself to repair the lock of a gun, and at Payta he began to teach himself all kinds of light work in metals, in which, being an ingenious fellow, he succeeded admirably, so that when some years later new revolutions recalled him to Ecuador, he opened at Ambato a flourishing business—what we should call that of a whitesmith—employing Indian smiths to do the heavier work. He resisted the most urgent solicitations of the Government to take a new and higher commission in the army, and resolved to maintain himself by the work of his hands and brain. Add to all this, that he is a man (like yourself) overflowing with the milk of human kindness, and you will comprehend how I came to regard him with great affection, and regretted much having to part from him.

Santander's lameness prevents his travelling much, but he knows the Cinnamon gatherers and sometimes trades with them. I will give him full instructions as to what we want. . . . If any one can get the Canelo—without going to the spot where it grows, which is a good month's journey out and in from Ambato—I believe it is Santander.

[A year later he has a reply from Santander, a translation of which he sends to Mr. Hanbury.]

Translation of Santander's Letter

To Ricardo Spruce

Ambato, *June* 30, 1867.

[The letter begins: "My never forgotten friend," and after two pages giving a full description of the box of specimens (also asked for) he has sent to Mr. Hanbury, and his prospects of getting the much-desired Canelo, he continues thus:]

I now pass on to my own affairs and those of my family. I wrote to Inez (his eldest daughter) with your salutations, and she replies saluting you most affectionately. She says that her first little boy already bears the name of Juan Elias, and that she reserves the name of Ricardo for her second. Her husband and her father-in-law (Don Rafael Paz y Miño), who both know you, salute you with many caresses. . . . As respects my family, we are all here at your orders, truly desiring to see you and embrace you, for even yet tears accompany the memory of our absent friend. Isabelito and Pachito (his younger children) are in despair to see you and embrace you, and say: "Oh that London was no farther off than Ambato to Lligna, that we might go to Señor Ricardo!" But as an immense distance separates us, there is no alternative but to console ourselves with your letters. Isabel (his wife) is ready to complain that she ever knew you, because she could not then have felt your loss; but consoles herself with the hope that one day you will return to Ambato, stout, healthy, and rich. This is what we all desire, and that, leading with us a simple and peaceful life, we may end our days together.

With this I await your reply, desiring that it may find you well. Your truest heart-friend sighs to see and to embrace you.

MANUEL SANTANDER.

Addition.—If convenient to you, and you consider that Pachito might be useful to you, and you will tell me how he may get there, I will give him to you, Señor Ricardo, that he may serve you as a companion and assist you in something.

(So endeth the epistle according to Santander.)

R. S.

[Nearly two years later, in a letter to Mr. Daniel Hanbury from Welburn (dated December 31, 1868), we have the conclusion of the long story of the repeated efforts to get flowers and fruits of the much-desired Canelo or Cinnamon tree of Quito. This tree and its spicy bark were known to the Spanish conquerors of Peru and Ecuador, and has been an article of commerce ever since; the great forest of Canelos was so named after it; many travellers and botanists have traversed this forest, including the enthusiastic Richard Spruce, yet no one had yet been able to obtain or even to see its flowers or fruit. Some of the causes of this failure are indicated in a letter from Santander, dated "Ambato, November 12, 1868." He therein describes the extraordinary series of accidents and misfortunes which made all his efforts of no avail; and as it also serves to illustrate further the character of the very interesting writer, and also that of Spruce himself, who could excite such enthusiastic affection (though this will surprise none who knew him intimately), I will here give the more interesting portions of it.]

Señor Santander to Señor Don Ricardo Spruce

MY MUCH-THOUGHT-OF AND NEVER-FORGOTTEN FRIEND—The receipt of your much-desired and consolatory letter has filled me and my family with joy, especially on seeing the portrait that accompanied it. But what a notable difference it presents from that you sent us in 1864, which showed you much the same as we had known you, whereas this last shows you with a beard as white as the snow of Chimborazo, and a stoutness that (for you) is extreme. What changes time makes in the features—one would think from this portrait you were seventy years old!

[Then follows an account of his own domestic troubles: the death of his eldest son, the dangerous illness of his wife, and the loss of a fine mastiff, "our old and faithful friend and the guardian of our house!" He then continues:]

Notwithstanding these calamities, I did all I could to procure the specimens of Cinnamon for Mr. Hanbury, but I have found it impossible.

In the first place, I availed myself of Padre Fierro, our friend, and in effect he sent the desired specimens by Pacho Gallegos and José Torres. But see what happened. The Padre's nephew ran away from Canelos and carried off all his uncle's clothes, some ounces of gold, a gun, etc. He reached Baños along with the two men, and as the branches, etc., of the Canelo were stowed in two baskets of Ishpingo, he sold the Ishpingo (which was then at 22 reals the pound) and threw away the branches, which were of no value to him. Nobody knows what has become of him; but I was almost at my wits' end.

The second time I made a treaty with Pedro Andicho, the Governor of Pindo (a few Indian huts in the middle of the Forest of Canelos), who was going there to make lance-shafts for the war that menaced us at that epoch, and I gave him three frascos (large square bottles) prepared according to your directions. I paid him in advance four dollars, and made him several presents—a gun among the rest—that he might deliver the frascos filled with specimens to Padre Fierro. He had scarcely reached Pindo when he died, and though I have again and again solicited Padre Fierro to recover the frascos, he has found it impossible.

I wrote to him also asking him to send me the branchlets in paper (as you used to prepare them), which indeed he took the trouble to do, and sent them by some Indians who were going to Baños, but who threw them into the river, so that they never reached me. How unfortunate I have been!

On the 8th of December last year I gave four dollars to Pedro Valladares, with a written agreement that he should obtain for me the desired objects. He goes to Canelos, starts on the return journey, and is stopped by death, and none of his effects have been recovered!

After this I made a treaty with Manuel Meneses for two dollars—as can be proved by my books—

and ever since I have neither seen nor heard tell of him!

I have tried to treat with the traders who go from Pelileo to Canelos, such as Hilario Flores and others; but not one has been willing to undertake to bring the Cinnamon fruits, etc., on account of its exceeding difficulty.

By Padre Fierro, who himself has just come out of the forest, I certainly hoped to obtain it, but he has only brought two young living plants, which (as they were beginning to wither) he has left behind him (planted) at St Iné's, on the farm of Dr. Lizarzaburo. He thought he was doing the best he could for me in bringing the live plants. He brought also seeds, but they got them from him at St. Iné's.

All that I have been able to obtain is some loose leaves and calyces with their fruits, but not of their original colour. Tell me, may I send Mr. Hanbury these dried leaves? The young plants will prosper, but the difficulty is how to send them.

As to getting the flowers, that is the most impossible of all, for Padre Fierro tells me that no sooner does the young calyx appear than it already contains the young fruit [here Spruce remarks—" hence the tree appears to be diœcious"], so that the calyx with the fruit ought to suffice for the identification of the Cinnamon tree. I hope to give these things to Mr. Seckel, who is on his way to Quito, that he may send them to Mr. Hanbury.

How I wish this affair had depended on me alone, and that I could have gone to Canelos; but that is impossible because of the precipitous ways. If I could have gone myself, even although I had perished, my death would have been praiseworthy, and my friends could not have been dissatisfied with me. Have the goodness to salute Mr. Hanbury for me, and to explain to him all the obstacles that have opposed the execution of his commission. If he is not satisfied that I have done my best, I must return him the £5—there is no other alternative.

What a pleasure it has been for me to learn something of your actual position, and it has been the same for my family, who charge me to embrace you with a thousand tender caresses; for they say the lapse of time only makes them remember you and regret your absence the more. For me, what shall I say? I preserve in my heart the image of Señor Ricardo, but this my joy is troubled by the hopelessness of ever seeing him again. What happiness it would be for us to have you at Ambato just now, in the most agreeable season of the year. The time of ripe pears and peaches is near; our friend Mantilla, with his accustomed kindness, is waiting for us to go and eat them. Miraflores is now planted with poplars all along the avenue where we used to walk. Tamatamas[1] are ready for our innocent games. Isobel is at the gate waiting for you. Frank and I are ready to accompany our dear friend. But—sweet dream—delusive hopes—where is he?

Adieu, my beloved friend, adieu! Thus your sincere friends bid you farewell!

MANUEL SANTANDER.

[In sending this translation to Mr. Hanbury, Spruce writes: "You will read about the disasters

[1] Sticky fruits with which children pelt each other.

that have attended the quest of the Quito Cinnamon. I know nearly all the people whose names are mentioned, and I have no doubt his relation is exact, for I know well the simple and truthful character of the man. . . .

"After all the time, etc., Santander has lost, I do not think we can ask him any more."

Thus ends the quest for botanical specimens of one long-known tree whose scented bark is still an object of commerce, but which grows only in a limited district of the great forests at the foot of the Andes of Ecuador.

The following interesting paper was sent to the Linnean Society in 1867, and published in the Society's Journal, vol. ix. (pp. 346-367), under the following title: "Notes on some Insect and other Migrations observed in Equatorial America. By Richard Spruce, Esq. Communicated by the President."

This title, however, does not convey an idea of its whole subject, which is almost as much botanical as zoological, the first portion of it containing an admirable sketch of the broader aspects of the vegetation of the Great Amazon Valley and adjacent regions. I have therefore subdivided the paper under separate headings, and have omitted a few of the less interesting details.]

The Broader Characteristics of Amazonian Vegetation

In endeavouring to trace the distribution of plants in the Amazon valley, and to connect it with that of animals, I have been struck with the fact that there are certain grand features of the vegetation which prevail throughout Cisandine America, within the tropics, and even beyond the southern tropic—features independent of the actual distribution of the running waters, partly also of the geological constitution, and even of the climate—to which the range of the larger species of Mammals and Birds corresponds in a considerable degree, but not that of any other class or tribe of animals, and especially not of lepidopterous Insects. These features depend on the prevalence of certain groups, or even of single species, of plants over vast areas: one set prevailing in the Virgin or Great Forests (Caa-guaçú of the Brazilians, Monte Alto of the Venezuelans) which clothe the fertile lands beyond the reach of inundations, and constitute the great mass of the vegetation; another in the Low or White Forests (Caa-tinga, Monte Bajo)—those curious remnants of a still more ancient and humbler but surpassingly interesting vegetation, which (especially on the Rio Negro and Casiquiari) are being gradually hemmed in and supplanted by the sturdier growth of the Great Forests, wherein they are interspersed like flower-beds in a shrubbery; another in the Riparial Forests (Ygapú or Gapó of the Brazilians, Rebalsa of the Spaniards), on lowlands bordering the rivers, and laid under water for several months in the year, where the trees when young, and the bushes throughout their existence, must have the curious property of being able to survive complete and prolonged submersion, constituting for them a species of hybernation; a fourth in the Recent Forests (Caa-puéra, Rastrojo), which spring up to replace the Primitive Forests destroyed by man, and, notwithstanding their weedy character, consist chiefly of shrubs and trees; a fifth in the savannas or campos — grassy or scrubby knolls, or glades, or hollows (dried-up lakes), which bear a very small proportion indeed to the vast extent of woodland in the Amazon valley proper, but towards its northern and southern borders compete with the woods for the possession of the ground, and in the centre of Venezuela enlarge to interminable grassy llanos or plains.

From an elevated site that should embrace the landscape on all sides to the extreme limit of vision, as, for instance, from the heights at the confluence of the Rio Negro and Amazon, or, better still, from one of the steep granite rocks that overlook the noble forests of the Casiquiari, a practised eye would distinguish the various kinds of forest by their aspect alone. The Virgin Forests are distinct enough by the sombre foliage of the densely-packed, lofty trees, out of which stand, like the cupolas, spires, and turrets of a large city, the dome-shaped or pyramidal or flat-topped crowns of still loftier trees, overtopping even the tallest palms, both palms and trees being more or less interwoven with stout, gaily-flowering lianas; the White Forests by the low, neat-growing, and thinly-set trees and bushes, with scarcely any lianas — the Palms few, but peculiar, and often odd-looking—on a near view by the greater abundance of Ferns, especially on the trees, and sometimes of terrestrial Aroids and Cyclanths; the Recent Forests by their low, irregular, tangled growth, paler foliage, and general weedy aspect; the Riparial Forests, even where the water is not visible, by the varied tints of the foliage, and by the trees rarely equalling those of the Virgin Forest in height—sometimes, indeed, beginning on the water's edge as low bushes, thence gradually growing higher as they advance inland, until at the limit of inundations they mingle with the primeval woods, and are almost equally lofty—by the greater proportion of herbaceous lianas which drape the trees and often form a curtain-like frontage—and by the abundance of Palms, whereof the taller kinds usually surpass the exogenous trees in height, and (the Fan palms especially) often stretch in long avenue-like lines along, or parallel to, the shore. On some black-water rivers, such as the Pacimoni, the Atabapo, and the Rio Negro in some parts of its course, the breadth of inundated land is entirely clad with bushes and small trees of very equable height, on the skirts of which the Virgin Forest rises abruptly to a height more than twice as great. This is called by the natives "caatinga-gapó."

Besides these differences of aspect, the natives will tell you there are other more intrinsic ones; for instance, that the riparial trees have softer and more perishable timber, as well as inferior fruits; while the caatingas, with a far greater show of blossom, have hardly any edible fruit at all, and very few indeed of the trees rise to the magnitude of timber trees. And yet, when the constituent plants of the different classes of forest come to be compared together, they are found to correspond to a degree quite unexpected; for although the species are almost entirely diverse, the differences are rarely more than specific. It is only in the caatingas that a few genera, each including several species, seem to have taken up their exclusive abode: such are Commianthus among Rubiaceæ, Pagamea among Loganiaceæ, Myrmidone and Majeta among Melastomaceæ; and there are a few other peculiar genera, chiefly monotypic. But, of the riparial plants, nearly every species has its congener on terra firme, to which it stands so near that, although the two must of right bear different names, the differences of structure are precisely such as might have been brought about by long exposure even to the existing state of things, without supposing them to date from widely different conditions in the remote past; and this is especially true of such genera as Inga, Pithecolobium, Lecythis, and of many Myrtles and Melastomes, Sapotads, etc.

As an illustration of the features which tend to impress a certain character of uniformity on the vegetation of the Amazon region, I will take the case of a single tree, *Bertholletia excelsa* (H. and B.)—perhaps the noblest tree of the Amazon region, and the most characteristic of its Virgin Forests—and briefly sketch its distribution. In aspect and foliage it is not unlike a gigantic Chestnut tree; and the seeds (the Pará nut of commerce), if not much like chestnuts in their trigonous bony shell, are not very different in taste, whence the Brazilian name of the tree, "Castanheira," and of the seeds "castanhas." This tree is found almost throughout the Amazon valley, both to north and south, chiefly wherever there is a great depth of that red loam which it pleases M. Agassiz to call "glacial drift." About Pará itself there is no lack of it, especially in the fine woods of Tauaú; and 1200 miles farther to the west it may be seen in some abundance on the very banks of the Amazon, between Coary and Ega, at a part called Mutúncoará (Curassow's Nest), where steep red earth-cliffs border the river and forest; while it extends many hundred miles up the Purús and other southern affluents. North of the main river I have seen it at many points—for instance, in the forests of the Trombetas and at the falls of the Aripecurú; in various places along the Rio Negro, where one village (Castanheiro) takes its name from it; and on the Casiquiari and Upper Orinoco, where it was first seen and described by Humboldt and Bonpland.

A magnificent palm, *Maximiliana regia* (Mart.)—Inajá of the Amazon, Cocurito of the Orinoco—frequently accompanies the Bertholletia, and is still more widely and generally dispersed. I have seen it as far to the south as in 7° lat.; and in 5½° N. lat., at the cataracts of the Orinoco, it is still as abundant as on the Amazon. It even climbs high on the granite hills. On one which I ascended near the falls of the Rio Negro, an Inajá palm occupied the very apex, at 1500 feet above the river; and with the telescope I have distinctly recognised this Palm at a much greater elevation on Duida and other mountains. Both the tree and the Palm range to northward and southward beyond the limits of my own explorations; and there are a few other arborescent plants which stretch all through South America, from the base of the coast-range of Caracas (or even in a few cases from the West India Islands) to the region of the river Plate; but these are chiefly trees such as sprinkle the savannas, or are gathered into groves, along both the northern and southern borders of the great Amazonian forest-belt, wherein they now barely exist on the bits of campos that at wide intervals break the monotony of the woodland—although they probably at some antecedent period ranged continuously from north to south.

In other cases, closely allied species occupy distinct areas. One of the finest fruits of Equatorial America, the Cocura (Pourouma of Aublet), is borne in large grape-like bunches on trees of the Bread-fruit tribe, having large, palmatifid, hoary leaves, quite like those of their near allies the Cecropias. Now the Cocura of the mid-region of the Rio Negro, of the Japurá, and of the Upper Amazon or Solimoẽs is *one* species (*Pourouma cecropiæfolia*, Mart.), while that of the mouth of the Rio Negro and adjacent parts of the Amazon is a very distinct and smaller-fruited species (*P. retusa*, Spruce), and that of the Uaupés is a third species (*P. apiculata*, Spruce), all three being so plainly diverse that the Indians distinguish them by adjective names, although that diversity or divergence, as in a great many parallel instances, is but a measure of the time that has elapsed since their derivation from a single stirp.

But the most general cause of resemblance lies in this fact, that there are many orders and families of plants whereof many of the species are confined to limited areas, and yet, throughout the Amazon valley, each order, or family, will be everywhere represented by about the same number of individuals and species, having to each other nearly the same correlation, as regards aspect and sensible properties, provided always that the conditions of growth (as above defined) be the same; so that a plant which serves as food for any particular animal or tribe of animals in a given locality is pretty certain to have its congener (or at least its co-ordinate) in any other locality of the same region.

The riparial plants of the Amazon (such, namely, as grow between ebb- and flood-mark, or within the limits to which the annual inundations extend) range in many instances from the very mouth of the river up to the roots of the Andes; and I do not yet know of a single tree which is not found both on the northern or Guayana shore and on the southern or Brazilian.[1] The most notable example of this extensive range is the Pao Mulatto or Mulatto tree (Enkylista, Benth.), a tall, elegant tree allied to the Cinchonas, and conspicuous from its deciduous brown bark, which grows everywhere on lands flooded by the Amazon, and, from its accessibility and the readiness with which its wood burns while green, supplies a great part of the fuel consumed by the steamers that navigate the Amazon. It is almost equally common on some of the white-water tributaries; I have seen it, for instance, far away up the Huallaga to the south, and up the Pastasa to the north. Two of the commonest river-side Ingas of the Amazon (*I. splendens*, W., and *I. corymbifera*, Benth.) reappear together on the Upper Casiquiari and Orinoco; and similar instances might be multiplied indefinitely.

Streams of black or clear water have also their proper riparial vegetation, some species being

[1] Hence I suspect that those insects of the south side of the Amazon which have been identified with Guayana species belong chiefly to the riparial forests.

apparently repeated on all of them. For example, many of the trees of the inundated margins of the Tapajoz (some of them undescribed when I first gathered them) I found afterwards on the Rio Negro up to its very sources—although none of them inhabit the shores of the Amazon, either between the mouths of those two affluents or elsewhere. A few recur on the Teffé and other black-water streams entering still farther to the west, and even on similar affluents of the Orinoco.

Here, at least, would seem to be a case of the vegetation depending on the distribution of the running waters; but in reality both the kind of water and the vegetation nourished by it depend entirely on the nature of the soil, those rivers which run chiefly through soft alluvial bottoms being turbid, while those that have a hard rocky bed run clear; and the two classes of rivers are repeated over and over throughout the length and breadth of the Amazon region. Into the black Rio Negro runs that whitest of rivers, the Rio Branco, and imparts to the vegetation of the former, for a little way below their confluence, quite an Amazonian character.[1] The two largest tributaries of the Casiquiari, namely, the Pacimoni and the Siapa, run nearly parallel through a longish course, and at rarely more than 15 miles apart; yet the former has clear dark water and the latter is excessively muddy. Moreover, when I explored the Pacimoni to its very sources, I found it divide at last into two nearly equal rivulets, whereof the one had white and the other black water. The true riparial vege-

[1] Here, for instance, is the only locality throughout the Rio Negro for *Bombax Munguba*, a fine Silk-Cotton tree abounding on the Amazon.

tation in all these and in hundreds of other cases is invariably modified after the same fashion by the colour of the waters. How it became what it is, and how it came there at all, are questions not to be discussed here.

After what has been said, it is scarcely necessary to add that many species of plants which grow down to the very coast in Guayana exist also in the Peruvian province of Maynas—that is, at the eastern foot of the Andes, and even up to a height of a few thousand feet in those mountains—*e.g.* Humboldt's Willow (*Salix humboldtiana*, W.) and the Cannon-ball tree (*Couroupita guianensis*, Aubl.), called Aia-uma or Dead Man's Head in Maynas; while the proportion of Orinoco plants repeated on the Amazon is much greater than that of the plants of South Brazil. Nor does this uniformity of character, and the constant recurrence of certain species, preclude the possibility of the flora being wonderfully rich; for I have calculated that by moving away a degree of either latitude or longitude I found about half the species different; while in the numerous caatingas I have explored I always found a few species in each that I never saw again, even in other caatingas.

The Relations of Plants and Animals

The importance of inquiries of this class is obvious, even from a zoological point of view; for that an animal should flourish in any region it must there find suitable food; and there is perhaps no part of the world where so large a proportion of the animals is so directly vegetarian in its diet. I have reason to believe that there is no carnivorous animal on the Amazon and Orinoco which does not occasionally resort to vegetables, and especially to fruits, for food—not always of necessity, but often from choice. When, however, we come to consider and compare the distribution of the various classes and subordinate groups of animals, we see that the range of a fruit-eating species or tribe can rarely correspond to that of one which feeds on leaves, and similarly of other pairs of differences or contrasts in the nature of the food—that, in short, the only animals which can be expected to range from sea to sea in a wide continent are a few general feeders and their parasites, the larger beasts of prey, and the scavengers, such as Vultures among birds (and perhaps Termites among insects).

As to the distribution of the Lepidoptera in the Amazon valley, it is plain that it can rarely correspond to the grander features of the vegetation, for the simple reason that the food of caterpillars is scarcely ever the foliage, etc., of the loftier forest trees, but chiefly of soft-leaved undershrubs and low trees (1) which grow under the shade of the forest and have, many of them, a restricted range; or (2) which spring up where the primeval woods have been destroyed, and in waste places near the habitations of men, and whose range in many cases is coextensive at least with Cisandine Tropical America. The bushy trees and the luxuriant herbs which border savannas and caatingas and broad forest paths, and sometimes those which grow on the very edge of streams, are also apt to be infested by caterpillars. Of about two thousand forest trees I have had cut down in the Amazon region for the sake of their flowers and fruits, very few indeed have been infested by caterpillars. A tall Leguminous (tree or liana) or Bombaceous species would sometimes have caterpillars on it; more rarely a Laurel or a Nutmeg; but a Fig or a Guttifer never. A vast number of trees and lianas of all sizes are, indeed, excluded from serving as food to caterpillars by their strongly resinous or else acrid and poisonous juices, and many more on account of their hard, leathery leaves, which are untouched except, rarely, by minute caterpillars that eat themselves galleries in the parenchyma.

Of plants which afford food for caterpillars, Leguminosæ hold decidedly the first place; next to these rank Mallow-like plants (including Malvaceæ proper, Sterculiaceæ, Büttneriaceæ, and Tiliaceæ); then Melastomaceæ and Solanaceæ. Caterpillars armed with stinging hairs seem peculiarly partial to Leguminosæ, as I know to my cost, the bushy Inga trees in some parts being scarcely approachable when with flowers and young leaves. In the neighbourhood of Guayaquil children that stray under the Tamarind trees sometimes get severely stung by the hairy caterpillars that drop on them from the trees.

Other orders of plants on which I have encountered caterpillars are chiefly the following:—Among Endogens: Grasses, Sedges, Palms, and Aroids—on all rather rarely; on Scitamineæ and Musaceæ more frequently. Among Exogens: Euphorbiaceæ (principally on those with aromatic foliage); Samydeæ; Bixaceæ; Vochysiaceæ; Sapindaceæ (few); Malpighiaceæ; Anonaceæ and Myristiceæ (rarely); Anacardiaceæ; Ochnaceæ (on very young leaves only, the adult foliage being hard and vitreous); Podostemeæ; Polygoneæ; Amarantaceæ; Piperaceæ; Lauraceæ (few); Chrysobalaneæ (often much infested); Combretaceæ; Myrtaceæ (rarely on true Myrtles, but a great pest to the large handsome flowers of the sub-orders Barringtonieæ and Lecythideæ); Passifloreæ; Cucurbitaceæ; Rubiaceæ (few out of the vast number of Amazon species); Compositæ (all weeds); Boragineæ; Verbenaceæ; Bignoniaceæ. Besides these, there are other orders which contain a few species with mild juices, and leaves (and even wood) not too tough for a caterpillar's jaws, which are doubtless chosen by certain species of butterflies as food for their progeny; and nearly all the very large flowers are apt to be plagued by caterpillars, as well as by the grubs of flies and beetles.[1]

Some caterpillars seem to have a decided taste for bitters; and narcotics are rarely objected to; indeed, I should say that most insects are decidedly partial to them, while bees and wasps seem to have a positive pleasure in getting drunk. The very few phyllophagous beetles whose habits have come under my notice feed on narcotic plants. At the falls of the Rio Negro, just south of the Equator, a common weed in the village of Saõ Gabriel is *Solanum jamaicense*, Sw., growing (when not disturbed) to the size of a currant-bush, and bearing large, angular, soft, woolly leaves. In February 1852 there appeared swarms of a large black beetle whose corpulent abdomen was barely half-covered by the elytra (whence I suppose it an ally of our Meloës),

[1] The above list has no further value than that of indicating, so far as my notes and recollections serve me, the kinds of plants which I have seen most maltreated by caterpillars in the Amazon region.

and whose sole food was this Solanum. Their feeding-times were the dusk of evening and morning, when they would arise, as it were, out of the earth, hover over the plants like a swarm of bees, and then settle down in such numbers that the plants were black with them.

. . . . .

For myself, I am free to confess that I, too, generally looked on the insect world as enemies to be avoided or destroyed. Mosquitoes and ticks sucked my blood; cockroaches ate and defiled my provisions; caterpillars mutilated the plants when growing; and ants made their nests among the dried specimens and saturated them with formic acid, or even cut them up and carried them away bodily. I recollect my horror at coming home and finding my house invaded by an army of Arriero or Saúba ants who had fallen on a pile of dried specimens and were cutting them up most scientifically into circular disks whose radius was just equal to the artist's own longest diameter. The few notes on insects scattered through my journals relate, indeed, chiefly to ants, who deserve to be considered the actual owners of the Amazon valley far more than either the red or the white man. In fine, when I venture to offer these imperfect jottings to the notice of zoologists, I feel that I can at best be considered only an interloper in a province not my own.

Some Cases of Insect Migration

Having above indicated the kinds of plants apparently most in request with the larvæ of the Lepidoptera, I wish now to recall the attention of naturalists to certain transits or migrations of the adult insects across the Amazon, such as have already been noticed by Messrs. Edwards, Wallace, and Bates, and perhaps by other travellers. The first time I fell in with such a migration was in November 1849, near the mouth of the Xingú, when I was travelling up the Amazon from Pará to Santarem; and it is thus sketched in my Journal:—

". . . As we returned to the brig we saw a vast multitude of Butterflies flying across the Amazon, from the northern to the southern side, in a direction about from N.N.W. to S.S.E. They were evidently in the last stage of fatigue: some of them attained the shore, but a large proportion fell exhausted into the water, and we caught several in our hands as they passed over the canoe. They were all of common white and orange-yellow species, such as are bred in cultivated and waste grounds, and having found no matrix whereon to deposit their eggs to the northward of the river (the leaves proper for their purpose having probably been already destroyed, or at least occupied, by caterpillars), were going in quest of it elsewhere."

The very little wind there was blew from between E. and N.E.; therefore *the butterflies steered their course at right angles to it*; and this was the case in subsequent flights I saw across the Amazon, although when the wind was strong the weaker-winged insects made considerable leeway, and would doubtless most of them succumb before reaching land. But the most notable circumstance is that *the movement is always southward*, like the human waves which from the earliest times seem to have surged one after the other over the whole length of America, generating after a time a reflux northwards, as in the case of the empire of the Incas. . . .

Since my return to England I have read Mr. Bates's graphic description of a flight of butterflies across the Amazon below Obidos, lasting for two days without intermission during daylight. These also all crossed in one direction, from north to south. Nearly all were species of Callidryas, the males of which genus are wont to resort to beaches, while the females hover on the borders of the forest and deposit their eggs on low-growing, shade-loving Mimosæ. He adds, "The migrating hordes, so far as I could ascertain, are composed only of males."[1] It is possible, therefore, that in the flights witnessed by myself the individuals were all males—in which case the flights should probably be looked upon not as migrations but dispersions, analogous to those of male ants and bees when their occupation is done, and they are doomed by the workers to banishment, which means death. In the case I am about to describe, however, the swarms certainly comprised both sexes, although I know not in what proportion; and their movements were more evidently dependent on the failure of their food.

In the year 1862 I spent some months at Chandúy, a small village on the desert coast of the Pacific northward of Guayaquil, where one or two smart showers are usually all the rain that falls in a year; but *that* was an exceptional year, such as there had not been for seventeen years before—with heavy rains all through the month of March, which brought out a vigorous herbaceous vegetation where almost unbroken sterility had previously prevailed.

[1] *Naturalist on the Amazons*, vol. i. p. 249.

In April swarms of butterflies and moths appeared, coming from the east, sucking the sweets of the newly-opened flowers, and depositing their eggs on the leaves, especially of a Boerhaavia and of a curious Amaranth (Fröhlichia, sp. n.) not unlike our common Ribgrass in external aspect—until caterpillars swarmed on every plant. New legions continued to pour in from the east, and finding the field already occupied, launched boldly out over the Pacific Ocean, as Magellan had done before them, there to find a fate not unlike that of the adventurous navigator.[1] No better luck attended most of the offspring of their predecessors, especially those who fed on the Boerhaavia, which was much less abundant than the Fröhlichia. The shoal of caterpillars advanced continually westward, eating up whatever to them was eatable until, on nearing the seashore and the limit of vegetation, I used to see them writhing over the burning sand in convulsive haste to reach the food and shelter of some Boerhaavia which had haply escaped the jaws of preceding emigrants; but, failing this, thousands of them were scorched to death, or fell a prey to the smaller seaside birds, to whom they were doubtless a rare dainty.

The explanation of this continual westward movement is not difficult. A few leagues inland, instead of the sandy coast-desert with here and there a tree, we find woods, not very dense or lofty, but where there is sufficient moisture to keep alive a few remnants of the above-mentioned herbs all the year round, and doubtless also of the insects that feed

[1] Here also the course attempted to be steered by the insects was across the strong southerly breeze that was blowing.

upon them. There are also cattle-farms; and around the wells from which water is drawn and served to the cattle the same weeds are continually springing up; while the seeds, even of those that grew on the desert, remain embedded in the sand and retain their vitality during all the years of drought. When the rains come on, therefore, they cause, as it were, a unilateral development of the vegetation from the forest across the open grounds, and a corresponding expansion of the insect-life which breeds and feeds upon it.

Results the same in principle, but diverse in mode, would take place under different local circumstances. Thus, if we suppose an oasis in the midst of a desert exposed to the same exceptional access of moisture as the desert of Chandúy with its forest skirt, there would be generated an extension of organic life radiating outwards in all directions.

Besides the migrations above recorded, I have many times in South America seen butterflies flying across rivers so wide that it is impossible to suppose they could be guided by any indication of sight or smell. Animals of higher organisation and stronger reasoning powers would probably turn aside along the shore of the river or ocean in quest of food for themselves and their offspring; but there are plainly cases where frail little creatures, such as butterflies, must go straight forward at a venture, and either attain their object or perish.

Migrating Ants

The movements of Ants registered in my journal are (as may be supposed) chiefly such as were hostile to myself, and they do not throw much additional light on their habits. Ecitons or Foraging Ants (called Cazadoras in Peru) seem to be true wandering hordes, without a settled habitation; for a certain number of them may always be seen carrying pupæ, apparently of their own species; but they sojourn sometimes for several days whenever they come upon suitable food and lodging. . . .

The first time I saw a house invaded by Cazadoras was in November 1855, on the forest slope of Mount Campana, in the Eastern Peruvian Andes. I had taken up my abode in a solitary Indian hut, at a height of 3000 feet, for the sake of devoting a month to the exploration of that interesting mountain. The walls of the hut were merely a single row of strips of Palm trees, with spaces between them wide enough to admit larger animals than ants. One morning soon after sunrise the hut was suddenly filled with large blackish ants, which ran nimbly about and tried their teeth on everything. My charqui proved too tough for them; but they made short work of a bunch of ripe plantains, and rooted out cockroaches, spiders, and other suchlike denizens of a forest hut. So long as they were left unmolested, they avoided the human inhabitants; but when I attempted to brush them away they fell on me by hundreds and bit and stung fiercely. I asked the Indian's wife if we had not better turn out awhile and leave them to their diversions. "Do they annoy you?" said she. "Why, you see it is impossible for one to work with the ants running over everything," replied I. Whereupon she filled a calabash with cold water, and going to the corner of the hut where the ants still continued to stream in, she devoutly crossed herself, muttered some invocation or exorcism, and sprinkled the water gently over them. Then walking quietly round and round the hut, she continued her aspersion on the marauders, and thereby literally so damped their ardour that they began to beat a retreat, and in ten minutes not an ant was to be seen.

Some years afterwards I was residing in a farm-house on the river Daule, near Guayaquil, when I witnessed a similar invasion. The house was large, of two stories, and built chiefly of bamboo-cane—the walls being merely an outer and an inner layer of cane, without plaster inside or out, so that they harboured vast numbers of cockroaches, scorpions, rats, mice, bats, and even snakes, although the latter abode chiefly in the roof. Notwithstanding the size of the house, every room was speedily filled with the ants. The good lady hastened to fasten up her fresh meat, fish, sugar, etc., in safes inaccessible even to the ants; and I was prompt to impart my experience of the efficacy of baptism by water in ridding a house of such pests. "Oh," said she laughingly, "we know all that; but let them first have time to clear the house of vermin; for if even a rat or a snake be caught napping, they will soon pick his bones." They had been in the house but a very little while when we heard a great commotion inside the walls, chiefly of mice careering madly about and uttering terrified squeals; and the ants were allowed to remain thus, and hunt over the house at will, for three days and nights, when, having exhausted their legitimate game, they began to be troublesome in the kitchen and on the dinner-table. "Now," said Doña Juanita, "is the time for the water cure"; and she set her maids to sprinkle water over the visitors, who at once took the hint, gathered up their scattered squadrons, reformed in column, and resumed their march. Whenever their inquisitions became troublesome to myself during the three days, I took the liberty to scatter a few suggestive drops among them, and it always sufficed to make them turn aside; but any attempt at a forcible ejectment they were sure to resent with tooth and tail; and their bite and sting were rather formidable, for they were large and lusty ants. For weeks afterwards the squeaking of a mouse and the whirring of a cockroach were sounds unheard in that house.[1]

Migrations of Birds and Mammals

The most remarkable migration that I have myself witnessed in South America is that of the great Wood-Ibis (*Tantalus loculator*), called Jabirú in Brazil, Gauán in Venezuela, between the Amazon and the Orinoco, a distance of from 300 to 500 miles in a straight line, but a thousand or more following the course of the rivers. The migrations are so timed that the birds are always on the one river or the other when the water is lowest and there is most sandy beach exposed, affording the greatest extent of fishing-ground. In the years 1853 and

[1] The ants called Carniceras or Butchers in Maynas are probably of a tribe distinct from the Foragers; for they are burrowing ants, and are said to prefer the flesh of human carcasses to any other food. Padre Velasco, in his *History of Quito*, assures us that they will make a perfect skeleton of a corpse the very day it is buried, and that they devour any disabled animal, however large, they find in the forest.

1854, when I was at San Carlos del Rio Negro (lat. 1° 53½′ S.), I saw them going northward in November and returning southward in May, and had the pleasure of having some of them stay to dine with me. One of their halting-places on their way to the Orinoco was on islands near the mouth of the Casiquiari, at only a few hours' journey above San Carlos. There I have seen them roosting on the tree-tops in such long close lines, that by moonlight the trees seemed clad with white flowers. They descend to sandy spits of islands to fish in the grey of the evening and morning, *i.e.* before betaking themselves to their eyrie, and before resuming their journey on the following day. The scarcity of fish in rivers of clear or black water is well known; and even were they more abundant, this very clearness of the water would render it difficult for fish-eating fowls to catch them, unless when there was little light; hence, perhaps, the Ibis's choice of hours for fishing; and the turbid water poured into the Rio Negro by the Casiquiari dulls its transparency at that point, which makes it eligible for a fishing-station, leaving probably only a single day's stage for the travellers to reach the Orinoco. The Ibises, however, did not, as one might have supposed, turn up the Casiquiari, but held right on to the north, crossing the isthmus of Pimichin, and descending the Atabapo to the Orinoco. Some of them, I was told, would halt on the Guaviare, whose turbid waters, alligators, turtles, etc., quite assimilate it to the Solimoẽs or Upper Amazon; and others push on to the Apuré; the former lot, however, are said to travel chiefly by way of the Japurá from the Amazon. Those that frequent the Upper Orinoco return in May; and their halting-place near San Carlos is not at the mouth of the Casiquiari, but on islands a day's journey below the village, so that they are at that season less persecuted by the Indians. If they went all the way down the Rio Negro in May, they would reach the Amazon long before its beaches began to be exposed; but it has been ascertained that they sojourn awhile on the Rio Branco, whose beaches are earlier uncovered. Flocks of Wild Ducks sometimes accompany the Ibises; and it is quite possible that some of the smaller aquatic and riparial fowls make similar migrations.

When the Ibises are roosting, a shot or two from a gun is enough to make the whole caravan take to flight and remove to some distance; but the Indians of San Carlos know better than to scare them away with firearms. They get into their canoes a little after midnight, creep silently up the river, and under cover of the night disembark beneath the trees where the Ibises are roosting. Then, when at break of day the birds wake up and begin to stir and to be visible, the Indians pick them off with poisoned darts from their blowing-canes, in great numbers, before the bulk of the flock takes alarm; so that they mostly return to the village with great piles of dead Ibises; and although this lasts only three or four days, the quantity killed is so great that, what with fresh and what with barbecued game, everybody feasts royally for a fortnight; whereas throughout the rest of the year the dearth of provisions exceeds what I have experienced elsewhere in South America.

The Ibises doubtless undertake these voyages from the testimony and under the guidance of the elders, far more than from any inherited knowledge or instinct; whereas the flights of butterflies one would think must be directed by instinct alone, without any aid from experience.

Many mammals wander far in search of food; and some that go in bands, such as wild Pigs and some Monkeys, have known feeding-places at certain times of the year, when some particular kind of fruit is in season there; so that the experienced Indian hunter often knows in what direction to bend his steps to fall in with a certain class of game. It is well known how fond all animals are of the Alligator pear, which is the fruit of a large Laurel (*Persea gratissima*). I have seen cats prefer it to every other kind of food; and the wild cat-like animals are said to be all passionately fond of it. I have been told by an Indian that in the forests between the Uaupés and the Japurá, he once came on four Jaguars under a wild Alligator pear tree, gnawing the fallen fruits and snarling over them as so many cats might do. I have gathered flowers of at least four species of Persea, but was never fortunate enough to find one of them with ripe fruit; so that I have missed seeing the concourse of animals of many kinds which I am assured assemble in and under those trees, attracted by the fruit. While speaking of fruit-eating carnivora, it is worth mentioning that dogs in South America often take naturally to eating fruit. I had in Peru a fine Spanish spaniel who, so long as he could get ripe plantains, asked for no better food. He would hold them between his paws and pull off the skin in strips with his teeth so delicately as not to foul them in the least; so that I have occasionally eaten a plantain of his peeling.

I fancy Monkeys sometimes go on day after day along the banks of a river, their rate of progress depending on the quantity of food they find to eat and waste. I have watched them at this in a strip of Mauritia palms, which stretched for a distance of some days' journey along the banks of a river. The Chorro (Barrigudo of Brazil), a monkey of the hot plain, sometimes ascends the slopes of the Andes to 5000 or 6000 feet, apparently to eat the fruit of the Tocte or Quitonian walnut (an undescribed species of Juglans), which is frequent at that elevation; but it is said never to pass a night there.

An Indian will tell you at what time of year certain fruit-eating fowls are to be met with on the banks of a river, and at what time they must be sought for deep in the forest. I remember coming on a flock of one of the small Turkeys called Cuyubi (*Penelope cristata*, or an allied species), on the banks of the Uaupés, feeding on the fruit of so deadly a plant as a Strychnos (*S. rondeletioides*, Benth.); but the succulent envelope of the fruit is innocuous, like that of our poisonous Yew. I had been forewarned that we might expect to find them at that particular spot, and thus occupied; so that we had our guns ready, and knocked several of them over. Indeed, they were so tame, or so gluttonous, that when a shot was fired and one of them fell, the rest either took no heed or only hopped on to another branch and recommenced feeding; and it was not until we had fired and reloaded three or four times that the survivors took wing and flew off.

On the slopes of the volcano Tunguragua, the steepest and most symmetrical cone, though not the loftiest, of the Quitonian Andes, I have seen flocks of another Turkey (allied to, but distinct from, the Uru-mutún of Brazil) feeding on the plum-like drupes of the Motilon,[1] and on the berries of an undescribed Melastome. Besides these fruit-trees, there were also numerous fruit-bearing bushes near, including some true Brambles, Whortleberries, and a Hawthorn, all of which probably afforded food to the turkeys. This species seems to inhabit a zone, between 6000 and 10,000 feet, on the wooded flanks of Tunguragua, and within those limits to make the perpetual round of the mountain, being always found on that side where there is most ripe fruit to be had; and the birds are so tame and sluggish when feeding that the Indians easily kill them with sticks.

I should suppose that these and other gallinaceous birds have their fixed centres of resort (breeding- and roosting-places), from which they never stray far. Many Parrots and Macaws, I know, have. On the western slopes of the Quitonian Andes, immense flocks of Parrots ascend by day to a height of 8000 or 9000 feet, where they ravage the fields of maize and other grain, but always descend to certain warm wooded valleys, at 2000 to 4000 feet, to roost. The flights of vast multitudes of garrulous parrots and macaws to and fro between their roosting- and feeding-places, in

[1] This name is given to *Symplocos cernua*, H. B. K., and also to two (or more) species of Hieronyma, all bearing edible drupes.

the grey of the evening and morning, is one of the first things that strikes the attention of the voyager on the Amazon.

The periodical appearance of certain birds in a district has been supposed by the inhabitants to have some mysterious connection with the Christian festivals. Thus there are two beautiful little birds in Maynas, apparently belonging to different genera, for one of them is a Seven-coloured Tanager (Pajaro de siete colores), and the other (which I have not seen) is said to be of a bright blue colour and differently shaped; but both are called by the Indians Huata-pisco (Bird of the Year), because they make their appearance together, in large flocks, about the end of the year (people will tell you, precisely on Christmas Day), and remain throughout January, when they are seen no more until the same epoch comes round again. Mr. Bates has given a capital account of the movements of these hunting-parties of frugivorous and insectivorous birds, and of the superstition of the Papauirá or Patriarch Bird, who is supposed to head them (vol. ii. p. 333 *et seq.*). I suspect that this is something more than mere superstition, and that the Patriarch leaders are not one but several to each predatory band.

Distribution of Fishes

The abundance of fish in rivers of white water, and their scarcity in black-water rivers, may easily be shown to depend chiefly on the luxuriant littoral vegetation of the former and its scarcity or utter absence in the latter; for on the Rio Negro there are (with one notable exception[1]) no aquatics and no shore grasses. Compare this with the broad fringe of tall, succulent, amphibious grasses on the shores of the Amazon, or detached and floating down it in the shape of large islands, and of luxuriant aquatics, some fixed by roots, others floating (Victoria, Jussiæa, Pontederia, Frogbits, Azolla, Salvinia, Pistia, etc.), in deep still bays, but especially in lakes and channels communicating with the main river.

Some of the tributaries of the Rio Negro, however, have plenty of fish, namely, those of more or less turbid water, of which the Rio Branco holds the first rank, and after it come the Marariá and Cauaborís, all entering on the left bank. In these rivers many Amazon fish are said to be repeated. About the mouth of the Rio Branco is the only place in the Rio Negro where the Pirarucú is found—that noble and remarkable fish, so characteristic of the Amazon. With the exception of the Pirarucú, most of the larger fish of the Amazon recur on the Upper Orinoco, above the cataracts; at least the Indians assert them to be the same, and to unskilled eyes they are undistinguishable. The Valenton or Lablab of the Orinoco, for instance, is surely the same as the large Pirahyba of the Amazon; the Pavon as the Tucunaré; the Rallado as the Surubím; the Murucútu as the Tambaquí; the Cajáru as the Pirá-arára, and so on.

. . . . . .

Many of the fishes of the Rio Negro travel up it to spawn, and especially up some of its tributaries; but the wanderings to and fro of fish in quest of

[1] That of the Podostemons on granite rocks in the falls and rapids.

food may be compared to that already noted of wild turkeys; for the principal subsistence of fish in the Rio Negro is on the fruits of riparial trees, some of which seem scarcely touched by either bird or monkey. A small laurel-like bush (*Caraipa laurifolia*, S.) lines the banks in many places, and bears damson-like drupes, which are the favourite food of that delicious fish the Uaracú or Aracú. When the ripe drupes are dropping into the water they attract shoals of Uaracú. Then the fisherman stations his canoe at dawn of day in the mouth of some still igarapé, overshaded by bushes of Uaraçú-Tamacoarí (the native Indian name of the tree), and with his arrows picks off the fish as they rise to snatch the floating fruits. It ought to be mentioned that the fish of the Negro, if much fewer, are some of them perhaps superior in flavour to any Amazon fish, whereof the Uaracú is an example, and the large Pirahyba is another, the latter being so luscious that it is difficult to know when one has had enough of it, whereas the same or a very closely allied species of the Amazon is often scarcely edible.[1]

I have, in what precedes, purposely avoided speaking of the way in which animals prey on each other, because the ultimate measure of the amount of animal life must always depend on that of vegetable life, and not because I shut my eyes to the fact.

Concluding Remarks

I leave these *disjecta membra* in the hands of naturalists, hoping that they may find among them

[1] For further information on the fishes of the Rio Negro I must refer to Mr. Wallace's interesting account of that river (*Travels*, chaps. ix., x., and xvi.), and to Schomburgk's *Fishes of Guiana*.

some bone to pick. They bear on many problems for which there do not yet exist materials, nor do I possess the skill requisite to arrive at a correct solution. On one point only I am pretty clear, viz. that almost every kind of animal now existing in Cisandine Tropical America might find suitable food and lodging on any parallel between the southern tropic and the mouth of the Orinoco; which is as much as to say that they would find everywhere either the one plant they most delighted to feed on or others which might suit them almost or quite as well. The continual substitution of new forms encountered as we advance in any direction does not, on a superficial view, show much correspondence between animals and plants—a fact which may be put otherwise, thus: Suppose on a given area at the foot of the Andes every species of some class of animals to be distinct from those of the same class on an equal area at the mouth of the Amazon, it does not therefore follow that every plant is different on the two areas; we know, indeed, that such is not the case. Yet the modifications that have been and are still in progress among vegetable forms must have some correspondence with those that take place in animals; for all the realms of Nature act and react on each other. The atmosphere and the earth (with its productions, animal and vegetable) are continually giving and taking; and as their actual relations to each other vary more widely at different points along the equatorial belt than elsewhere on the earth's surface, it is plain that what seems equilibrium is either oscillation or progress in some direction. If plants were the only organic existences, and there were no animals to aid in their reproduction, to feed upon them, to dispose of their dead carcasses, etc., the dominant forms would doubtless be quite different from what they are now. Darwin has shown by an admirable series of observations how necessary insect agency is to the fertilisation of the flowers of many plants. Hence the organs of those insects and the parts of the flowers have been (and are being) continually modified, or moulded, the one on the other. I can conceive that if certain Orchids were henceforth entirely freed from the visits of insects, their flowers, notwithstanding the apparent permanence of inherited (though now useless) peculiarities, would immediately *tend* to revert to the symmetry which no doubt they possessed in the remote types. I have a good deal of evidence to show that in tropical countries many peculiarities of structure in the leaves and other parts of plants (prevailing through large suites of species and genera) have been brought about, and are still in part maintained, by the unremitting agency of insects, especially of Ants. These and many other matters require the fullest investigation before the precise relations of the changes, in animals and plants, that are taking place under our eyes, can be properly understood and appreciated.

CHAPTER XXIV

ANTS AS MODIFIERS OF PLANT-STRUCTURE

[THE paper which forms the greater part of this chapter was written during the first few years after Spruce's return to England, and at a time when he had probably not seen, and had certainly not carefully read, the *Origin of Species*, the teachings of which at a later period he fully appreciated. At this period he accepted—as did almost all naturalists, including Darwin himself—what is termed the heredity of acquired characters, such as the effects on the individual of use or disuse of organs, of abundant or scanty nutrition, of heat and cold, excessive moisture or aridity, and other like agencies. But in the paper here given he went a step beyond this, and expressed his conviction that growths produced by the punctures and gnawings of ants, combined perhaps with their strongly acid secretions, continued year after year for perhaps long ages, at length became hereditary and thus led to the curious cells and other cavities on the leaves and stems of certain plants, which are now apparently constant in each species and appear to be specially produced for the use of the ants which invariably frequent them.

This paper Spruce sent to Darwin, asking him to send it to the Linnean Society if he thought it worthy of being read there. I will here give some passages from Darwin's reply, dated April 1, 1869.

"The facts which you state are extraordinary, and quite new to me. If you can prove that the effects produced by ants are really inherited, it would be a most remarkable fact, and would open up quite a new field of inquiry. You ask for my opinion; if you had asked a year or two ago I should have said that I could not believe that the visits of the ants could produce an inherited effect; but I have lately come to believe rather more in inherited mutilations. I have advanced in opposition to such a belief, galls not being inherited. After reading your paper I admit, Firstly, from the presence of sacs in plants of so many families, and their absence in certain species, that they must be due to some extraneous cause acting in tropical South America. Secondly, I admit that the cause must be the ants, either acting mechanically or, as may perhaps be suspected from the order to which they belong, from some secretion. Thirdly, I admit, from the generality of the sacs in certain species, and from your not having observed ants in certain cases (though may not the ants have paid previous visits?), that the sacs are probably inherited. But I cannot feel satisfied on this head. Have any of these plants produced their sacs in European hot-houses? Or have you observed the commencement of the sacs in young and unfolded leaves which could not *possibly* have been visited by the ants? If you have any such evidence, I would venture strongly to advise you to produce it. . . .

"I may add that you are not quite correct (towards the close of your paper) in supposing that I believe that insects directly modify the structure of flowers. I only believe that spontaneous variations adapted to the structure of certain insects flourish and are preserved."

The paper was read on April 15, 1869, and then, as usual, was submitted to the Council to decide as to its publication. After full consideration, their decision was communicated to Spruce by the secretary as follows :—

"I am requested to communicate to you their opinion that the paper will require modification before they can recommend its publication. It is considered that the evidence adduced is insufficient to overcome the improbability of the sacs in the course of ages having become inherited, and that although there would be no objection to a statement that the author has been led to suspect that the structures in question are now inherited (which might lead to further investigations), it would be inadvisable for the Society to publish positive statements on the subject of inheritance without much fuller evidence. The Council wish me to say that if you do not object to alter the title of the paper, and to strike out some short passages, marked in pencil on the margin, they will be glad to undertake the publication of the paper, as they think it highly desirable that the facts recorded should be made known."

The paper was returned to him to make the alterations required if he wished to do so, but nothing more was heard of it, and it has remained among his papers till now. Spruce was very sensitive to criticisms of his writings by persons who had not the same knowledge that he possessed; but in this case I think it probable that he himself, later on, recognised the incompleteness of the evidence. A year and a half later he corresponded with Mr. Hanbury on the subject, and he was evidently seeking for more information. I therefore now print his paper in full, with a few omissions of unimportant details or digressions, giving the passages objected to within square brackets. It will be seen that they involve very slight alterations, in no way affecting the facts or observations of the paper itself. That he intended to modify and enlarge the paper may perhaps be concluded from the fact that the paper cover in which the MSS. was kept contains in pencil two alternative titles, both less dogmatic than that on the paper itself. They are as follows:—

(1) "On Changes in the Structure of Plants produced by the Agency of Ants."

(2) "On Structures formed in Living Plants by Ants, which apparently become permanent in the Species."

The paper here follows, and I shall at the end adduce a few additional facts which will serve as a partial reply to the questions put by Darwin.]

ANT-AGENCY IN PLANT-STRUCTURE; Or the Modifications in the Structure of Plants which have been caused by Ants [by whose long-continued Agency they have become Hereditary and have acquired sufficient Permanence to be employed as Botanical Characters].

In the forests of the Amazon and Orinoco, and elsewhere in Tropical America, there are numerous plants belonging to very distinct orders, which have singular dilatations of the tissues and membranes, in the form of sacs on the leaves, or of hollow fusiform nodes on the petioles or branches (becoming tubers on the rhizomes), or of slender inordinately-elongated fistulose branches. I have reason to believe that all these apparently abnormal structures have been originated by ants, and are still sustained by them; so that if their agency were withdrawn, the sacs would immediately tend to disappear from the leaves, the dilated branches to become cylindrical, and the lengthened branches to contract; [and although the inheritance of structures no longer needed might in many cases be maintained for thousands of years without sensible declension, I suppose that in some it would rapidly subside and the leaf or branch revert to its original form].

§ 1. *Of Sac-bearing Leaves*

These exist chiefly in certain genera of Melastomes, whereof one (Tococa) is very numerous in species and individuals throughout the Amazon valley, growing in the form of slender weak bushes, 8 to 12 feet high, chiefly in that part of the forest which is adjacent to and inundated by the rivers and lakes, but sometimes deep in the virgin forest, wherever the land is so low that the water of rains may accumulate thereon to a slight depth. All the species have the unmistakable aspect of their order —the ribbed opposite leaves, the polypetalous flowers with beaked porose anthers, etc.; but they are distinguished at sight from most others of the order by the large, thin, lanceolate or ovate acuminate leaves, very sparsely set with long hairs, and having a hollow sac or a pair of sacs at the base either of all the leaves, or (more frequently) of only one of each pair when that one is much larger than the other. The leaves in the majority of the species have but three ribs; a few species, however, have five- or even seven-ribbed leaves; but, in all, the origin of the innermost pair of ribs is an inch or so up the midrib from the base of the leaf; and it is this portion of the leaf, from the insertion of the inner ribs downwards, which is occupied by the sac. The latter sometimes takes up only a part of the breadth of the leaf, when it is technically considered to be seated *on* the leaf (Epiphysca); in other cases the sac in its lower half absorbs the whole breadth of the leaf, when it seems to be seated half on the leaf, half on the petiole (Anaphysca); or, lastly, throughout its length it absorbs the whole breadth of the leaf, and then seems seated entirely on the petiole (Hypophysca). That it is really formed in all cases at the expense of the lamina, and not of the petiole, is proved by the occasional occurrence of imperfectly-developed sacs in the hypophyscous form, bordered by a narrow wing continuous with the leaf, and giving to the latter a panduriform outline. Sometimes there is a pair of sacs, one on each side of the midrib, but in most cases the two sacs are confluent into one, which has a medial furrow along the upper side.

I proceed to describe a few forms of sacs in various species of Tococa. In one species (*T. disolenia*, MSS. hb. 1412) which grows by forest-streams entering the lower part of the Rio Negro, the leaves of each pair are very unequal, and the larger of the two (11 by 3½ inches) is alone sacciferous. The axils of the inner pair of ribs are perforated, giving entrance to two tubes or fistulæ—one on each side of the midrib—which conduct to a large basal sac, inhabited by small brownish ants, which pour out of the tubes and patter over the leaves to attack any animal that disturbs their domicile.

In most species, however, the sac springs at once from the base of the inner ribs, through whose perforated axils the ants have access to it without any intervening tubular way.

T. bullifera, Mart., grows in moist forests about the mouth of the Rio Negro, and is of humbler growth than the other species of the genus, reaching barely 5 feet; but the berries are more juicy and better flavoured than in any other Tococa, although so scanty and perishable that they cannot possibly serve as food for ants except for a very short period, and can hardly have influenced them in the choice of an abode. The leaves are long-lanceolate, either subequal and then with a large fusiform sac at the base of each of the pair, or very unequal and then the smaller leaf esaccate. The sacs afford refuge to multitudes of minute reddish ants which are fragrant when crushed. Most species of Tococa, however, are inhabited by ants of medium size, with a blackish or brownish abdomen and pale thorax, and a milky fluid exudes from them when crushed; they bite but do not sting.

T. macrophysca, Benth. (Spruce, 2188), grows in moist caatingas of the Rio Negro and Uaupés, and has leaves sometimes a foot long, not very unequal, and all of them usually bearing a stout elongato-cuneiform sac, an inch long, at the top of the petiole.

Tococas are scattered over the Amazon region from the sea-coast to the roots of the Andes, and two species (*T. pterocalyx*, sp. n., and *T. parviflora*, sp. n.) ascend the Peruvian Andes to 2500-3000 feet. I gathered altogether twenty-four or twenty-five species of Tococa, and all but one or two (*T. planifolia*, Benth., and a closely-allied species or variety) have sacs on the leaves inhabited by ants. An examination of the circumstances of growth of the esaccate *T. planifolia* seems to throw light on the origin of sacs on the leaves of the other species.

Tococa planifolia grows here and there along the shores of the Rio Negro, at least as far up as to the foot of the cataracts, or say for about 700 miles. From the cataracts upwards, on the main river, on its tributary the Uaupés, and on some clear-water affluents of the Casiquiari, it is replaced by an allied non-sacciferous species or possibly a mere variety. Wherever it grows, it always occupies the very edge of the riparial forest, to which it forms an inner fringe, along with various Rubiaceæ, Apocyneæ, etc., of similar humble growth, all of which are *completely submerged* in the time of flood;

so that even if the leaves of this Tococa were sacciferous, they could not afford a permanent refuge to ants. But all the other sub-riparial species grow so far away from the real shore that the periodical inundations never overwhelm them completely, but leave at least the tops of the branches out of water; and it is noticeable that not only are the first leaves of young plants of every Tococa often esaccate, but that also the lowest leaves of each ramulus of the adult plant have either no sac or only the slightest rudiment of one. I suppose, then, that the primeval Tococa—the ancestor of all the existing species—had no sac at all on the leaves, but that a few ants having sheltered in the deep narrow angles formed by the junction of the prominent lateral ribs with the midrib, found the axils perforable, and having thereby reached the interior of the leaf, scooped out the parenchyma between the two surfaces. The leaves of any plant, when its juices are sucked away by insects (Aphides, for example) or otherwise diverted from their usual course on the one surface, are apt to become bullate on the opposite surface; hence it is easy to understand that, when mined by ants, the cuticular tissue of both surfaces should expand outwardly and contract laterally so as to form a sac, whose further enlargement would be effected by the continual crowding in of ants. [This process repeated on the plants for many generations would induce an hereditary tendency to the production of sac-bearing leaves.] It is natural that the ants should select the largest leaves, as affording most room for their operations; but that one leaf of each pair should be often larger than the other depends on some cause anterior to any action of ants, for it is a very common thing all through the order of Melastomes. In species which have the leaves of each pair nearly equal, it is usual to see some of the smaller ones saccate and others altogether esaccate on the same plant. [I have often examined *half-grown plants and have seen that sacs begin to be developed (by inheritance) long before any ants touch them*, but that when the sacs are taken possession of by ants they speedily became much enlarged.]

Seeing, then, how the sacs on the leaves have originated, and what purpose they serve, it is plain that a species of Tococa, like *T. planifolia*, inhabiting the very river's brink, and liable to be completely submerged for several months of every year, could never serve as a permanent residence for ants, nor consequently have any character impressed on it by their merely temporary sojourn; even if their instinct did not teach them to avoid it altogether, as they actually seem to do; whereas the species of Tococa growing far enough inland to maintain their heads above water even at the height of flood are thereby fitted to be permanently inhabited, and are consequently *never destitute of saccate leaves*, nor at any season of the year clear of ants; as I have reason to know from the many desperate struggles I have had with those pugnacious little creatures when breaking up their homes for the sake of specimens.

In one species (hb. 3477) with seven-ribbed leaves, growing by the Rio Negro near the mouth of the Casiquiari, the leaves on some plants have a small distorted sac at the base inhabited by ants, and on others are nearly all esaccate; and] I noted of this species that the plants grow sometimes where they are totally overwhelmed by the periodical floods, rendering them a precarious dwelling-place for the ants. This leads to the suspicion that some of the sacciferous species, growing far away in the forest, may have sprung originally from *T. planifolia*, which grows on the river-banks; and even that some of the epiphyscous, anaphyscous, and hypophyscous species may be mere varieties of one another, or may have had a common progenitor at no very remote epoch. This and many other interesting problems can only be solved when naturalists shall become permanent members of the fauna of Equatorial America, and not as now have to be classed among "occasional visitants"; for their solution would require observations to be carried on through many consecutive years on the same spot.

Besides Tococa, there are other allied genera of Melastomes, viz. Myrmidone, Mart., Majeta, Aubl., and Calophysa, DC., which have sac-bearing leaves infested by ants. They are all found in the forests of humble sparse growth called "caatingas," and especially where the soil of white sand, or the granite floor almost bare of herbs, lies low and is liable to get transformed into a shallow lake in the time of heavy rains, thus driving ants and other insects to take refuge in the trees and bushes. Of Myrmidone I gathered four species, including the original *M. macrosperma* of Martius. They are low-growing, sparingly-branched shrubs of 3 to 8 feet; the leaves of each pair are very unequal in size, the smaller one sometimes even obsolete, the larger saccate, as in the *Tococa Anaphyscæ*, but the sac always rugose as well as unisulcate;

flowers solitary, rather large, terminal or axillary, rose (turning red); hairs of stem, leaves, etc., spreading, more copious than in Tococa, and red or crimson, corresponding curiously with the colour of the minute ants—of that viciously-stinging tribe called "Formiguinhas de fogo" (Little Fire-Ants)—which inhabit the sacs, and also make covered ways of intercommunication along the outside of the stem and branches—a precaution I have rarely noted among the Tococa-dwellers.

Myrmidone rotundifolia, sp. n., grows in caatingas in the lower angle of the confluence of the Rio Negro and Casiquiari. It is only 3 feet high, and has crowded, subunequal leaves, the larger of each pair $3\frac{1}{2}$ inches long, orbiculari-panduriform, cordate at the base, where there is a large sac; while the smaller leaf is orbiculari-cordate and mostly (but not always) has no sac.

Majeta guianensis, Aubl., has very much the habit of the Myrmidones, but it has also fistulose branches swollen at the nodes, so that the inhabitants have an inner way of communication between the sacs at the base of the larger of each pair of sessile leaves.

Calophysa tococoida, DC., is a slender shrub with thin hairy leaves, the larger leaf of each pair having a large bifid sac at the base of the petiole; but the frequent presence of a narrow wing connecting the leaf with the sac proves that the latter belongs really to the lamina (as in the Tococas) and that the leaf is sessile.

Examples of sac-like ant-dwellings exist in the leaves of plants of other orders, so like those already described in Melastomes, that it is scarcely worth while to do more than indicate some of the species. The solitary instance known to me in Chrysobalans is that of *Hirtella physophora*, Mart., a slender arbuscle growing just within reach of inundations in the forests about the mouth of the Rio Negro. The distichous, oblong, apiculate leaves are nearly a foot long, and at the cordate base have a pair of compresso-globose sacs tenanted by ants. On cutting open the sacs I was rather surprised to find them lined with cuticular tissue and hairs, just like the underside of the leaf; which seems to show that they have been produced by a recurvation of the alæ of the leaf, through the ants nestling at first (Aphis-like) under the leaf and causing it to become bullate, and that the recurved margins have at length reached and coalesced with the midrib so as to form a pair of sacs.

Rubiads afford a few instances of sac-bearing leaves, especially in the genus Amaiona (Aubl.). In caatingas of the Rio Negro, almost throughout its extent, grows *Amaiona saccifera*, Mart., a small bushy tree with leaves three together, above a foot long, obovate with a minute apiculus, tapering to the base, where there are two contiguous sacs inhabited by small red fire-ants. The fruit resembles a large plum (except that like the leaves it is harshly hairy), and when ripe is soft and edible; but long before it reaches that stage the ants crowd on it and seem to suck the juices through the pores of the cuticle.

To the same order belongs *Remijia physophora*, Bth., a remarkable tree found at the falls of the Uaupés, having the aspect of an Amaiona, but the dry capsules and other characters of Cinchona and its allies. The opposite leaves, 9 inches long, are oblong-oval, obtuse with a short apiculus, near the base abruptly panduriform, and bearing a small ant-sac on the midrib. All the other known species of this large genus have non-sacciferous leaves.

In all the plants I have seen bearing sacs on the leaves, to whatever order they belong, it is remarkable that the pubescence consists of long hairs having a tubercular base; and although I do not see what connection that peculiarity can have with the ants' choice of a habitation, it is probable they find some advantage in it.

§ 2. *Of Inflated Petioles*

A true swelling of the petiole, inhabited by ants, and (as I believe) owing its existence to their agency, I have seen only in two genera of Leguminosae Cæsalpinieæ, viz. Tachigalia and Sclerolobium. The Tachigaliæ are low-growing riparial trees, of black-water rivers, and have pinnate, often silky foliage, and small, yellow, sweet-smelling, nearly regular flowers disposed in panicles. All have trigonous petioles, which are mostly dilated at the base into a fusiform sac tenanted by ants. *T. caripes*, sp. n., grows abundantly on the banks, and on inundated islands, of the Uaupés. It is a spreading tree of 30 feet, and has the ramuli, petioles, and leaves clad with a fine, close, silky pubescence. The sacs of the petiole are inhabited by small black ants, whose entrance is by a little hole on the underside of the sac. *T. ptychophysca*, sp. n., grows in moist sandy caatingas by the same river, and has a similar sac on the petiole.

The species of Sclerolobium are not usually riparial, but one species (*S. odoratissimum*, sp. n.) is eminently so, constituting a great ornament of the shores and islands of the Rio Negro towards the mouth of the Casiquiari, and perfuming the whole breath of the river with the abundance of its pale yellow honey-scented flowers; and it is notable that this is the only species of the genus in which I have found sacciferous petioles. The sac is large, extending upwards from the knee of the petiole to the base of the second pair of leaflets, and it has a furrow along the upper face.

I presume the ants have been induced to take up their residence on these particular trees on account of the abundance and long persistence of their honied flowers. On other species of Sclerolobium, inhabiting dry lands solely, such as *S. tinctorium*, Benth., and *S. paniculatum*, Vog., I have seen the flowering panicles infested with little fire-ants, which, however, seemed to have their permanent habitation in the ground, about or near the tree-roots, and never to perforate the leaf-stalks. Many other Leguminosæ, especially the woody climbing Phaseoleæ, are visited by ants when in flower, and knobs or galls caused by the perforation of those insects are frequent on the panicles of Dioclea and allied genera; [but I have not remarked any instance of such knobs having become hereditary, except in *Pterocarpus ancylocalyx*, Benth., a small tree on the banks of the Solimoẽs or Upper Amazon, which has the rachis of the racemes thickened in the middle, the swelling being sometimes (but not always) tenanted by ants].

In the shrubby Cassias, which are common weeds

of tropical America, the knee of the petiole may sometimes be seen hollowed and enlarged by ants; [but the action of these insects has not been maintained with sufficient constancy to render the swelling a permanent character in any species of Cassia I have met with].

Ants congregate on the pods of some Cassias and other plants which have seeds in sweet pulp; and on those parts of any plant where they find suitable food, in the shape of mucilaginous exudations, etc.; but they mostly sojourn there just so long as that food lasts, and no longer; or otherwise they merely visit the plants for the sake of collecting their products and carrying them off at once to a permanent storehouse elsewhere.

§ 3. *Of Inflated Branches*

Ants' nests in swellings of the branches are found chiefly in soft-wooded trees of humble growth, which have verticillate or quasi-verticillate branches and leaves, and especially where the branches put forth at the extremity a whorl or fascicle of three or more ramuli; then, either at each leaf-node or at least at the apex of the penultimate (and sometimes of the ultimate) branches, will probably be found an ant-house, in the shape of a hollow swelling of the branch; communication between the houses being kept up, sometimes by the hollowed interior of the branches, but nearly always by a covered way along their outside.

The genus Cordia (Boraginaceæ) affords many examples of this structure. One of the rather artificial sections into which Cordia is divided in the

"Prodromus," viz. Physoclada, is characterised by "rami sub foliis congesto-verticillatis inflati cavi," the hollow inflation being tenanted by ants, whence *C. nodosa*, the type-species of the group, is known to the South Americans as "Ant tree" (Pao de formiga). *C. formicarum*, Hoffmans, and *C. callococca*, Aubl., are supposed to be synonyms of *C. nodosa.*

. . . . .

Cordia gerascantha, Jacq., differs from the Physocladæ in the structure of its rather showy white flowers. It rises to a stoutish tree of 30 to 40 feet, and is throughout fasciculately branched (branches 3–5-nate). At the point where the branches divide there is mostly a sac, inhabited by very vicious ants of the tribe called "Tachí" by the Brazilians. The preceding species are usually tenanted by the small fire-ant, but sometimes by the Tachí. Probably the former was in all cases the original occupant, and the Tachí is an intruder.

All these sacciferous Cordiæ have fascicled or whorled branches, and are beset (not often densely) with long coarse hairs arising from tubercles, much as in the Amaiona and the Melastomaceæ above described; but of the numerous other Cordiæ I have gathered, with vague ramification and often short soft pubescence, not one was seen with saccate branches, or any other structure serving as a permanent residence to ants.

Some of the aromatic shrubby Crotons, with trichotomous branches, have occasionally the branch-axils perforated by ants and swollen; but the process does not seem to have been carried on long enough to make the character permanent in any species I have met with.

To this category belong the creeping rhizomes of some ferns which are often beaded with globose swellings inhabited by ants; *e.g.* of *Phymatodes Schomburgkii*, J. Sm., a not uncommon fern on shady rocks and trees by the Rio Negro. [In a small Polypodium, found by Dr. Jameson on the river Napo, the moniliform character of the rhizomes seems to have become permanent, for he did not see a single specimen wanting it; but the presence of ants in all the swellings revealed the origin of the latter.]

A curious epiphytal genus of Solanaceæ, Marckea, whereof I gathered two species on the Rio Negro and Uaupés, is singularly affected by ants. The stem is reduced to a large tuber—sometimes as big as a child's head—and attains that size through the agency of ants, who inhabit its hollow interior and cover it outwardly with paper of their own manufacture. From the tuber radiate several branches, simple or sparingly forked. The leaves are very like those of *Acnistus arborescens*, save that they are verticillate (or at least approximated) in one species (*M. ciliata*, Benth.) in threes and in the other species in fives; but the large hypocrateriform corollas, with a tube 3 inches long, are more like those of some Gesnerea. There are perforated swellings at the forks of the branches, and sometimes also at the leaf-nodes, which serve the ants as detached apartments. I did not see a single plant wanting the basal tuber.

§ 4. *Of Elongated and Fistulose Stems and Branches*

There is an order of plants, whereof several genera and species inhabit Equatorial America, and all, with the exception of the herbaceous species, are infested by ants. The order is Polygoneæ; the ant-infested species belong to the genera Triplaris, Coccoloba, Campderia, Symmeria, and Rupprechtia; and the exceptions are species of Polygonum, some of them closely resembling common European species. All, both trees and herbs, grow in moist situations, and most of them on lands subject to periodical inundations. Not only is every lignescent Polygonea a habitation for ants, but the whole of the medulla of every plant, from the root nearly to the growing apex of the ramuli, is scooped out by those insects. The ants make a lodgment in the young stem of the tree or shrub, and as it increases in size and puts forth branch after branch, they extend their hollow ways through all its ramifications. They appear to belong all to a single genus, and are long and slender, with a fusiform, very fine-pointed, dark-coloured, shining abdomen, and they all sting virulently. They are known in Brazil by the name of "Tachí" or "Tacýba," and in Peru by that of "Tangarána"; and in both countries the same name is commonly applied to any tree they infest as to the ants themselves.

A few trees and shrubs of other orders are similarly infested by Tachí ants; such as Platymiscium (Vog.) in Leguminosæ, Tachia (Aubl.) in Gentianeæ, and Mabea (Aubl.) in Euphorbiaceæ.

Triplaris surinamensis, Camb., a Polygoneous tree of very rapid growth, reaching at maturity a hundred or more feet in height, and conspicuous from afar when in fruit from the abundance and bright red colour of its enlarged shuttlecock-like calyces, is common all along the Amazon, both on the river banks and in marshy inland sites; and solitary trees of it are often seen standing out above the Cacao plantations. *T. Schomburgkii*, Benth., a smaller tree, grows in the same way on the Upper Orinoco and Casiquiari. These trees, as well as the other arborescent Polygoneæ, have slender elongated tubular branches, often geniculate at the leaf-nodes, and nearly always with perforations, like pinholes, just within the stipule of each leaf, which are the sallyports of the garrison, whose sentinels are besides always pacing up and down the main trunk, as the incautious traveller finds to his cost when, invited by the smoothness of the bark, he ventures to lean his back against a Tachí tree. I suspect that the remote progenitors of these ants have at first sheltered in the ocrea (sheathing stipule) which is so characteristic a feature of the Polygoneæ; but, having found the wood soft and thin and the pith easy to scoop out, have made their more secure abode within the stem and branches.

Some Tachí trees seem as if they were actually trying to run away from the ever-encroaching ants. *Coccoloba parimensis*, Benth., found by Schomburgk in British Guayana and by myself on the river Uaupés, is an arbuscle with a stem 15 feet long, that tapers upwards and arches over so as finally to touch the ground, the ants all the while hollowing it out, as it stretches away apparently in the hopeless attempt to escape their invasion. Some slender Coccolobas climb high into the adjacent trees, not by twining but by crooking their branches and thereby hoisting themselves up; others are self-standing bushy trees, but still have the same slender geniculate branches.

The pretty Gentianeous shrubs of the genus Tachia have long, slender, hollowed branches, that either hang down or support themselves on the branches of adjoining shrubs and trees; [yet although this character is (as I suppose) an undoubted inheritance of the effects of ant-agency, it is singular that Tachias are nowadays often found entirely free from ants; while the name, taken by Aublet from the Tupí language, distinctly implies that in his day they were notoriously ant-infested.] The genus Tachigalia, spoken of above, also doubtless owed its name to the same peculiarity, which it still enjoys unabated. Aublet tells us he got these and other Tupí names from a colony of Indians from Pará, who had crossed the Amazon and established themselves in Cayenne.

Some Mabeas are still more remarkable, the long sarmentose branches stretching away to a great length among the adjacent vegetation, although never actually twining. All Mabeas of the section Taquari have this habit, and all are infested by Tachí ants. The slender but tough twigs, hollowed and polished interiorly by ants, are a favourite material for tobacco-pipes with the Indians of the Amazon, who strip off the bark and paint and varnish the surface of the wood. These "Taquaris," as they are called, are commonly sold in the shops at Pará. A bundle of them which I purchased there is now in the Kew Museum. The arborescent Mabeas, however, with tall erect trunks and paniculate inflorescence, are apparently never touched by ants.

None of these fistulose trees and shrubs have any sacs or swellings on the branches, except the leguminous genus Platymiscium, which has the pinnate leaves usually in whorls of three, and the tubular branches sometimes dilated at the leaf-nodes; so that this genus has almost as much right to be placed in the preceding section as here.

All the plants above named belong to the eastern side of the Andes and the Amazonian plain; but when I crossed over to the western side of the Andes I saw a Triplaris in the Red Bark forests of Chimborazo, and *Rupprechtia Jamesoni*, Meisn., and a Coccoloba on the inundated savannas of Guayaquil, with just the same long, slender, geniculate branchlets—infested by the same class of ants—as their congeners east of the Andes.

A few other plants with long-drawn-out stems and branches, such as some species of Remijia, may be supposed to owe at least the exaggeration of that feature to the ants which still continue to infest them.

Nearly all tree-dwelling ants, although in the dry season they may descend to the ground and make their summer-houses there, retain the sacs and tubes above-mentioned as permanent habitations; and some kinds of ants appear never to reside elsewhere, at any time of year. The same is probably true also of ants which build nests in trees, of extraneous materials, independent of the growing tissues of the tree itself. There are some ants which apparently must always live aloft; and the Tococa-dwellers continue to inhabit Tococas where there is never any risk of flood, as in the case of the *T. pterocalyx*, which grows on wooded ridges of the Andes. Their case is parallel to that of the lake-dwellers of the mouth of the Orinoco and the inundated savannas of Guayaquil, whose descendants must needs elevate their houses on stages six feet or more in height, although nowadays erected on rising ground far beyond the reach of river floods or ocean-tides. We call this "instinct" in the case of ants, "inherited custom" in the case of men; yet there is obviously no difference.

There are numerous instances of the effects of Ant-agency in the plants of Tropical America, not reducible to any of the foregoing sections. At Tarapoto, in the Andes of Maynas, a prickly suffruticose Solanum, with pinnate leaves, is frequent in sandy ground. The fruit is a small scarlet edible berry, tasting like that of Physalis. The very prickly calyx persists with the fruit, and is dilated into a wide cup which holds the water of rains, for whose sake it is visited by fire-ants that have their burrows in the sand. The contained water is slightly mucilaginous, and possibly, after standing a while, partakes of the flavour of the berry that is partially immersed in it. After a shower, the ants may be seen crowding on the inner edge of the calyx and sipping the liquid; but in dry weather they fill the calyx, bent apparently on extracting the last drop. The consequence of this crowding into the calyx is to sustain and augment the inflation. The bulging, gummy, water-holding leaf-bases of many epiphytal Bromels seem to owe those properties to the same influence, for they are commonly infested by ants, whose papery nest, indeed, often envelops the root of the plant.

[When I compare these and similar instances with the Pitchers of the Nepenthes, in which (as I learn from the accounts of travellers) ants as well as water are nearly always found, I cannot doubt that those curious appendages have attained their actual dimensions through the deepening and widening which they have undergone from ants through untold ages.]

We have a curious example, in the genus Cinchona, of the supposed correlation of a minute structural peculiarity with chemical and medical properties. Eminent botanists, such as Weddell and Karsten, who have studied that genus in its native forests, have thought they had found a character in the leaves always associated with a bark rich in alkaloids, viz. the presence of a small pit or scrobicule in the axil of each vein on the underside of the leaf. But when good specimens of *C. succirubra,* the richest of all the barks in alkaloids, came to be examined, the leaves were found entirely destitute of scrobicules! See now how this comes about. The leaves of the Hill Barks—those, namely, that grow at an elevation of 8000 feet and upwards —are liable to be infested by a small mite which nestles in the scrobicules—has caused them, in fact—its remote ancestors having at first sheltered in the vein-axils; but *C. succirubra* grows always below that elevation—indeed, as low down as 2400 to 6000 feet—and is the only quinine-producing Cinchona that descends so low, the other species of Cinchona that grow at a low elevation having all medically worthless bark. But as all these species, *C. succirubra* included, are equally destitute of scrobiculate leaves and of mites, the reasonable inference is that

that kind of mite is confined to a higher and cooler zone, and never descends to the warm zone of the Red Bark.

Let it be observed that these scrobicules, although I have no doubt of their origin by insect-agency, are quite as good and permanent a botanical character as many others—as the sacciferous leaves of Tococa, for example. [What a vast length of time, compared with man's brief life, it must have taken to impress a character of permanence on the latter character and render it hereditary! Probably a period far longer than those we choose to designate "historical" or "bronze" or "stone." The inimitable researches of Mr. Darwin have rendered it (to my mind) almost certain that many of the deviations from symmetry in the form and direction of the parts of a flower have been brought about by the direct mechanical agency of insects; and that the origin of every obliquity, unequal-sidedness, and so forth, in any organ of a plant, is to be sought in the action of forces not only internal, but also external to the plant itself.] In this wonderful "life," which exists only through perpetual change, every equilibrium is unstable, and even what we call "permanence" is but a transitory state.

In fine, the list of structures which I have above assigned to Ant-agency might no doubt be very much extended, and perhaps more satisfactorily classified. I have described only what I have seen with my own eyes and noted down on the spot; and corroborative specimens of all the plants mentioned exist in the Royal Herbarium at Kew, by means of which the accuracy of my account of the structures inhabited by ants may at any time be tested.

REMARKS BY THE EDITOR

[The Director of the Kew Gardens, Lieutenant-Colonel Prain, informs me that the genus Tococa in cultivation produces the inflated bladders, but he does not know that the plant has ever been raised from seed, which is not produced in Europe. Prof. James W. H. Trail, who has observed these plants and the ants that infest them in Amazonia, informs him that in one or two cases plants which had no ants on them, though possessing the ant-dwellings moderately developed, were being damaged by herbivorous pests. This important observation indicates the "utility" to the plant itself, which is always needed to bring natural selection into play for the purpose of modifying and rendering permanent any special adaptation in plant- or animal-structure.

Much light is thrown on this question by the observations of Mr. Henry O. Forbes, recorded in his *Naturalist's Wanderings in the Eastern Archipelago* (pp. 79-82). He found the strange tuberous Myrmecodia and Hydnophytum abundant in Sumatra and Amboyna (as they are all over the Archipelago), and raised many young plants from seed, which, though completely isolated from the ants that make their homes in the wild plants, grew vigorously and developed the internal branching cells and galleries from the very first. These chambers are formed by the shrivelling up of a delicate pith with which they are at first filled, and as they grow rapidly and form irregular tuberous masses as large as a man's head, it seems probable that this pith, as well as the watery liquid secreted in a large central chamber, are the primary attraction to the ants, which are always of one species and sting virulently.

I find that I had myself given a short account of these ant-infested plants of both hemispheres in my volume on *Natural Selection and Tropical Nature* (p. 284), in which I refer to Mr. Forbes's observations, and also to those of the late Mr. Belt on the Bull's-Horn Acacia, which has the thorns in a young state filled with a sweetish pulpy substance which at first serves as food for the ants, while later on they are supplied by honey-glands upon all the leaves. He also notices and figures in his *Naturalist in Nicaragua* (p. 223) the leaves of one of the Melastomæ with swollen petioles, and he states that, besides the small ants always infesting them, he noticed, several times, some dark-coloured Aphides. He also suggests that these small virulently-stinging ants are of use to the plants by guarding them from leaf-eating enemies such as caterpillars, snails, and even herbivorous mammals, but above all from the omnipresent Saúba or leaf-cutting ant, which he declares he observed to be much afraid of these small species.

I think the facts that have now been observed in both the western and eastern tropics are really sufficient to enable us to understand the probable origin of the various remarkable structures that have been developed in many different groups of plants and are utilised by ants. There is clearly "utility" on both sides. The ants obtain dwellings, protection from floods, a safe shelter for their eggs and larvæ, and a portion of their food—in some cases perhaps all—from the plant they inhabit; while the plant derives protection to its foliage, and perhaps also in some cases to its flowers—as shown by Kerner—by the presence of whole armies of virulently-stinging ants whose very minuteness renders them the more formidable. In the most remarkable plant-formicaria known—those of the Myrmecodia and Hydnophytum of the Malay Archipelago—the whole structure has been proved to be hereditary, and we may therefore conclude that in the Tococas of the Amazon, and other cases in which the cavities inhabited by the ants are constantly present, they are also hereditary. In other cases, as Spruce himself states, they are not so, being directly formed by the ants or being abnormal growths due to their irritations.

Spruce's error was in not recognising that the ever-present variability in all the parts and organs of plants furnished *the material*, and the survival of the fittest *the agency*, by which these, as well as all other specific modifications of plants, have been brought about; and that this is a far more powerful, as well as a more exact and certain, mode of doing so than the hereditary transmission of mutilations, the effects of which would in many cases be the reverse of beneficial.

In my recent work, *My Life* (vol. ii. p. 64), I give a letter from Spruce written shortly after the paper was rejected, in which he explains his reasons for refusing to alter his paper. Three years later he wrote me another letter on an allied subject—the purport of aromatic leaves (printed at p. 65), at the commencement of which he says: "Every structure, every secretion of a plant is (before all) beneficial to the plant itself. That is, I suppose, an incontrovertible axiom."

This is a great advance on the views stated in the earlier letter, in which he wrote: "The ants cannot be said to be useful to the plants, any more than fleas and lice are to animals; and the plants have to accommodate to their parasites as best they may." The evidence, however, now shows that, in all probability, they are always useful, in which case their becoming hereditary is merely a question of variability in the plant, and the continued preservation of those whose variations were in the direction of utility to the ants.

The whole of these very interesting phenomena, so well described by Spruce, are thus seen to be in complete accordance with those of the modification of flowers by insect-agency, which are now admitted to depend upon a mutual adaptation for the benefit of both plant and insect.

They lead, I think, to the establishment of the general principle, that no special adaptation of one organism to another can become fixed and hereditary unless it is of direct utility to both.]

CHAPTER XXV

INDIGENOUS NARCOTICS AND STIMULANTS USED BY THE INDIANS OF THE AMAZON

[THIS chapter consists of a carefully written account of the above subject, compiled by Spruce about 1870 from his notes and observations, and printed in the short-lived *Geographical Magazine.* Fortunately, he presented the beautifully written manuscript to his Yorkshire friend and fellow-botanist, Mr. G. Stabler, of Milnthorpe, Westmoreland, who has kindly lent it me for reproduction here, and I feel sure that it will be both new and interesting to the great majority of readers of this volume. Besides its main subject, it touches upon the beliefs and customs of the Indians who use these narcotics, and on the proceedings of their "pajés" or medicine-men; and incidentally it narrates the occurrence of rare and mysterious sounds in the forest, and their very curious explanation, which I believe he was the first, and probably still the only, traveller to obtain. The whole essay affords a good example of the writer's style and of his power of making even technical details interesting, and of introducing bright descriptive flashes and touches of human nature in what might otherwise be a rather dry exposition of botanical and pharmaceutical facts. Two paragraphs only have been omitted as unsuitable for the present work. The rest is printed verbatim, and will, I think, even to the non-botanical reader, prove not one of the least interesting chapters of this volume.]

ON SOME REMARKABLE NARCOTICS OF THE AMAZON VALLEY AND ORINOCO

In the accounts given by travellers of the festivities of the South American Indians, and of the incantations of their medicine-men, frequent mention is made of powerful drugs used to produce intoxication, or even temporary delirium. Some of these narcotics are absorbed in the form of smoke, others as snuff, and others as drink; but with the exception of tobacco, and of the fermented drinks prepared from the grain of maize, the fruit of plantains, and the roots of *Manihot utilissima*, *M. Aypi*, and a few other plants, scarcely any of them are well made out. Having had the good fortune to see the two most famous narcotics in use, and to obtain specimens of the plants that afford them sufficiently perfect to be determined botanically, I propose to record my observations on them, made on the spot.

The first of these narcotics is afforded by a climbing plant called Caapi. It belongs to the family of Malpighiaceæ, and I drew up the following brief description of it from living specimens in November 1853.

I. BANISTERIA CAAPI, Spruce

(*Pl. Exsicc.* No. 2712, *Anno* 1853)

Description.—Woody twiner; stem = thumb, swollen at joints. Leaves opposite, 6.4 × 3.3, oval acuminate, apiculato-acute, thinnish, smooth above, appresso-subpilose beneath; on a petiole 0.9 inch long. Panicles axillary, leafy. Umbels 4-flowered. Pedicels appresso-tomentose, bracteolate only at base. Calyx deeply 5-partite; segments ligulate, eglandulose, or with only rudimentary glands, appresso-tomentose. Petals 5, on longish thick claws; lamina pentagonal, fimbriate, the fimbriæ clavate. Stamens 10, subunequal; anthers roundish. Styles 3, subulate; stigmas capitate. Capsules muricato-cristate, prolonged on one side into a greenish-white semiobovate wing (1.7 × 0.6 inch).

Habitat.—On the river Uaupés, the Içanna, and other upper tributaries of the Rio Negro, where it is commonly planted in the roças or mandiocca-plots; also at the cataracts of the Orinoco, and on its tributaries, from the Meta upwards; and on the Napo and Pastasa and their affluents, about the eastern foot of the Equatorial Andes. Native names: Caapi, in Brazil and Venezuela; Cadána, by the Tucáno Indians on the Uaupés; Aya-huasca (*i.e.* Dead man's vine) in Ecuador.[1]

The lower part of the stem is the part used. A quantity of this is beaten in a mortar, with water, and sometimes with the addition of a small portion of the slender roots of the Caapi-pinima.[2] When sufficiently triturated, it is passed through a sieve, which separates the woody fibre, and to the residue

[1] Caapí (the Portuguese have made it Caapîm) is the Tupí or Lingoa Geral name for "grass." It means simply "thin leaf," and in that sense may correctly be applied to the *Banisteria Caapi*. In the same language the Maté of Paraguay (*Ilex Paraguayensis*) is called Caamirîm, *i.e.* "small leaf," which is certainly not so truly said of it. The Brazilian Indians accent the last, the Venezuelan the first, syllable of Caapi.

[2] Caapi-pinima, *i.e.* "painted Caapi," is an Apocyneous twiner of the genus Hæmadictyon, of which I saw only young shoots, without any flowers. The leaves are of a shining green, painted with the strong blood-red veins. It is possibly the same species as one I gathered in flower, in December 1849, at an Indian settlement on the river Trombetas (Lower Amazon), and has been distributed by Mr. Bentham under the name of *Hæmadictyon amazonicum*, n. sp. It may be the Caapi-pinima which gives its nauseous taste to the caapi drink prepared on the Uaupés, and it is probably poisonous, like most of its tribe; but it is not essential to the narcotic effect of the Banisteria, which (so far as I could make out) is used without any admixture by the Guahibos, Zaparos, and other nations, out of the Uaupés.

The Tucáno Indians call this plant Cadána-pirá, which means the same as the Tupí name. They are the most powerful tribe on the Uaupés, and the greatest consumers of caapi; but all the other tribes on that river—and they are about a dozen—use it in the same way.

enough water is added to render it drinkable. Thus prepared, its colour is brownish-green, and its taste bitter and disagreeable.

The Use and Effects of Caapi

In November 1852 I was present, by special invitation, at a Dabocurí or Feast of Gifts, held in a mallóca or village-house called Urubú-coará (Turkey-buzzard's nest), above the first falls of the Uaupés; the village of Panuré, where I was then residing, being at the base of the same falls, and about four miles away from Urubú-coará, following the course of the river, which during that space is a continuous succession of rapids and cataracts among rocky islands. We reached the mallóca at nightfall, just as the botútos or sacred trumpets began to boom lugubriously within the margin of the forest skirting the wide space kept open and clear of weeds around the mallóca.[1] At that sound every female outside makes a rush into the house, before the botútos emerge on the open; for to merely see one of them would be to her a sentence of death. We found about 300 people assembled, and the dances at once commenced. I need not detail the whole proceedings, for similar feasts have already been described by Mr. Wallace (*Travels on the Amazon and Rio Negro*, pp. 280 and 348). Indeed, there

[1] Some of the trumpets used at this very feast are now in the Museum of Vegetable Products at Kew. To get them out of the river Uaupés, when I left for Venezuela in March 1853, I wrapped them in mats and put them on board myself at dead of night, stowing them under the cabin floor, out of sight of my Indian mariners, who would not one of them have embarked with me had they known such articles were in the boat. The old Portuguese missionàries called these trumpets juruparís or devils—merely a bit of jealousy on their part; the botúto being the only fetish—not worshipped, but held in high respect—throughout the whole Negro-Orinoco region. (See figures opposite.)

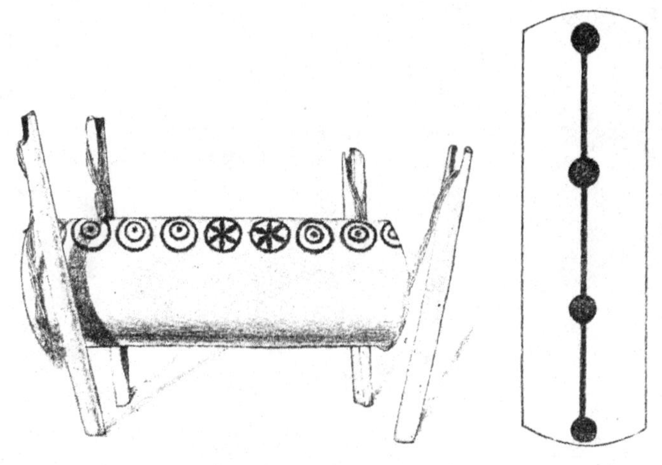

Side view. Back view.

End view.

FIG. 15.—INDIAN SACRED DRUM OR TRUMPET.

The upright outline shows the holes at bottom of the drum (as suspended) through which the inside has been scooped out. The patterns are in red and blue.

is such a family likeness in all the Indian festivities of Tropical America that, allowing for slight local variations, the description of one might serve for all. There is no more graphic account of a native feast than that by old Wafer, of one he saw on the Isthmus of Darien (*New Voyage and Description of the Isthmus of America*, p. 363).

In the course of the night, the young men partook of caapi five or six times, in the intervals between the dances; but only a few of them at a time, and very few drank of it twice. The cup-bearer—who must be a man, for no woman can touch or taste caapi—starts at a short run from the opposite end of the house, with a small calabash containing about a teacupful of caapi in each hand, muttering "Mo-mo-mo-mo-mo" as he runs, and gradually sinking down until at last his chin nearly touches his knees, when he reaches out one of his cups to the man who stands ready to receive it, and when that is drunk off, then the other cup.

In two minutes or less after drinking it, its effects begin to be apparent. The Indian turns deadly pale, trembles in every limb, and horror is in his aspect. Suddenly contrary symptoms succeed: he bursts into a perspiration, and seems possessed with reckless fury, seizes whatever arms are at hand, his murucú, bow and arrows, or cutlass, and rushes to the doorway, where he inflicts violent blows on the ground or the doorposts, calling out all the while, "Thus would I do to mine enemy (naming him by his name) were this he!" In about ten minutes the excitement has passed off, and the Indian grows calm, but appears exhausted. Were he at home in his hut, he would sleep off the remaining fumes, but now he must shake off his drowsiness by renewing the dance.

I had gone with the full intention of experimenting the caapi on myself, but I had scarcely dispatched one cup of the nauseous beverage, which is but half a dose, when the ruler of the feast—desirous, apparently, that I should taste all his delicacies at once—came up with a woman bearing a làrge calabash of caxirí (mandiocca-beer), of which I must needs take a copious draught, and as I knew the mode of its preparation, it was gulped down with secret loathing. Scarcely had I accomplished this feat when a large cigar, 2 feet long and as thick as the wrist, was put lighted into my hand, and etiquette demanded that I should take a few whiffs of it—*I*, who had never in my life smoked a cigar or a pipe of tobacco. Above all this, I must drink a large cup of palm-wine, and it will readily be understood that the effect of such a complex dose was a strong inclination to vomit, which was only overcome by lying down in a hammock and drinking a cup of coffee which the friend who accompanied me had taken the precaution to prepare beforehand.

White men who have partaken of caapi in the proper way concur in the account of their sensations under its influence. They feel alternations of cold and heat, fear and boldness. The sight is disturbed, and visions pass rapidly before the eyes, wherein everything gorgeous and magnificent they have heard or read of seems combined; and presently the scene changes to things uncouth and horrible. These are the general symptoms, and intelligent traders on the Upper Rio Negro, Uaupés, and

Orinoco have all told me the same tale, merely with slight personal variations. A Brazilian friend said that when he once took a full dose of caapi he saw all the marvels he had read of in the *Arabian Nights* pass rapidly before his eyes as in a panorama; but the final sensations and sights were horrible, as they always are.

At the feast of Urubú-coará I learnt that caapi was cultivated in some quantity at a roça a few hours' journey down the river, and I went there one day to get specimens of the plant, and (if possible) to purchase a sufficient quantity of the stems to be sent to England for analysis; in both which objects I was successful. There were about a dozen well-grown plants of caapi, twining up to the tree-tops along the margin of the roça, and several smaller ones. It was fortunately in flower and young fruit, and I saw, not without surprise, that it belonged to the order Malpighiaceæ and the genus Banisteria, of which I made it out to be an undescribed species, and therefore called it *Banisteria Caapi.* My surprise arose from the fact that there was no narcotic Malpighiad on record, nor indeed any species of that order with strong medicinal properties of any kind. Byrsonima—a Malpighiaceous genus that abounds in the Amazon valley—includes many species, all handsome little trees, with racemes of yellow or rose-coloured flowers, followed by small edible but rather insipid drupes. Their bark abounds in tannin, and is the usual material for tanning leather at Pará, as also, by the Indians, for dyeing coarse cotton garments a red-brown colour. Another genus—Bunchosia—grows chiefly on the slopes of the Andes, at from

7000 to 9000 feet elevation, and the species are trees of humble growth, bearing large yellowish-green edible drupes known as Ciruelas de fraile (Friar's plums). In cultivation the fruits are mostly seedless, and in that state are sometimes brought for sale to Ambato and other towns. The seed is described in books as poisonous, and if it be really so, then it is the only instance, so far as I know, of the existence of any hurtful principle in the entire family of Malpighiads, always excepting that of the Caapi. Yet strong poisons may lurk undiscovered in many others of the order, which is very large, and (the twining species especially) of great sameness of aspect; and the closely-allied Soapworts (Sapindaceæ) contain strong narcotic poisons, especially in the genus Paullinia.

I obtained a good many pieces of stem, dried them carefully, and packed them in a large box, which contained botanical specimens, and dispatched them down the river for England in March 1853. The man who took that box and four others on freight, in a large new boat he had built on the Uaupés, was seized for debt when about half-way down the Rio Negro, and his boat and all its contents confiscated. My boxes were thrown aside in a hut, with only the damp earth for floor, and remained there many months, when my friend Senhor Henrique Antonij, of Manáos, whom I had advised by letter of the sending-off of the boxes, heard of the mishap, and succeeded in redeeming them and getting them sent on to the port of Pará. When Mr. Bentham came to open them in England, he found the contents somewhat injured by damp and mould, and the sheets of specimens near the bottom of the boxes quite ruined. The bundle of Caapi would presumably have quite lost its virtue from the same cause, and I do not know that it was ever analysed chemically; but some portion of it should be in the Kew Museum at this day.

Caapi is used by all the nations on the river Uaupés, some of whom speak languages differing *in toto* from each other, and have besides (in other respects) widely different customs. But on the Rio Negro, if it has ever been used, it has fallen into disuse; nor did I find it anywhere among nations of the true Carib stock, such as the Barrés, Banihuas, Mandauacas, etc., with the solitary exception of the Tarianas, who have intruded a little way within the river Uaupés, and have probably learnt to use caapi from their Tucáno neighbours.

When I was at the cataracts of the Orinoco, in June 1854, I again came upon caapi, under the same name, at an encampment of the wild Guahibos, on the savannas of Maypures. These Indians not only drink the infusion, like those of the Uaupés, but also chew the dried stem, as some people do tobacco. From them I learnt that all the native dwellers on the rivers Meta, Vichada, Guaviare, Sipapo, and the intervening smaller rivers, possess caapi, and use it in precisely the same way.

In May 1857, after a sojourn of two years in the North-Eastern Peruvian Andes, I reached, by way of the river Pastasa, the great forest of Canelos, at the foot of the volcanoes Cotopaxi, Llanganati, and Tunguragua; and in the villages of Canelos and Puca-yacu—inhabited chiefly by tribes of Zaparos—I again saw Caapi planted. It was the identical species of the Uaupés, but under a different name, in the language of the Incas, Aya-huasca, *i.e.* Dead man's vine. The people were nearly all away at the gold-washings, but from the Governor of Puca-yacu I got an account of its properties coinciding wonderfully with what I had previously learnt in Brazil. Dr. Manuel Villavicencio, a native of Quito, who had been some years governor of the Christian settlements on the Napo, published the following year, in his *Geografia de la Republica del Ecuador* (New York, 1858), an interesting account of the customs of the natives of that river, and amongst others of their drinking the aya-huasca; but of the plant itself he could tell no more than that it was a liana or vine. The following is a summary of what I learnt at Puca-yacu and from Villavicencio of the uses and effects of the aya-huasca or caapi, as observed on the Napo and Bombonasa.

Aya-huasca is used by the Zaparos, Angutéros, Mazánes, and other tribes precisely as I saw caapi used on the Uaupés, viz. as a narcotic stimulant at their feasts. It is also drunk by the medicine-man, when called on to adjudicate in a dispute or quarrel—to give the proper answer to an embassy—to discover the plans of an enemy—to tell if strangers are coming—to ascertain if wives are unfaithful—in the case of a sick man to tell who has bewitched him, etc.

All who have partaken of it feel first vertigo; then as if they rose up into the air and were floating about. The Indians say they see beautiful lakes, woods laden with fruit, birds of brilliant plumage, etc. Soon the scene changes; they see savage beasts preparing to seize them, they can no longer hold themselves up, but fall to the ground.

At this crisis the Indian wakes up from his trance, and if he were not held down in his hammock by force, he would spring to his feet, seize his arms, and attack the first person who stood in his way. Then he becomes drowsy, and finally sleeps. If he be a medicine-man who has taken it, when he has slept off the fumes he recalls all he saw in his trance, and thereupon deduces the prophecy, divination, or what not required of him. Boys are not allowed to taste aya-huasca before they reach puberty, nor women at any age: precisely as on the Uaupés.

Villavicencio says (*op. cit.* p. 373): "When I have partaken of aya-huasca, my head has immediately begun to swim, then I have seemed to enter on an aerial voyage, wherein I thought I saw the most charming landscapes, great cities, lofty towers, beautiful parks, and other delightful things. Then all at once I found myself deserted in a forest and attacked by beasts of prey, against which I tried to defend myself. Lastly, I began to come round, but with a feeling of excessive drowsiness, headache, and sometimes general *malaise*."

This is all I have seen and learnt of caapi or aya-huasca. I regret being unable to tell what is the peculiar narcotic principle that produces such extraordinary effects. Opium and hemp are its most obvious analogues, but caapi would seem to operate on the nervous system far more rapidly and violently than either. Some traveller who may follow my steps, with greater resources at his command, will, it is to be hoped, be able to bring away materials adequate for the complete analysis of this curious plant.

Niopo Snuff and the Mode of using it

II. Piptadenia Niopo, Humboldt

Synonyms—*Acacia? Niopo*, Humb., *Rel. Hist.* ii. p. 620; ejusdem *Nov. Gen. Amer.* vi. p. 282; *DC. Prodr.* ii. p. 471. *Inga Niopo*, Willd.

Description.—Tree, 50 feet by 2 feet, with muricated bark, otherwise unarmed. Leaves bipinnate; pinnæ twenty-four pairs; pinnules very numerous, minute, linear, mucronato-apiculate, ciliated, sparsely sub-pubescent. An oblong gland on petiole above base; another between terminal pinnæ. Racemes axillary and terminal; pedicels twin, each bearing a small globose head of white flowers. Corolla slightly emersed from 5-angled calyx. Stamens 10; anthers tipped with a gland. Pod linear, sub-compressed, apiculate, 7–12-seeded, sub-constricted between seeds. Seeds flattish, green.

Habitat.—In the drier forests of the Amazon, and along its tributaries, both northern and southern; on the Rio Negro, throughout its course; also at the cataracts of the Orinoco; both wild and planted near villages. (Santarem, fl. Amazonum, Spruce, *Exsicc.* No. 828, etiam Janauarí, fl. Negro, No. 1786.) Native names: Paricá in Brazil; Niópo in Venezuela.

We owe our first knowledge of Niopo snuff, and of the tree producing it, to Humboldt and Bonpland, whose brief account of it is thus condensed by Kunth: "Ex seminibus tritis calci vivae admixtis fit tabacum nobile quo Indi Otomacos et Guajibos utuntur" (*Synopsis*, iv. p. 20). In the modern niopo, as I saw it prepared by the Guahibos themselves, there is no admixture of quicklime, and that is the sole difference. My specimens of the leaves, flowers, and fruit agree so well with Kunth's description of *Acacia Niopo* that I cannot doubt their being the same species; especially as I have traced the tree all the way from the Amazon to the Orinoco, and found it everywhere identical, although it bears a different name on the two rivers, as is commonly the case where the same plant or animal occurs on both. Mr. Bentham believes my plant to be the old *Mimosa peregrina* of Linnæus (*Acacia peregrina*, Willd.); and if both opinions be correct, then the species must be called *Piptadenia peregrina* (L.), Benth.; and *Acacia Niopo*, Humb., will stand as a synonym.

I first gathered specimens of the Parica (or Niopo) tree in 1850, near Santarem, at the junction of the Tapajoz and Amazon, where it had apparently been planted. In the following year I gathered it on the little river Jauauarí—one of the lower tributaries of the Rio Negro—where it was certainly wild. But I did not see the snuff actually prepared from the seeds and in use until June 1854, at the cataracts of the Orinoco. A wandering horde of Guahibo Indians, from the river Meta, was encamped on the savannas of Maypures, and on a visit to their camp I saw an old man grinding Niopo seeds, and purchased of him his apparatus for making and taking the snuff, which is now in the Museum of Vegetable Products at Kew. I proceed to describe both processes.

The seeds being first roasted, are powdered on a wooden platter, nearly the shape of a watch-glass, but rather longer than broad (9¼ inches by 8 inches). It is held on the knee by a broad thin handle, which is grasped in the left hand, while the fingers of the right hold a small spatula or pestle of the hard wood of the Palo de arco (Tecomæ sp.) with which the seeds are crushed.

The snuff is kept in a mull made of a bit of the leg-bone of the jaguar, closed at one end with pitch, and at the other end stopped with a cork of marima bark. It hangs around the neck, and from it are suspended a few odoriferous rhizomes of a sedge (*Kyllingia odorata*). Rhizomes of the same sedge, or of an allied species, are in use among the Indians throughout the Amazon and Orinoco. They render the wearer secure from the bad wish and evil eye of his enemies.

For taking the snuff they use an apparatus made of the leg-bones of herons or other long-shanked birds put together in the shape of the letter Y, or something like a tuning-fork, and the two upper tubes are tipped with small black perforated knobs (the endocarps of a palm). The lower tube being inserted in the snuff-box and the knobs in the nostrils, the snuff is forcibly inhaled, with the effect of thoroughly narcotising a novice, or indeed a practised hand, if taken in sufficient quantity; but this endures only a few minutes, and is followed by a soothing influence, which is more lasting.

The Guahibo had a bit of caapi hung from his neck, along with the snuff-box, and as he ground his niopo he every now and then tore off a strip of caapi with his teeth and chewed it with evident satisfaction. "With a chew of caapi and a pinch of niopo," said he, in his broken Spanish, "one feels so good! No hunger—no thirst—no tired!" From the same man I learnt that caapi and niopo were used by all the nations on the upper tributaries of the Orinoco, *i.e.* on the Guaviare, Vichada, Meta, Sipapo, etc.

I had previously (in 1851) purchased of a Brazilian trader at Manáos an apparatus for taking niopo snuff rather different from that of the Guahibos. He had brought it from the river Purús, where it had been used by the Catauixí Indians. My note on it (as taken down from his account) is as follows:—

The Catauixís use niopo snuff as a narcotic stimulant, precisely as the Guahibos of Venezuela, and as the Muras and other nations of the Amazon, where it is called paricá. For absorbing paricá by the nose, a bent tube is made of a bird's shank-bone, cut in two, and the pieces joined by wrapping, at such an angle that one end being applied to the mouth, the other reaches the nostrils. A portion of snuff is then put into the tube and blown with great force up the nose. A clyster-pipe is made, on the same principle, of the long shank-bone of the tuyuyú (*Mycteria americana*). The effect of paricá, taken as snuff, is to speedily induce a sort of intoxication, resembling in its symptoms (as described to me in this instance) that produced by the fungus *Amanita muscaria*. Taken in injection, it is a purge, more or less violent according to the dose. When the Catauixí is about to set forth on the chase, he takes a small injection of paricá, and administers another to his dog, the effect on both being (it is said) to clear their vision and render them more alert!

Herndon (*Valley of the Amazon*, p. 318) gives the following account of the use of paricá among the Mundrucús, on the river Tapajoz, which he derived from an intelligent Frenchman (M. Maugin) who had traded among them. They powder the seeds of paricá, make the powder into a paste, and repulverise a portion whenever they want to take it as snuff. Two quills of the royal heron, joined side by side, make a double tube, which is applied to the nostrils and the powder snuffed up with a strong inspiration. M. Maugin thus describes its effects on an Indian whom he saw take it. "His eyes started from his head, his mouth contracted, his limbs trembled. It was fearful to see him. He was obliged to sit down or he would have fallen. He was drunk, but only for about five minutes; he was then gayer."

"Medicine-Men" and their Customs

Among the native tribes of the Uaupés and of the upper tributaries of the Orinoco, niopo or paricá is the chief curative agent. When the payé is called in to treat a patient, he first snuffs up his nose such a quantity of paricá as suffices to throw him into a sort of ecstasy, wherein he professes to divine the nature of the evil wish which has caused the sickness, and to gather force to counteract it. He next lights a very thick cigar of tobacco, inhales a quantity of smoke, and puffs it out over the sick man, over the hammock in which he is laid, and over everything he habitually uses, but especially over the food he is to eat. This done, the payé professes to suck out the ill, by applying his mouth to the seat of pain, or as near to it as practicable; and he spits out the morbid matter—most likely tobacco or coca juice—and sometimes produces from his mouth thorns and other substances, previously hidden there, but which he pretends to have extracted from the sick man's body. If the sickness ends fatally, he denounces the enemy whose evil wish has caused it, and not infrequently it is some rival payé, of the same or another nation. Hence I was told that the payés never travel without an accompaniment of at least four or five well-armed men, their lives being in continual jeopardy from such denunciations.

I have never been so fortunate as to see a genuine payé at work. Among the civilised Indians the Christian padre has supplanted the pagan payé, who has besides been discountenanced and persecuted by the civil authorities; so that if any now exist, he must exercise his office in secret. With the native and still unchristianised tribes I have for the most part held only passing intercourse during some of my voyages. Once I lived for seven months at a time among them, on the river Uaupés, but even there I failed to catch a payé. When I was exploring the Jauarité cataracts on that river, and was the guest of Uiáca, the venerable chief of the Tucáno nation, news came to the mallóca one afternoon that a famous payé, from a long way up river, would arrive that night and remain until next day, and I congratulated myself on so fine a chance of getting to know some of the secrets of his "medicine." He did not reach the port until 10 P.M., and when he learnt that there was a white payé (meaning myself) in the village, he and his attendants immediately threw back into the canoe his goods, which they had begun to disembark, and resumed their dangerous voyage down the river in the night-time. I was told he had with him several palm-leaf boxes, containing his apparatus. (There is a similar box now in the Kew Museum, sent by me from the Uaupés.) I could only regret that his dread of a supposed rival had prevented the interview which to me would have been full of interest; the more so as I was prepared to barter with him for the whole of his materia medica, if my stock-in-trade would have sufficed.

Rochefort (*Histoire Naturelle et Morale des Isles Antilles*, Rotterdam, 1665), says: "Their Boyés or medicine-men practise both medicine and devilry. They are resorted to: 1, to cause punishment to fall on some one who has wronged or injured the applicant; 2, to cure some disease; 3, to foretell the advent of a war; 4, to drive out the Maboya or Evil Spirit" (p. 472).

Their functions are very much the same at the present day among the native tribes of the mainland as they were two or three hundred years ago in the isles of the Caribbean Sea. I propose, in what follows, to review briefly the use made by the payés of their materia medica in the treatment of disease.

The apparatus and materia medica of the medicine-men of the region lying adjacent to the Upper Rio Negro and Orinoco, and extending thence westward to the Andes, are chiefly the following:—

The Maraca or Rattle.

Tobacco, juice and smoke.

Niopo (or Paricá), powdered seeds in snuff.

Caapi (or Aya-huasca), stems in infusion.

1. *The Maraca or Rattle.*—This is the hard globose or oval pericarp of the *Crescentia Cujete*, or sometimes of a gourd, tastefully engraved and perforated in geometrical or fantastic designs, and the lines usually coloured. To make it rattle, a few small bright-red or red-and-black beans are put into it; those most used on the Uaupés are seeds of *Batesia erythrosperma* (Spruce) and of *Ormosia coccinea* (Jack). I have seen the maraca used in

dances, but it is also employed by the payés in their divinations, and Bancroft's account of its use in Guayana corresponds so nearly with what was told to me on the Uaupés, that I cannot do better than transcribe it here.

"The medicine-men, called Peiís [Stedman says Peiis or Pagayers], unite in themselves the sacerdotal and medicinal functions. One of the implements of the peii is a hollowed calabash (cuya) through the centre of which an axis is passed projecting about a foot on each side, the thick end forming a handle, the thin end decorated with feathers; it is also carved and painted and perforated with small holes—some long, some round—and several quartz pebbles and red-and-black beans are put inside it, so that it forms a rattle. When the peii is called to a patient, he begins his exorcism at night, the lights being put out and he left alone with the patient. He rattles his maraca by turning it slowly round, singing at the same time a supplication to the Yawahoo. This goes on for say a couple of hours, when the peii is heard conversing with the Yawahoo—at least there are two distinct voices. Afterwards the peii makes a report in an ambiguous style, on what will be the event of the disorder. The exorcisms are repeated every night until after a favourable turn, when the peii pretends to extract the cause of the disorder by sucking the part affected, after which he pulls out of his mouth fish-bones, thorns, snake's teeth, or some such substance, which he has before concealed therein, but pretends to have been maliciously conveyed into the affected part by the Yawahoo. The patient then fancies himself cured, and the influence of imagination helps his recovery. If the patient dies, the peii attributes it either to the implacable Yawahoo or to the influence of some inimical peii." (*An Essay on the History of Guayana*, by Dr. Edward Bancroft, 1769, p. 310.)

Long before Bancroft's time the use of the maraca and of tobacco by Brazilian payés was described by Thevet, as follows: "Existimant enim, cum hunc fructum (quem Maraka et Tamaraka nuncupant) manibus pertractant, crepitantemque ob Mayzi grana injecta audiunt, cum suo se Toupan, id est, Deo sermones conferre atque ab eo quodam responsa accipere, sic a suis Paygi (divinatorum genus est, qui suffitu herbae Petun, et quibusdam obmurmurationibus illorum Tamaraka divinam facultatem attribuunt tribuere perhibent) persuasi."[1]

The accounts given by the early missionaries of the doings of the payés are seldom full or reliable. Those pious men regarded them as the great obstacle to the reception of the Christian faith by the natives, and always wrote of them with a certain impatience and disgust, under the belief (no doubt sincere) that the payés had direct dealings with the devil. But the cure of disease by suction is alluded to by missionaries in every part of South America. In the *Lettres Édifiantes et Curieuses*, consisting of selections from the correspondence of missionaries in various heathen countries, published with the sanction of the holy see, there is this note about the medicine-men of the Moxos Indians: "L'unique soulagement qu'ils se procurent dans

[1] Thevetus, as quoted by Chusius, in *Aromatum et Simplicium aliquot . . . Historia.* Auctore Garcia ab Horto, Medico Lusitanico. Ed. Clusio. Antverpiæ, 1579.

leurs maladies, consiste à appeler certains enchanteurs, qu'ils s'imaginent avoir reçu un pouvoir particulier de les guérir ; ces charlatans vont trouver les malades, récitent sur eux quelque prière superstitieuse, leur promettent de jeûner pour leur guérison, et de prendre un certain nombre de fois par jour du tabac en fumée ; ou bien, ce qui est une insigne faveur, ils sucent la partie mal affectée, après quoi ils se retirent, à condition qu'on leur payera libéralement ces sortes de services" (tome viii. p. 83). And at p. 339 of the same volume, speaking of the enchanters of the Chiquitos, it is said : "Le médecin suce ensuite la partie mal affectée, et au bout de quelque temps il jette par la bouche une matière noire : Voilà, dit-il, le venin que j'ai tiré de votre corps."

It is not necessary to be a payé to "suck out a pain." Among the Barrés it is commonly practised, and I have seen a fellow hang on to his comrade's shoulder for half an hour together, "sucking out the rheumatism." But as they know the whites ridicule the practice, they avoid as much as possible being surprised in it. Formerly they had professional chupadores or suckers ; but in my time there were none such, besides the payés, who were found only among the unchristianised tribes.

2. *Tobacco.*—This was possibly the first narcotic ever used in South America, and is likely to be the last. In one form or another it is a prime ingredient in the medicine of the payés. Rochefort says : "Each Boyé has his familiar demon, whom he evoques by a chant, accompanied by the smoke of tobacco, whose perfume is supposed to be attractive to devils" (*loc. cit.* p. 473). And it is

essential to the *making* of payés. Bancroft says: "The order of Peiis is inherited by the eldest sons. A young Peii is initiated with superstitious ceremonies lasting several weeks. Among other things, he is dosed with tobacco till it no longer operates as an emetic" (*loc. cit.*).

Tobacco-smoke is blown on the sick person by the payé in almost all methods of cure, whether the maraca, niopo, or caapi be the primary agent. In lieu of the two latter it would seem that in some nations the enchanters narcotised themselves by chewing tobacco and swallowing the juice. The large cigar used on the Uaupés is smoked in the ordinary way, and the smoke blown from the mouth; but in the country bordering the Pacific coast of Equatorial America the cigar—two or three feet long, but slenderer than that of the Uaupés—was held in the mouth *at the lighted end*, and the smoke blown from the opposite end upon the sick person, or, at a feast, in the faces of the guests, whereof Wafer has an amusing account and a rude picture (p. 327, *loc. cit.*). He calls the payés pawawers, evidently the same name, with a merely dialectic difference. It is curious that at the present day the Indians and negroes along that coast frequently hold the lighted end of a cigar in their mouths, as any one who has sojourned at Panama or Guayaquil may have observed.

The uses of niopo (or paricá) and of caapi (or aya-huasca) I have already indicated above. The former is the chief "medicine" of the payés on the affluents of the Amazon, both northern and southern, and on the Orinoco; but the latter in the roots of the Equatorial Andes. I have not learnt that they are ever used in conjunction, except as an occasional stimulant, and in small quantity.

On Spirits or Demons among the Indians

I have never heard any mention among the native races with whom I have sojourned of a Spirit or Demon the payé was supposed to invoke, but there has been so much testimony to that effect, that it can hardly fail to be true. This demon—the Maboya of the Antilles, the Yawahoo of Guayana (according to Bancroft and Stedman)—is surely the Yamádu of the Casiquiari and Alto Orinoco. But when I made inquiry about the latter, I was always assured that it had a bodily, and not merely a ghostly existence. It is, in fact, a Wild Man of the Woods or Forest Devil—the Curupira or Diabo do mato of the Amazon, the Munyía of the eastern foot of the Equatorial Andes —a little hairy man, not more than four to five feet high, but so strong and wiry that no single Indian can cope with him. His great peculiarity is that, although his tracks are often met with, no one can tell which way he has gone. Either, as on some parts of the Amazon, he has a perfectly human foot, but set on the contrary way; or else, as on the Casiquiari, Uaupés, Napo, etc., he has two heels on each foot and never a toe. This little devil plays many pranks, of which the most serious is his carrying off women who venture alone into the forest; but he never attacks two people together, so that in some parts a man or woman will take a little child into the forest rather than go alone. If an Indian loses his way in the forest, he blames the

Curupira, and to find it again he twists a liana into a ring—or, if he be a Christian, into the form of a cross—in such a way that the points of the liana are completely hidden; he then throws it behind him, taking care not to look which way it goes, and afterwards picks it up and follows the direction in which it has fallen. I cannot here recount all the tales I have heard about this mysterious being, but I suppose they point to the former existence in the regions of some *homo primordialis*, and that the fact has come down by tradition from untold ages, coupled with the belief that the species is even yet not extinct. Meanwhile, until the animal, or its skeleton, be found—which I do not look on as impossible—I suppose we must consider the Curupira, or Munyía, or Yamádu, the analogue of the Barghaist of the north of England and Scotland, the Loup-garou of France, the Lobishomem of Portugal, and other similar mythical creatures.

A Strange Occurrence and its Explanation

In my voyage to the Upper Orinoco, by way of the Casiquiari, in 1853-54, when the river was so low at Christmas that I had great difficulty in getting my piragoa up as far as Esmeralda, and it was quite impossible to ascend farther, as I had at first intended, I afterwards explored its northern tributary, the Cunucunúma, and re-entered the Casiquiari, intending to go as far down as Lake Vasiva. The dry season should have held all through the months of January and February, and Vasiva was described to me as having at that time broad sandy beaches, sprinkled with curious little plants, and bordered with flowering bushes, so that I reasonably hoped to make a fine collection there. But the first night of our downward voyage (Jan. 7) the rains came on, out of their time, and continued daily for many days, until the river had risen to its winter level, and the forest-margin was mostly flooded. There are only two small pueblos on the Casiquiari above the outlet of Lake Vasiva, and at the lower of these I halted nine days, hoping the floods might subside. This pueblo was of only recent formation, and was peopled by Pacimonari Indians, who had named it Yamádu-bani, that is, Wild Man's Land, because the adjacent forests were said to be haunted by the Yamádu. I explored them as much as the heavy rains permitted, and never encountered any Yamádu; but on the very first day I was myself taken for it by two girls whom I met suddenly at the turning of a large buttressed tree, on a forest trail, and who threw down their baskets, laden with manioc, and fled affrighted. At length the weather seemed to take up a little, although the river was still high, and I determined to go on to Vasiva. We accordingly re-embarked early on the 21st, and eight oars, aided by a strong current, brought us to the lake at 4 P.M.; but in vain we coasted along to find a bit of dry land whereon to encamp, for the trees and bushes were all in water up to 4 or 5 feet; so that we had to return to the narrow winding channel forming the outlet of the lake, where there was a scanty strip of terra firme and a rancho left by a party that had gathered turtles' eggs there the previous year. Here we remained four days, but the weather was dreadfully rainy, the sun never once appeared, and all I could do was to creep about the margin of the lake and up its tributary creeks in my curiara, and gather specimens of the few trees that were in flower. On the 22nd, at 4 P.M., when we were cooking our dinner, we were startled by hearing the report of a musket in the forest on the opposite bank of the river, there not more than 80 yards wide. It is scarcely possible to conceive the strangeness of such a sound in savage, desolate forests which scarcely any human being could penetrate, especially one accustomed to firearms. A region of at least 10,000 square miles, of which we were the centre, had scarcely 400 inhabitants, and those chiefly half-wild Indians, whose weapon was the blowing-cane. The nearest settlement was that of Yamádu-bani, but we knew that none of their hunting tracks extended to Vasiva; and the half-dozen adult males had neither guns nor ammunition when we left them only the day before. There had been no inhabitants on Vasiva for very many years, and there were no traders or other travellers on the Casiquiari at that season beside ourselves. I was completely puzzled. The report was not exactly like that of either musket or rifle, nor was it any one of the accustomed sounds which at rare intervals break the silence of those vast solitudes, and with which I had become familiar. The crash of a huge tree falling from sheer age—the explosion, like distant cannon, of an old hollow Sassafras or Capivi tree, burst by the balsam accumulated in the cavity—the solitary thunderclap in an apparently cloudless sky—the roar of cataracts, and of the approaching hurricane—all these sounds I had previously heard, and had learnt to distinguish. My Indians, however, although even more startled than myself, soon made up their minds about the origin of the unwonted sound. It was the Yamádu, *in propria persona*, hunting near us, and he would infallibly send us terrible rain or some other calamity to warn us off his territory. The soughing of the approaching tempest was already heard, and presently it burst upon us, with thunder and lightning and deluging rain that lasted until midnight. The two following days were dull and dropping, and a little later on in the day—that is, towards nightfall—we each day heard a single report, not quite so near at hand, and then we had heavy rain from 7 P.M. throughout the night. My people became silent and gloomy, were afraid, they said, to hunt or fish, and I believe if I had remained another night would have every one deserted me. So in the afternoon of the 25th I gave the order for resuming our voyage down the Casiquiari, to their very great content. When I came on deck shortly afterwards to see if everything was in readiness for starting, I saw some of the men in a tree that overhung our encampment, fastening to the branches a couple of scarecrows they had rigged up out of old shirts and trousers. "What does this mean, Antonio?" said I to one of them who was fond of talking to me in Lingua Tupí. "Yánerangáua" (our effigies), said he. "Oh, I see," said I. "You think to cheat the Yamádu. Seeing us up the tree, he will fancy we are still here, and will not pursue us down the river!" But I had a quiet laugh over it in the recesses of my cabin. It reminded me of a fellow pursued by a bull, who throws off hat and coat to detain the savage brute until he himself can gain a place of safety.

For years afterwards the solitary shots in the sombre forests of Lake Vasiva used to haunt my memory and my dreams. They were as mysterious to me, although not so alarming, as the single footprint was to Robinson Crusoe. My ears were always open to some repetition of the sound which might lead to detecting its origin. In April 1857, I was on my voyage up the lonely Pastasa, at the eastern foot of the Andes. My companions were two Spaniards, two whitish lads who acted as our servants, and fourteen Cucáma Indians who paddled our two canoes. Five months before, there had been an uprising of the savage Jibaros and Huambisas, who had laid waste the Christian villages on the Amazon, below the Pongo de Manseriche, and the only village (Santander) on the Lower Pastasa. We travelled, therefore, in constant risk of being attacked, and were on the alert day and night. The Indians would never go on shore to cook until we had first landed with our arms and ascertained that the adjacent forest was clear. One morning we had cooked our breakfast, and were just squatting down, Turkish fashion, around the steaming pots, when what sounded like a gunshot—quite near—brought us all to our feet. But the Jibaros, we knew, had no firearms, and it at once struck me that it was the identical sound heard on Lake Vasiva. "What and where is that?" I exclaimed. "I will take you straight to it, if you like," said the old pilot of my canoe; and accepting his offer, I plunged into the bush with him, and in three minutes reached a heap of débris, like a huge haycock, the remains of a decayed Palm-trunk whose sudden fall it was that had startled us. It had been a very tall, stout Palm, 80 or 100 feet high at the least. When the vitality of a Palm is exhausted, the crown of fronds first withers and falls, and then the soft interior of the trunk gradually rots and is eaten away by termites until nothing is left but a thin shell; and when that can no longer bear its own weight, it collapses and breaks up in an instant, with a crash very like a musket-shot.[1]

A few weeks later, I had to make my way on foot through the forest of Canelos, and it sometimes happened that when we had to cook our supper, after a day of soaking rain, we could find no wood that would burn but these shells of Palm-trunks. (The Palm was the curious *Wettinia Maynensis*, which abounded there.) A single stroke of a cutlass would often suffice to cause them to collapse and fall, in a mass of dust and splinters, repeating each time the report of the weapon of the mysterious hunter of Vasiva, and not without risk to the operator of being buried in the ruins.

Sometimes when I have been deep in the virgin forest, and could not see through the overarching foliage any sign of rain in the sky, or was heedless of it—when not a sound or a breath of air disturbed the solemn calm and stillness—a shiver would all at once pass through the tree-tops, and yet no wind at all be sensible below. Then all would be still again, and it was not until a few minutes later that a distant soughing announced the coming tempest. The preliminary shudder would bring down dead leaves and twigs, and such a one might have

[1] This strange sound is briefly described in Spruce's Journal. See vol. i. p. 423.

prostrated the decayed Palm on Lake Vasiva. Other dead Palms might fall when the full force of the squall caught them, but the crash of their fall would be drowned in the general roar of the tempest, and especially in the continuous roll of the thunder. The truth seems to be that it is nearly always during a storm such Palms do fall, and that their prostration during a season of calm is the rarest possible occurrence; which accounts for my having passed four years and a half in the forest before I ever heard it, and for others having lived the best part of their lives there either without noticing it, or without caring to ascertain the origin of the sound caused by it. It hardly needs mention that perfectly vigorous Palm trees, and trees of all kinds, may fall during a violent storm. Hurricanes that open out long lanes in the forest are only too frequent towards the sources of the Orinoco, but are exceedingly rare on and near the Amazon.

Rarity of Curative Drugs among the Indigenes

From what was said above, it will have been seen that, although the medicine-man doses himself with powerful narcotics, no drug whatever is administered to the patient; nor could I learn that it was ever done by a "regular practitioner." The Indians have a few household remedies, but by far the greater portion of these have come into use since tne advent of the white man from Europe and the negro from Africa. Von Martius remarks nearly the same thing in the introduction to his *Systema Materiae Medicae vegetabilis Brasiliensis*

(1843, p. xvii.): "At valde fallerentur, qui putarent, Brasiliae plantas medicas omnes per autochthones colonis esse oblatas; potius multa me movent, ut dicam, totidem, quae nunc adhibentur, a nigris et albis incolis esse detectas et usu cognatas, quot ab illis." Of external applications, I have seen only the following. For a wound or bruise or swelling, the milky juice of some tree is spread thick on the skin, where it hardens into a sort of plaster, and is allowed to remain on until it falls of itself. Almost any milky tree may serve, if the juice be not acrid; but the Heveas (India-rubbers), Sapotads, and some Clusiads are preferred. Such a plaster has sometimes an excellent effect in protecting the injured part from the external air.

At Tarapoto, in the Eastern Peruvian Andes, where the people are all Christians, and some of them almost pure white, where there are churches and priests and schools, such medicine as they have is little more than necromantic practices of their curanderos. In all sicknesses the first curative operation is to sobar el espanto (rub out the fright), which is done thus: Chew a piece of the gum-resin called "sonitonio," place it in the hollow of the hand, and with it rub the legs of the sick person, from the knees downwards, and end by whistling between all the toes. There are other ridiculous and useless operations, but in some cases the rubbing is really beneficial. Take this mode of "rubbing out colic" as an example. Put a little fowl's grease in the hand, and rub it over the body of the patient, round and round, over the course of the colon, making every now and then a forcible twist and pressure on the navel, para soltar el empacho (to loosen the indigestion). Rubbing with a dry hand is still better, and for lumbago and other forms of rheumatism has sometimes an excellent effect. There are persons who, by long practice, acquire what is called "a good hand," and are much sought after as sobadores or shampooers.

Nervous Stimulants used by the Indians

Several plants are used in South America as nervous stimulants, and all are more or less narcotic. Of these, the foremost place must be assigned to *Erythroxylon Coca* (Lam.)—Coca of the Peruvians, Ipadú of the Brazilians. Of its use in Peru, chiefly by miners and cargueros, Poeppig has already given an excellent account. There the entire leaf is chewed, with a small admixture of lime. But in North Brazil, where also its use is almost universal, I have always seen it used in powder. The plant itself, a slender shrub, with leaves not unlike tea-leaves, except that they are entire at the margins, is frequently planted near houses. In Peru, as is well known, there are large plantations of it, called cocales. I have gathered it truly wild on the rocky banks of the Rio Negro, near Tomo in Venezuela (hb. 3565); and an Erythroxylon (*E. cataractarum*, n. sp. hb. 2614), which I found growing abundantly on rocks in the cataracts of the Paapurís, a tributary of the Uaupés, which has small dark-green leaves only an inch and a half long, is considered by Mr. Bentham a variety of the same species.

In January 1851 I saw ipadú prepared and used on the small river Jauauarí, near the mouth of the

Rio Negro, and I sent a quantity of it to Kew for analysis. My account of it was published in Hooker's *Journal of Botany* for July 1853, and I here reproduce it. The leaves of ipadú are pulled off the branches, one by one, and roasted on the mandiocca-oven, then pounded in a cylindrical mortar, 5 or 6 feet in height, made of the lower part of the trunk of the Pupunha or Peach Palm (*Guilielmia speciosa*), the hard root forming the base and the soft inside being scooped out. It is made of this excessive length because of the impalpable nature of the powder, which would otherwise fly up and choke the operator; and it is buried a sufficient depth in the ground to allow of its being easily worked. The pestle is of proportionate length, and is made of any hard wood. When the leaves are sufficiently pounded, the powder is taken out with a small cuya fastened to the end of an arrow. A small quantity of tapioca, in powder, is mixed with it to give it consistency, and it is usual to add pounded ashes of Imba-úba or Drum tree (*Cecropia peltata*), which are saline and antiseptic. With a chew of ipadú in his cheek, renewed at intervals of a few hours, an Indian will go for days without food and sleep.

In April 1852 I assisted, much against my will, at an Indian feast in a little rocky island at the foot of the falls of the Rio Negro; for I had gone down the falls to have three or four days' herborising, and I found my host—the pilot of the cataracts—engaged in the festivities, which neither he nor my man would leave until the last drop of cauim (coarse cane- or plantain-spirit) was consumed. During the two days the feast lasted I was nearly famished, for, although there was food, nobody would cook it, and the guests sustained themselves entirely on cauim and ipadú. At short intervals, ipadú was handed round in a large calabash, with a tablespoon, for each one to help himself, the customary dose being a couple of spoonfuls. After each dose they passed some minutes without opening their mouths, adjusting the ipadú in the recesses of their cheeks and inhaling its delightful influences. I could scarcely resist laughing at their swollen cheeks and grave looks during these intervals of silence, which, however, had two or three times the excellent effect of checking an incipient quarrel. The ipadú is not sucked, but allowed to find its way insensibly into the stomach along with the saliva. I tried a spoonful twice, but it had little effect on me, and assuredly did not render me insensible to the calls of hunger, although it did in some measure to those of sleep. It had very little of either smell or taste, and in both reminded me of weak tincture of henbane. I could never make out that the habitual use of ipadú had any ill results on the Rio Negro; but in Peru its excessive use is said to seriously injure the coats of the stomach, an effect probably owing to the lime taken along with it.

The Use of Guaraná as a Tonic

Another powerful nervous tonic and subnarcotic is cupána or guaraná, which is prepared from the seed of a twining plant of the family of Sapindaceæ. The first definite information about it was obtained by Humboldt and Bonpland in the south of

Venezuela. Humboldt says: "A missionary seldom travels without being provided with some prepared seeds of the Cupána. The Indians scrape the seeds, mix them with flour of cassava, envelop the mass in plantain-leaves, and set it to ferment in water, till it acquires a saffron-yellow colour. This yellow paste, dried in the sun and diluted in water, is taken in the morning as a kind of tea. This beverage is bitter and stomachic, but appeared to me to have a very disagreeable taste." (*Personal Narrative*, v. 278, Miss Williams's translation.)

It was at Javita, near the head of the Atabapo, that Humboldt made trial of cupána. I first tasted the cold infusion, prepared nearly in the same way, except that no cassava had been added to the grated seeds, I think at Tomo, on the Guainia, only two days' journey from Javita, in 1853; and I afterwards drank it frequently on the Atabapo and Orinoco, where the inhabitants still take it commonly the first thing in a morning, on quitting their hammocks, and consider it a preservative against the malignant bilious fevers which are the scourge of that region. It is as bitter as rhubarb, and is always drunk unsweetened, so that at first one finds it absolutely repulsive; but it soon ceases to be so, and those who use it habitually get to like it much, and to find it almost a necessary of life. When the bowels are relaxed and coffee taken in the morning, fasting excites too much peristaltic action, then cupána is decidedly preferable, for it is less irritating than coffee and has quite the same stimulating effect on the nervous system.

Long before I saw cupána in Venezuela—indeed, ever since the end of 1849—I had been familiar with it in Brazil, but under another name and prepared in a different way. There it is called guaraná, and is largely cultivated in the mid-Amazon region, especially on the river Mauhés, which is a little west of the Tapajoz, whence it is exported to all other parts of Brazil. Single plants of it may be seen in gardens and roças all the way up the Amazon, as far as to the Peruvian frontier; and throughout the Rio Negro. Martius's excellent account of the Guaraná of the Mauhés has been translated by Mr. Bentham in Hooker's *Journal of Botany* for July 1851. Martius called the plant *Paullinia sorbilis*, apparently not suspecting it to be the same as Humboldt's *Paullinia Cupana* ; yet the two are absolutely identical, and Humboldt's name, being the elder, must stand.

The specimens distributed by Mr. Bentham in my Plantae Exsiccatae (No. 2055) were gathered at Uanauacá, a farm on the Rio Negro, a little below the cataracts. I subjoin the brief description I drew up on the spot.

PAULLINIA CUPANA, H. B. K., *Nov. Gen. Amer.* v. p. 117; *DC. Prodr.* i. 605.

Synon. *Paullinia sorbilis*, Mart., *Reise*, ii. p. 1098; ejusdem *Syst. Mat. Med. Brasil.* p. 59; Th. Mart. in Buchner's *Repert. d. Pharm.* xxxi. p. 370.

Description.—Stout woody twiner, kept down in cultivation to the size of a compact currant bush. Ramuli and petioles sub-pubescent. Leaves alternate, pinnate; leaflets two and a half pairs, $5\frac{3}{4} \times 2\frac{3}{8}$ inches, oval, sub-acuminate, grossly and obtusely serrate, the apical tooth retuse, nearly smooth. Racemes axillary, with small white flowers in stalked clusters. Fruit (capsule) yellow, passing to red at the top, obovato-pyriform, tapering below into long neck (quasi-stipitate), at apex shortly rostrate, $1\frac{7}{16}$ inch long (neck $\frac{3}{8}$ inch, beak $\frac{1}{4}$ inch); pericarp thinnish, soft, glabrous externally, densely tomentose on the inner surface, 3-valved, but dehiscing along only two of the sutures, the third remaining closed, by abortion 1-celled, 1-seeded. Seed ovato-globose,

$\frac{11}{16}$ inch in diameter, black, polished, nearly half-immersed in a cupuliform white aril, with undulato-truncate mouth, which is seated on an obconical torus.

Humboldt's description of his *Paullinia Cupana* (*loc. cit.*) tallies with the above as to number, form, and cutting of leaflets, and the only difference is that the fruits are called "ovate," having probably been described from immature dried specimens, in which the true form of the fruit is apt to be disguised by the shrinking of the soft, half-formed seed and of its enclosing pericarp. I have, besides, seen with my own eyes that the Guaraná of Brazil and the Cupána of Venezuela are one and the same plant, which is cultivated in villages and farms all the way up the Rio Negro, and is known as Guaraná in the lower, but as Cupána in the upper part of that river; while about the line of demarcation between Brazil and Venezuela it is called indifferently by both names. The very same plant is cultivated also at Javita, and in the villages of the Atabapo and Orinoco, as far north as to the cataracts of the latter. I have nowhere seen it wild.

I gathered the following information about Guaraná at Santarem, on the Amazon, and at the mouth of the river Uaupés. The fruit is gathered when fully ripe, and the seeds are picked out of the pericarp and aril, which dye the hands of the operators a permanent yellow. The seeds are then roasted, pounded, and made up into sticks, much in the same way as chocolate, which they rather resemble in colour. In 1850, a stick of guaraná used to weigh from one to two pounds, and was sold at about 2s. 4d. the pound at Santarem; but at Cuyabá, the centre of the gold and diamond region, whither it was conveyed from Santarem and the Mauhés by the long and dangerous navigation of the Tapajoz, it was worth six or eight times as much. The usual form of the sticks was long oval or nearly cylindrical; but in Martius's time (1820) guaraná was "in panes ellipticos vel globosos formatum," and old residents at Santarem had seen it made up into figures of birds, alligators, and other animals. The intense bitterness of the fresh seed is almost dissipated by roasting, and a slight aroma is acquired. The essential ingredient of guaraná, as we learn from the investigations of Von Martius and his brother Theodore, is a principle which they have called guaranine, almost identical in its elements with theine and caffeine, and possessing nearly the same properties.

Guaraná is prepared for drinking by merely grating about a tablespoonful into a tumbler of water and adding an equal quantity of sugar. It has a slight but peculiar and rather pleasant taste, and it affects the system in much the same way as tea. I was told that at Cuyabá the thirsty miners used to resort to the tabernas, in the intervals of their toil, and call for a glass of guaraná, just as they would for one of lemonade, or of agoa doce. The brothers Martius strongly advocated the introduction of guaraná into the European pharmacopœias, and pointed out the maladies wherein its use seemed indicated. In South America I have frequently seèn it of late years exhibited in nervous affections, and it has even come to be regarded as a specific against the jaquéca (*i.e.* hemicrania) which is the fashionable ailment of a Peruvian lady. It has had the reputation of a remedy for diarrhœa, but I did not find it so, although I have tried it largely both on myself and others. The bitter unroasted seeds, as used in Venezuela, are probably more efficacious. The general notion on the Amazon was, however, that guaraná was rather a preventive of sickness, and especially of epidemics, than a cure for any, and Martius says of it "pro panacea peregrinantium habetur," which is precisely the estimate made of it in the south of Venezuela.

Guayúsa, a Tonic used in the Eastern Andes

Instead of Cupána or Guaraná, the Zaparos and Jibaros, who inhabit the eastern side of the Equatorial Andes, have Guayúsa, a plant of very similar properties, but used by them in a totally different way. The Guayúsa is a true Holly (Ilex), allied to the máte or Paraguay tea (*Ilex paraguayensis*), but with much larger leaves. I was unable to find it in flower or fruit, and cannot say if it be a described species. The tree is planted near villages, and small clumps of it in the forest on the ascent of the Cordillera indicate deserted Indian sites. The highest point at which I have seen it is at about 5000 feet above the sea, in the gorge of the Pastasa below Baños, on an ancient site called Antombós, a little above a modern cane-farm of the same name. There, in 1857, was a group of Guayúsa trees, supposed to date from before the Conquest, that is, to be considerably over 300 years old. They were not unlike old Holly trees in England, except that the shining leaves were much larger, thinner, and unarmed.

When I travelled overland through the forest

of Canelos, and my coffee gave out, I made tea of guayúsa leaves, and found it very palatable. The Jibaros make the infusion so strong that it becomes positively emetic. The guayúsa-pot, carefully covered up, is kept simmering on the fire all night, and when the Indian wakes up in the morning he drinks enough guayúsa to make him vomit, his notion being that if any food remain undigested on the stomach, that organ should be aided to free itself of the encumbrance. Mothers give a strong draught of it, and a feather to tickle the throat with, to male children of very tender age. I rather think its use is tabooed to females of all ages, like caapi on the Uaupés. Indians are not by any means so solicitous to empty the bowels early in the day as to clear out the stomach. On the contrary, all through South America I have noticed that when the Indian has a hard day's work before him, and has only a scanty supply of food, he prefers to go until night without an evacuation, and he has greater control over the calls of nature than the white man has. Their maxim, as an Indian at San Carlos expressed it to me in rude Spanish, is "Quien caga de mañana es guloso" (he who goes to stool in a morning is a glutton).

. . . . . .

From all that has been said, it may be gathered that the domestic medicine of the South American Indians is chiefly hygienic, as such medicine ought to be, it being of greater daily importance to preserve health than to cure disease. If their physicians be mere charlatans, their lack of skill may often be compensated by the ignorant faith of their patients; and their methods are scarcely more ridiculous—certainly less dangerous to the patient—than those of the Sangrados, Purgons, Macrotons, etc., portrayed by Lesage and Molière. If, to procure for himself fleeting sensual pleasures, the poor Indian's "untutored mind" leads him to sometimes partake of substances which are either hurtful in themselves or become so when indulged in to excess, examples of similar hallucination are not wanting even among peoples that boast of their high degree of civilisation.

This does not profess to be a treatise on all known South American narcotics, or I should have to speak of a vast number more, such as (for instance) the numerous plants used for stupefying fish. Some of these, but especially the Timbó-açú (*Paullinia pinnata*), are said to be also ingredients in the slow poisoning which some Amazonian nations are accused of practising; and on the Pacific side of the Andes the same is affirmed of the Yuca-ratón, which is the thick soft white root of a Leguminous tree (Gliricidiæ sp.) frequent in the plain of Guayaquil. The Curáre also would require a chapter to itself, and must be reserved for another occasion.

CHAPTER XXVI

THE WARLIKE WOMEN OF THE AMAZON: A HISTORICAL STUDY

[This essay was written by Spruce as an appendix to his chapter on the Trombetas river, near the mouth of which the early discoverers first encountered the fighting women. But as the evidence adduced by Spruce for their existence is spread over a large part of Amazonia, it seems better to give it here. By doing so I have been enabled to divide the present work into two volumes of nearly equal size, each dealing with a well-defined geographical area.]

The Women Warriors

I cannot dismiss the Trombetas without saying a few words about the warlike women whom Orellana affirmed that he encountered on his voyage down the Great River; the site of the encounter having been identified by subsequent travellers with the mouth either of the Trombetas or of the Nhamundá (called also the Cunurís), which is the next tributary of the Amazon to westward. It is of little moment to which river we assign it, when (according to Baena) the Nhamundá has two mouths, 14 leagues apart, and the lower mouth is but 6 leagues above the mouth of the Trombetas. That it was at no great distance above the mouth of the Tapajos is plain from Orellana's account that, two or three days after his fight with the "Amazons," he came to a pleasant country where there were Evergreen-oaks and Cork-trees (Alcornoques), the latter, as we have already seen, being the name the Spaniards still give to *Curatella americana*, and the former indicating probably the *Plumieria phagedænica*. (See vol. i. p. 67.) The country around Santarem is the only one which corresponds to this description throughout the whole course of the Amazon.

Orellana has been much ridiculed and called all sorts of hard names by people who have never taken the trouble to read his original Report to the Emperor Charles V., or the account of the voyage drawn up by F. Gaspar Carbajal, a Dominican friar who accompanied him. The voyagers heard rumours of the existence of the Amazons long before reaching them. Even before getting out of the Napo into the main river, we read that an Indian chief informed Friar Carbajal about the Amazons; and two hundred leagues below the mouth of that river, in the village where they built their brigantine, the friendly chief Aparia inquired of Orellana if he had seen the Amazons, whom in his language they called Coniapuyara (masterful women?). And when they actually encountered the real (or supposed) Amazons, what is their account of what befell them? That having landed at a place to traffic with the Indians, the latter attacked Orellana's party and fought bravely and obstinately. That ten or twelve women fought in front of the Indians, and with such vigour that the Indians did not dare to turn their backs. "These women appeared to be very tall, robust, and fair, with long hair twisted over their heads, skins round their loins, and bows and arrows in their hands, with which they killed seven or eight Spaniards." This is all that they profess to have seen with their own eyes of those warlike women; and, as Herrera remarks on it, "it was no new thing in the Indies for women to fight, and to use bows and arrows, as has been seen on some of the Windward Islands and at Cartagena, where they displayed as much courage as the men."

In the account of the return of Columbus from his second voyage we read that when he arrived at Guadeloupe (having started from Hispaniola), numbers of women, armed with bows and arrows, opposed the landing of his men. This is one instance, of many such, recounted by the Spanish historians.

I have myself seen that Indian women can fight. At the village of Chasuta, on the malos pasos of the river Huallaga, which in 1855 had a population of some 1800 souls, composed of two tribes of Coscanasoa Indians, the ancient rivalry of those tribes generally breaks forth when a large quantity of chicha has been imbibed during the celebration of one of their feasts. Then, on opposite sides of the village, the women pile up heaps of stones, to serve as missiles for the men, and renew them continually as they are being expended. If, as sometimes happens, the men are driven back to and beyond their piles of stones, the women defend the latter obstinately, and generally hold them until the men are able to rally to the combat. At that epoch there was no permanent white resident at Chasuta, and travellers who were so unfortunate as to be detained there during one of these fights were glad to keep themselves shut up until the stony storm had abated; and with reason, for there had been two instances, within a few years, of a white man being barbarously murdered by the Indians of Chasuta.

There is, therefore, no necessity for supposing that the Spaniards mistook men for women, either, according to the Abbé Raynal, because they were beardless, or, according to Wallace, because they were long-haired; for (1) American savages are generally beardless; and (2) the Spaniards had been for two whole years among Indians who wore their hair long, as they do to this day throughout the forest of Canelos, the scene of Orellana's wanderings with Gonzalo Pizarro; nay, the principal tribe among them, afterwards preached to by the most famous of the Quito missionaries and martyrs, F. Rafael Ferrer, were so notorious for the length to which they allowed their hair to grow as to have got the name of Encabellados. Moreover, on the Amazon itself, at the village of the chief Aparia, we read that "at this time four tall Indians came to the captain, dressed and adorned with ornaments, and with their hair reaching down to the waist."

As to the account given to Orellana by an Indian whom he captured some way farther down the river, about the whole country being subject to warlike women who were very rich in gold and silver, and had five houses of the sun plated with gold, while their own dwellings were of stone and their cities were fortified, Orellana merely repeats it as it was told to him, evidently, however, believing it himself; nor ought we to accuse him of credulity when we call to mind that he had lately left in Peru a reality in some respects more wondrous than this report. Herrera remarks very judiciously on it: "The tales of Indians are always doubtful, and Orellana confessed he did not understand those Indians, so that it seems he could hardly have made, in so few days, a vocabulary correct and copious enough to enable him to comprehend the minute details given by this Indian." I may add, too, that the Spaniards would probably ask as they went along for gold under its Peruvian name of cúri, and as curí (with merely a difference in the accent) is the Tupí term for coloured earth, it is not surprising that they should have received constant assurances of its abundance throughout the Amazon.

It is worthy to be noted that F. Carbajal, although he has left on record his dissatisfaction with the conduct of Orellana, confirms instead of contradicting the account of the combat with the Amazons, having, in fact, been himself one of the wounded in it. Besides, as is well remarked by Velasco (*Historia de Quito*, i. 167), "he (Orellana) did not go alone to the court, but with fifty companions, many of them so disgusted with his conduct that they refused to accompany him on his return. He was giving information to his sovereign, who might utterly ruin him if he detected him in a falsehood, and it ought to have been easy to detect him, with so many witnesses unfavourably disposed towards him. Besides, it is incredible that fifty persons, and amongst them a religious priest, should agree in guaranteeing the truth of a lie, especially when nothing was to be gained by it."

We have also a very good and independent

account of this voyage from Gonzalo Fernandez de Oviedo, who was in the Island of St. Domingo when Orellano touched there on his way to Spain, in the ship he had purchased in the Isle of Trinidad. Oviedo relates what he was told by Orellana's companions, and it corresponds in all essential points with the navigator's own narrative; with the important addition that the women fought naked to the waist, and that they had *not* one of the breasts cut off, like the Asiatic Amazons—a question Oviedo had particularly asked of the Spaniards.

The little I had read before leaving England about the existence of a nation of women living apart from men, somewhere in the interior of South America, threw ridicule on the notion, and attributed its origin to lying Spanish chroniclers, so that I confess to have not thought it worth while to make a single inquiry on the spot as to whether the tradition were still extant; but when I afterwards came to read carefully the relations of those authors who had bestowed most attention on the subject, I was surprised to find them all agreed on the tradition having been based on fact. I allude especially to Acuña, Feijoo, Condamine, Velasco, Southey, and Humboldt; but it is nowhere more fully discussed than in a small treatise by Van Heuvel entitled *El Dorado*, to which, and to the writings of the celebrated authors just mentioned, I must refer the reader.

The ways by which the country of those women might be reached, as related by travellers and missionaries, seem to converge not to one, but to two points; the one to northward of the Amazon, a good distance below the Rio Negro; the other to southward of it, above the Rio Negro, and somewhere between the rivers Coari and Teffé. In the very year of Orellana's encounter with the Amazons (1541), Cabeza de Vega headed an expedition which ascended the Plata and the Paraguay in search of gold. From the latter river he sent Hernando de Ribeiro ahead, in a brigantine, with fifty-two men, to explore the lake of Xarayes, a large tract of country periodically inundated, lying to eastward of what was afterwards the Province of Moxos. From the Xarayes Indians Ribeiro received information of the Amazons, whose country he was told lay two months' journey to the northward; and, disregarding the warning of the chief of the Xarayes, that it would be impracticable to traverse the forests at that season of floods, he and his party proceeded on foot for eight days, with the water up to their middle. This brought them to another nation, the Siberis; and a journey thence of nine days (the first four being still wading through the water) to the nation of Urtueses, who told them there was yet a month's journey to the Amazons, with much flooded ground to traverse. From this point they were compelled to regress by their provisions giving out; and the plantations of the Urtueses having been devastated for two successive years by some insect, no more food was to be had; but those Indians reiterated the assurance of the existence of a nation of women, governed by a woman, and possessing plenty of both white and yellow metal, their seats and utensils being made of them. They lived on the western (eastern ?) side of a large lake, which they called the Mansion of the Sun, because the sun sank into it (Southey's *History of Brazil*, pp. 156-159).

Towards the close of the sixteenth century, F. Cyprian Bazarre, a Jesuit missionary to the Tapacura Indians (a tribe of Moxos), heard accounts similar to those related by Ribeiro, tending to place the Amazons in the country lying southward of the Great River and westward of the Purús, or very nearly where Condamine many years afterwards (in 1741) heard such circumstantial accounts of them. This traveller spoke at Coari with an Indian whose grandfather had met a party of those women at the mouth of the river Cuchinará (now the Purús). " Elles venoient de celle de Cayamé, qui débouche dans l'Amazone du côté du Sud entre Tefé et Coari ; qu'il avoit parlé à quatre d'entr'elles, dont une avoit un enfant à la mamelle: il nous dit le nom de chacune d'elles ; il ajouta qu'en partant de Cuchinará elles traversèrent le Grand Fleuve, et prirent le chemin de la rivière Noire. . . . Plus bas que Coari, les Indiens nous dirent partout les mêmes choses avec quelques variétés dans les circonstances ; mais tous furent d'accord sur le point principal." For many other details, tending to the same conclusions, I must again refer the reader to the original.

The numerous missionaries on the Amazon during the seventeenth and eighteenth centuries all testify to the same traditions. It was no uncommon thing, they say, for Indians in confession to accuse themselves of having been of the number of those who were admitted to visit periodically the women living alone. Their testimony may be summed up in the words of an old Indian at San Regis de los Yameos (a village on the left bank of the Amazon above the mouth of the Ucayáli), as delivered to the priest F. Sancho Aranjo, who was Condamine's host when he passed that way, and who afterwards repeated them to F. Velasco.

1. That respecting the first combat the Spaniards had had with the warlike women, there was no one in all the missions who did not know of it by tradition from father to son.

2. That he had heard his forefathers say those women had retired far into the interior, across the Rio Negro.

3. That, according to common report, they still existed, and that some Indians visited them every year, but not in their proper country; for the women always met the men at some place previously agreed on a long way from their homes (whose site the men were not permitted to know), and, after conversing with them as long as they listed, dismissed them with presents of gold and green stones, and of the male children that had been born and had reached the age of two or three years.

4. That these women were always governed by one, chosen on account of her valour, and who always marched to battle at their head (Velasco, *loc. cit.* p. 173).

The green stones spoken of here and elsewhere —called also Amazon stones—were formerly met with among nearly all the Indians of Tropical America, but seem now to have totally disappeared from the Amazon. I, at least, never either saw or heard of one there in the hands of the Indians; nor is that to be wondered at when we recollect how eagerly they were at one time bought up by Europeans on account of their supposed medicinal virtues. At the beginning of the present century we learn from Humboldt that the price of a cylinder two inches long was from twelve to fifteen dollars in Spanish Guayana. He obtained a few of them from the dwellers on the Upper Rio Negro. According to Condamine they were once common articles on the site of the modern Santarem. "C'est chez les Topayos qu'on trouve aujourd'hui, plus aisément que partout ailleurs, de ces pierres vertes, connues sous le nom de Pierres des Amazones, dont on ignore l'origine, et qui ont été fort recherchées autrefois, à cause des vertus qu'on leur attribuoit de guérir de la Pierre, de la Colique néphrétique et de l'Épilepsie" (*Voyage*, p. 137). Even to this day their origin is doubtful, for it is said that no jade of the same kind as these stones has been found anywhere in South America, although it exists in Mexico. The notable thing about them is that the South American Indians in whose hands they have been seen by Europeans all agreed in asserting them to be obtained from the women without husbands, or, on the Orinoco, from the women living alone (Aikeambenanos in the Tamanac language, according to F. Gili).

Velasco cites also a conversation he had with a friar, F. José Bahamonte, who had been for forty years a missionary on the Marañon, to the effect that, being in 1757 in the village of Pevas, shortly after the Portuguese garrison of the fort of the Rio Negro had mutinied against their commandant, "those deserters, having left the major nearly dead and pillaged the warehouses and the royal treasury, fled up the Marañon, and reached Pevas a few at a time. Some of them remained in the mission; others went on to Quito. With one of those parties there arrived at my village a very good-looking Indian of about sixty, inquiring for the nation of the Pevas and speaking their language, and yet not known to anybody there. After a while he came to me and besought me to hear in secret the motive of his coming thither. Having taken him apart, where we could be overheard of no one, he prostrated himself at my feet, and earnestly entreated me to receive him into my village and make him anew a Christian. I asked him if, being baptized, he had denied the Christian faith. He said no, but that, although he was already a Christian, he had always lived like a heathen." The Indian then tells his story in full to the priest; how he was a Peva by birth, and had been baptized at the mission when young; but that, as he grew up, having taken a great dislike to the severe discipline of the mission, he had fled from it down the Amazon, and finally established himself in a village on the river Teffé. There he was recommended by an Indian to enter on the office of one lately deceased who used every year to visit the women without husbands. Having followed this employ for thirty years, and received from the women many presents of gold and green stones, he was obliged to relinquish it on account of an injury he received, and also (as he asserted) by a remorseful conscience which continually tormented him. "The death of this Indian," adds the good missionary, "a few months afterwards, having lived during that period a penitent and holy life, was one of the greatest consolations that befell me in the missions, for I felt convinced, from his good conduct, that he was predestinated" (Velasco, *loc. cit.* p. 175).

The accounts heard by Raleigh on the Orinoco, in 1595, of a nation of female warriors existing on the Amazon, seem to combine both the above-specified sites. "I made inquiry," says he, "among the most ancient and travelled of the Orinokoponi [the Indian inhabitants of the Orinoco] respecting the warlike women, and will relate what I was informed of as truth about them, by a Cacique who said he had been on that river [the Amazon], and beyond it also. Their country is on the south side of the river, in the province of Tobago [Topayos], and their chief places are in the islands on the south side of it, some 60 leagues from the mouth. They accompany with men once in a year for a month, which is in April. . . . Children born of these alliances, if males, they send them to their fathers; if daughters, they take care of them and bring them up,"[1] etc. Another report he heard was that "there is a province in Guyana called Cunurís, which is governed by a woman"—plainly a Cuñá-puyára. It is to be noted that these reports were heard near the mouth of the Orinoco, or some 2000 miles away from the supposed country of the Amazons, from Indians who had them from one another and not from the Spaniards; and that the Cunurís is for the first time indicated by name in this relation of Raleigh's. We have the most complete account of the river and district of Cunurís, and of the extant traditions respecting the Amazons, in Acuña's description of his voyage down the Amazon in 1639. He mentions four nations who inhabit on the river Cunurís, the Cunurís (Indians) being nearest the mouth, and the Guacarás the highest up; while

[1] Cayley's *Life of Raleigh*, pp. 194-195.

beyond the last were the Amazons. "These man-like women," he says, "have their abodes in great forests and on lofty hills, amongst which that which rises above the rest, and is therefore beaten by the winds for its pride with most violence, so that it is bare and clear of vegetation, is called Yacamiaba" (*Yacamí*, the Tupí name of the Trumpeter bird or Agamí; *Aba* or *awa*, people).

When I read this account of Acuña's, some years after I had left the Amazon, I was struck with the connection of the name of the hill Yacamiaba with that of an Indian dance I had seen on the Upper Amazon in 1851. The dance was called Yacamí-cuñá (Agami woman), and the performers in it moved to the rude music of a pipe and tambour; and to the words of a song, which I unfortunately neglected to take down at the time. A lot of young people joined hands to form a ring, in which males and females alternated, and danced round and round, singing the song of the Yăcamí. At the words "Yacamí-cuñá-cuñá!" the ring suddenly broke up—the partners turned tail to tail and bumped each other repeatedly, with such goodwill that one of the two (and as often the man as the woman) was frequently sent reeling across the room, amidst the uproarious laughter of the bystanders. The Yacamís or Agamís are, as is well known, birds without any tail-feathers, those appendages having diappeared from the birds continually rubbing their sterns together—so, at least, says Indian tradition, which has been embodied in the dance; and it is easy to understand its application to a rocky hill, shaggy below with woods, bare at the summit, such as I have seen many in both Brazilian and Spanish Guayana.

May not also both the names, Yacamí-women and Yacamí-people, allude to the women living alone?

Van Heuvel met with a Caribi chief at the head of the river Essequibo, who, when asked about the nation of women, said "he had not seen them, but had heard his father and others speak of them. That they live on the Wasa [the Ouassa of the French maps, a tributary of the Oyapock]. Their place of abode is surrounded with large rocks, and the entrance is through a rock" (*El Dorado*, p. 124).

Condamine was informed by a soldier in the garrison of Cayenne, that in 1726 he had accompanied a detachment which was sent to explore the interior of the country; in pursuance of which object they had penetrated to the country of the Amicouanes, a long-eared people, who dwell beyond the sources of the Oyapock, near to where another river takes its rise that falls into the Amazon [the Oyapock falling into the Atlantic in lat. about 4° N.]. The country lies high, and none of the rivers are navigable. There the soldier had seen on the necks of the women and girls certain green stones, which the Indians said they obtained from the women who had no husbands (*Voyage*, p. 102).

We have mention of the long-eared folk, and of the same kind of savage rocky country as all tradition assigns to the abiding-place of the Amazons, in Unton Fisher's relation of his voyage up the Mariwin (Marony). "The passage to the head of the Mariwin, from the men with long ears (which is the thirteenth town from the mouth), is very dangerous, by reason of the passage through hollow and concave rocks, wherein harbour bats of unreasonable bigness, which, with their claws and wings, do wound the passengers shrewdly; yea, and oftentimes deprive them of life."[1]

Van Heuvel cites various accounts which he found still current in Guayana, all tending to collocate the warlike women on a site just beyond the sources of the Essequibo, Marony, and Oyapock, which lie apparently very near to each other, and also to the sources of the Trombetas and Nhamundá, the two latter rivers running in a contrary direction to the three former, *i.e.* southwards, or towards the Amazon.

I might adduce a great deal more evidence to show the universality of the traditions in Tropical America of a nation of women, whose permanent habitation was from 1° to 2° north of the Equator, and in long. 54° to 58° W.; and whose annual rendezvous with their lovers was held on a site in lat. about 5° S., long. 65° W.

Those traditions must have had some foundation in fact, and they appear to me inseparably connected with the traditions of El Dorado. I think I have read nearly all that has been written about the Gilded King and his city and country; and, comparing it with my own South American experience, I can hardly doubt that that country was Peru—possibly combined (or confused) with Mexico. The lake called the Mansion of the Sun, because the

[1] The whole of this curious relation is given in Purchas's *Collection of Voyages*, Bk. vi. ch. xvii., and is placed immediately after that of the voyage made by Robert Harcourt to Guayana in 1608. Purchas says of it: "I found this fairly written among Mr. Hakluyt's papers, but know not who was the author." But Van Heuvel adduces ample proof of its having been written by Fisher, cousin of Harcourt, whom the latter left behind him at the third town on the Mariwin, with instructions to complete the exploration of the river, which he himself had unsuccessfully attempted.

sun sank into it, is plainly the Pacific Ocean; but some accounts seem to point to Lake Titicaca, and others to the lakes of Mexico; probably the general notion of such lake was made up of all three. It is scarcely necessary to remind the reader that most Indian nations call the ocean and a lake (and in some cases even a river) by one and the same name. The confusion of town (or city) and country is also universal among them. I have been gravely told by a Jibaro Indian in the Andes that France and England were two towns, standing on opposite banks of a river, the people on the left bank being Christians and those on the right heathens: a piece of ethnology derived from the teaching of Catholic missionaries, and not at all flattering to myself as an Englishman.

I think I can trace the progress of the fame of the riches of Peru quite across South America, to the Atlantic coast and islands, whence it surged back into the interior, so disguised and disfigured, that the Spaniards did not recognise it as indicating an El Dorado with which they were already familiar. Now the accounts of the real El Dorado of Peru (and of Mexico) would infallibly be accompanied by others of the Vestal communities dedicated to the worship of the sun, *i.e.* of women living alone, or women without husbands. If we deny the existence of a nation, or nations, of warlike women on the Amazon, then the tradition could only have had its origin in the Virgins of the Sun; and some accounts, such as that of Cabeza de Vega and Ribeiro, possibly point to them alone. But if we concede the fact of the existence of these warlike women, then may not the latter have been originally a community of Vestals, who, having fled in a body from their nunnery, carrying with them their ornaments of gold and green stones, established themselves in the forests of the plain? Or they may have accompanied one of those emigrations, led by chieftains who had revolted from the rule of the Inca, of which we read in the early historians. In either case they were probably at first respected by neighbouring savage tribes as a religious community; and they would gradually learn the use of the bow and other weapons, more as implements of the chase than of offence and defence; for we do not read that they were ever assaulted by other Indians. I put forward this as mere conjecture, my object in what precedes having been principally to vindicate the earlier travellers and historians, Spanish and English, from the charges of gross credulity, or even wilful falsehood, which have been wantonly brought against them. Is it to be wondered at that unlettered, or at best imperfectly educated, adventurers should have believed, and repeated as true, nearly every report they heard, when we find a man of so philosophic a turn of mind as Raleigh telling the most extravagant tales—just as they were told to him, no doubt, and not adding anything thereto, yet evidently believing them himself in the main?

No one has declared his convictions of the existence of a nation of Amazons more forcibly and eloquently than Acuña, and, without endorsing them fully myself, I close this long digression with his own words, recommending them to the candid consideration of my readers:—

"The proofs that give assurance that there is a

province of the Amazons on the banks of this river are so strong and convincing that it would be renouncing moral certainty to scruple giving credit to it. I do not build upon the solemn examinations of the sovereign court of Quito, in which many witnesses were heard, who were born in these parts and lived there a long time, and who, of all matters relating to the countries bordering on Peru, as one of the principal, particularly affirmed that one of the provinces near the Amazon is peopled with a sort of warlike women, who live together and maintain their company alone, without the company of men; but at certain seasons of the year seek their society to perpetuate their race. Nor will I insist on other information, obtained in the new kingdom of Grenada, in the royal city of Pasto, where several Indians were examined; but I cannot conceal what I have heard with my own ears, and concerning the truth of which I have been making inquiries from my first embarking on the Amazon; and am compelled to say that I have been informed at all the Indian towns in which I have been, that there are such women in the country, and every one gave me an account of them by marks so exactly agreeing with that which I received from others, that it must needs be that the greatest falsehood in the world passes through all America for one of the most certain histories."[1]

[1] *Voyages and Discoveries in South America*, by Christopher d' Acugna, London, 1698.

CHAPTER XXVII

INDIAN ROCK-PICTURES: ENGRAVED ROCKS ON THE RIO NEGRO AND CASIQUIARI (COMMONLY CALLED INDIAN PICTURE-WRITING)

[WHILE residing at Piura on the sea-coast of Peru in 1863, and being incapacitated by illness for outdoor work, Spruce wrote out a description of these curious works of art illustrated by the drawings he was able to make of some of them, and with an explanation of their meaning given him by the Indians who were with him and to whom they were familiar. He also gives his own view as to their probable age, and as to the causes that led to their production. In this paper he does not refer to the best known of these Picture-writings on the rocks of Pedra Island, near the mouth of the Rio Branco, which are briefly described in his Journal. (See vol. i. p. 260.) This paper refers solely to the examples of which he made drawings on the Casiquiari and Uaupés rivers.]

INDIAN PICTURE-WRITING[1]

When I ascended the Casiquiari in December 1853, I charged my pilot, an intelligent Indian of

[1] In his Journal (1851), when describing the figures on Pedra Island (Lower Rio Negro), he protested against the use of the term "picture-writing" as conveying the erroneous idea that they are in any sense writings or hieroglyphics. Twelve years later he uses the popular term, though showing that it is an incorrect one.

the Barré nation, to point out to me any engraved rocks which lay in our way. On reaching the Pedra de Culimacari, a bed of granite a little beyond the mouth of the Pacimoni, we found it still under water, so that the figures seen there and copied by Humboldt in the beginning of the century were not visible. The pilot consoled me by saying that when we reached the Laja de Capibara he would show me there ten times more figures than I had missed seeing at Culimacari. On the 9th of December we passed the mouth of Lake Vasiva, and on the 11th reached a modern Indian village called Yamádubani (Wild Man's Land), or more commonly Pueblo de Ponciano, having been founded by an Indian named Ponciano, who was not long dead. Early on the morning of the 13th we came upon the deserted site of another village called Capibara, being the *nom de guerre* of its founder, after whose death it has become depopulated. It is on the left (S.E.) side of the Casiquiari. Leaving here part of the crew to cook our breakfast, I took with me the rest, and under the guidance of the pilot struck into the forest in quest of picture-writing. After walking about half a mile, we came out on large flat sheets of granite rock, naked save where in fissures of the rock there were small oases of vegetation, the first plants to establish themselves there being a few lichens and mosses, and, rarely, some stunted shrubs. The bare places, one of which was an acre in extent, were covered with rude figures, the outlines of which were about half an inch wide, and were graven in the rock to nearly an inch deep. The figures were in perfect preservation except that in rare cases they were obliterated by the shaling of the rock, the granite of that region having often three or more thin coats comparable to those of an onion, as if the cooling down had not been equable.[1] I immediately set to work to copy, and the Indians of their own accord cleared out the earth and lichens which had filled up some of the lines. As it was impossible to copy all, I selected those figures which were most distinct, and those which, by their frequent repetition, might be considered typical. That marked A (Fig. 17), for instance, varying only slightly in the details, was repeated several times. It was not possible to draw all by hand to the same scale, but as I measured most of the figures, that defect can easily be remedied in recopying them.

In all the drawings which illustrate this chapter, the small figures give the dimensions in feet and inches. When underlined they show the entire length of the object copied, as 3/10 in the centre figure of Fig. 17 means that it is 3 feet 10 inches long; otherwise they indicate the length of the line at which they are written. Thus 2/5 on the right side of A shows that the longer side of the oblong is 2 feet 5 inches long, and the cross line on the right is 4 feet long.

As I sketched, I asked the Indians, "Who had made those figures, and what they represented?" but received only the universal reply of the Indian when he cares not to tell or will not take the trouble to recollect, "Quien sabe, patron?" ("Who knows?"). But I understood enough of Barré to note that in

[1] [For drawings of such onion-like rocks see Plate x. in my *Amazon and Rio Negro*. It occurs on every scale from that of moderate-sized boulders up to whole mountains. It is seen on a great scale in the huge domes of the Yosemite valley, and is now believed to be the result of a process of aerial decomposition due to the action of sun and rain.—ED.]

their talk to each other they were saying, "This is so-and-so, and this so-and-so." "Yes," I struck in, "and don't you think this is so-and-so?" Thus led on, I got them to give their opinion of most of the figures. About some they were quite certain; about

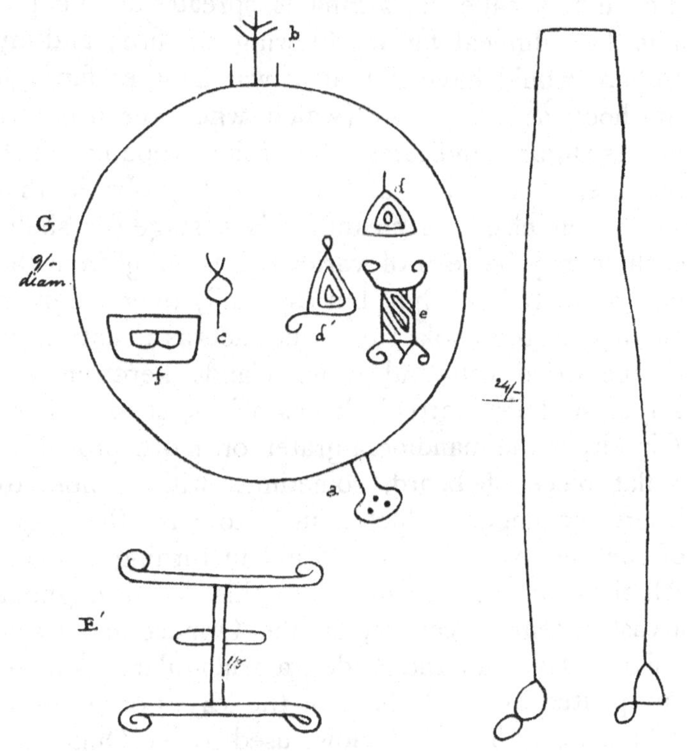

FIG. 16.—GROUP OF PICTURES AT LAJA DE CAPIBARA, RIVER CASIQUIARI.

others they would only speculate. Of all the figures the one marked G (Fig. 16) was that whose origin seemed clearest both to them and to me. It represents a mandiocca-oven (called budári in Barré)—a large circular dish of fireproof pottery, supported on a wall of mud-masonry, which has an opening on one side (rudely figured at *a*), into which fire is put, and another at the opposite (as at *b*), which serves as a flue. Of the articles laid on the budári, *c* is the brush of piassaba tied tightly round at midway, which serves for sweeping the oven before the cassava cake or farinha is spread out to bake; *d* is the palm-leaf fan for blowing the fire; and my Indians would have it that *d'* was another fan, but the hook at one corner (which, whenever it occurs in these figures, indicates a bit of liana-rope by which the utensil is hung up) renders it probable that something else was meant; *e* is a stage (or shelf) such as may be seen of various sizes hung from the roof of an Indian's hut, but especially over the oven and hearth, the smoke from which acts as an antiseptic to the dried fish and other viands kept on the stages, and also partially keeps off the cockroaches; *f* is either the mandiocca-grater or, more probably, a flat piece of board, sometimes with a hole to insert the fingers, which is used to raise the edges of the cassava cake and to aid in turning it over. All these articles are in use to this day throughout a vast extent of country on the Orinoco and Casiquiari. Even in the Andes, a triangular or square fan, plaited by the Indians of the leaves of maize or wild cane, is the only bellows used by the Quitonian housewife.

The figures marked B (Fig. 17) were declared by my Indians to be dolphins, whereof two species abound in the Amazon and Orinoco.

C they said was plainly the same sort of thing as the big papers (maps) I was continually poring over. For *a* is the town—often consisting of a single annular house, with a road from it leading down to the caño (or stream leading into the main river, *c*), while *b* is a track leading through the forest to another tributary stream which here and there expands into lakes, while other lakes send their waters to it. There were other figures apparently geographical, but the one I copied was the most complicated and perfect.

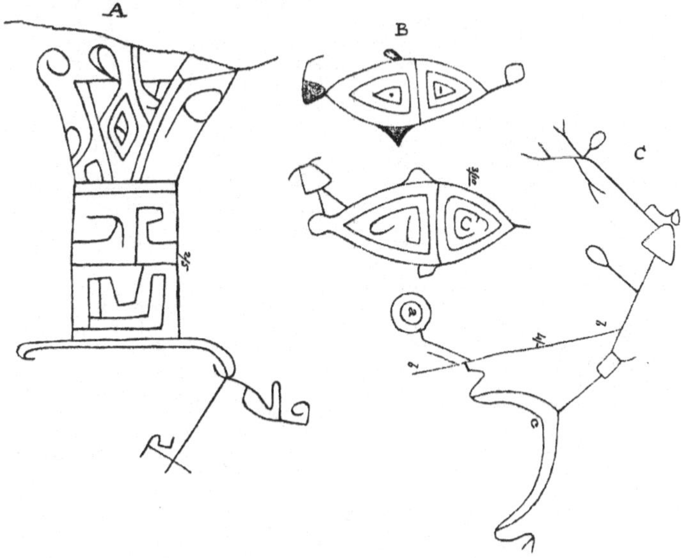

FIG. 17.—GROUP OF PICTURES AT LAJA DE CAPIBARA, RIVER CASIQUIARI.

D (Fig. 18) are ray-fishes, which are found of enormous size in the Casiquiari and Rio Negro, and sometimes inflict deadly wounds on incautious bathers.

E on Figs. 16 and 18 and perhaps A on Fig. 17 was thought by my companions to be the quiver for holding the darts of the blowing-cane.

By the time I had covered three sheets with figures, the sun began to beat hot on my head, protected by only a light cap, and although my pilot told me that farther away in the forest there were more granite sheets covered with pictures, I was obliged to content myself with what I had

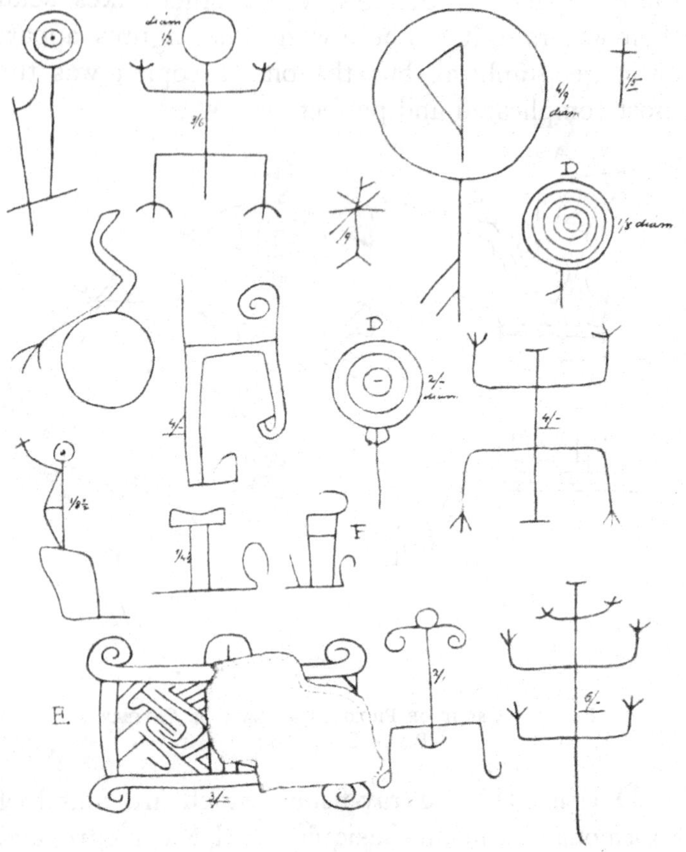

FIG. 18.—GROUP OF PICTURES AT LAJA DE CAPIBARA, RIVER CASIQUIARI.

already seen and done; for I had engaged to meet the Comisario of San Fernando at Esmeralda on Christmas Day, and to get there I had still a long voyage before me, going slowly along as I did in my large boat and gathering plants all the way.

A few miles from the upper mouth of the Casiquiari a stream called Calipo enters it where there is some picture-writing that was covered with water when I passed up; but when I returned (on January 6, 1854) the Casiquiari had lowered 2 feet, and at the mouth of the Caño Calipo a good many figures were laid bare, all of which I copied. The figures on Fig. 19 have the same relative positions and distances as on the rock, and apparently

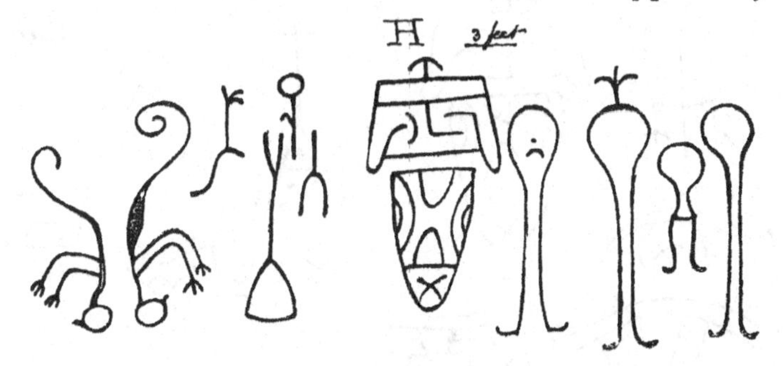

FIG. 19.—GROUP OF PICTURES ON RIGHT BANK OF THE CASIQUIARI, A LITTLE ABOVE THE CAÑO DE CALIPO.

represent a family group, whereof my interpreter assured me that H symbolised a chief, and that the figures on the right were his three wives and a child, the principal wife being distinguished by the plume worn on her head. The curious figures on the left may perhaps be meant for the prehensile-tailed Iguanas, which being very good food would be of especial interest.

The other group (Fig. 20) repeats the symbol of a chief (at H H), with some four-footed animal, perhaps a dog, on the left. The rest are probably household goods of some kind.

Picture-writing is frequent throughout the granite district of the Casiquiari, but I have nowhere seen so much of it together as at the Laja de Capibara. The best executed figures, however, I have met with, and the only ones about which I could make out any extant tradition, are in the river Paapurís, which enters the Uaupés from the south at Jauarité caxoeira, and is inhabited by Fish and Mosquito Indians (Pira-Tapuyas and Carapanás). The Paa-

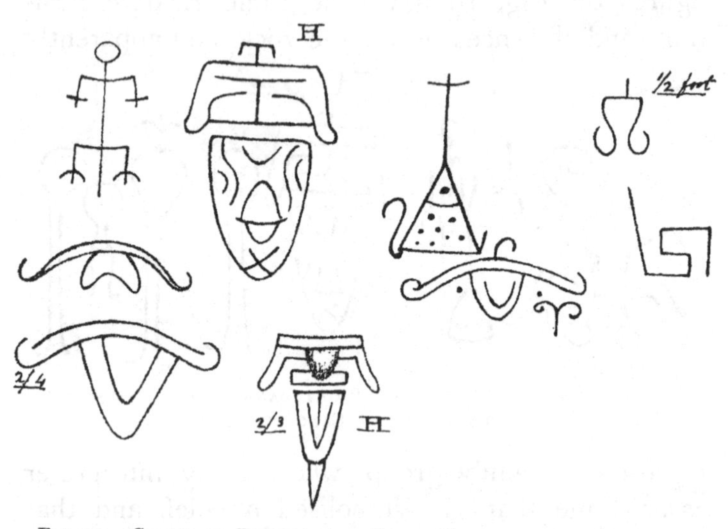

FIG. 20.—GROUP OF PICTURES ON RIGHT BANK OF THE CASIQUIARI, A LITTLE ABOVE THE CAÑO DE CALIPO.

purís in its lower part is an uninterrupted and dangerous rapid; and at Aracapá caxoeira, a few miles up, two islands divide it into three narrow channels, each of which is a nearly perpendicular cascade of about 15 feet high. At this point canoes have to be unladen and dragged over one of the islands, which are masses of granite having on them much picture-writing, where not clad with shrubs. The most distinct figures are on the top of a rock which rises perpendicularly by the highest fall, and cannot be reached without risk. They were engraved by a young woman who was lamenting the death of her mother, for whose epitaph they were probably intended. Day by day she sat on the rock engaged in her task, while her fast-falling tears ceased not to mingle with the cataract. Thus months passed away, until one day the maiden, worn with grief and fading almost to a shadow, fell over the rock and disappeared among the roaring breakers at its base.

I had not with me pencil or paper of any kind, and I was obliged to content myself with a hasty glance at the figures, some of which represented human beings; nor was I able to revisit the spot. On the top of the same rock there are shallow impressions, apparently the work of nature, which bear some resemblance to a human form, and are called by the Indians Tupana-rangaua (the figure of God). The damsels of the Paapurís visit the spot on stated occasions, and kneeling down on the knees of the figure, perform some kind of devotion —what, I could not learn.

I copied a few rude figures on the rocks near the village of Jauarité. Those on Fig. 21 seem to represent very rudely various types of trees, as seen in the three figures on the right. The two upper ones indicate a buttressed stem or aerial roots, with flowers or fruits on the three terminal branches; while the lower one has a tap-root, and diverging branches of a more usual type. The lower middle figure is probably the very rudest symbol of a human form; while the remainder seem to be merely fanciful geometrical patterns.

The large figure on Fig. 22 is called by the Indians

the buta or dolphin. On these and other rocks

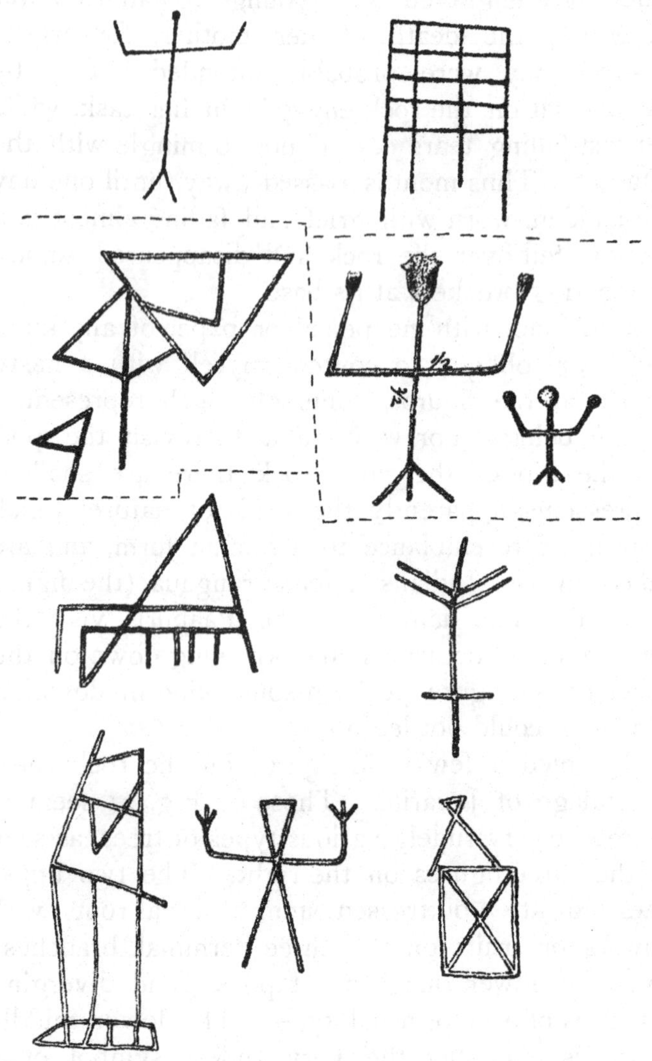

FIG. 21.—GROUP OF PICTURES AT JAUARITÉ CAXOEIRA, RIO UAUPÉS.

of the Uaupés there are impressions called Pé de Anta (Tapir's foot), which look as if some three-toed foot had trod on the rock while still soft; but they are scattered, not consecutive. It is not so

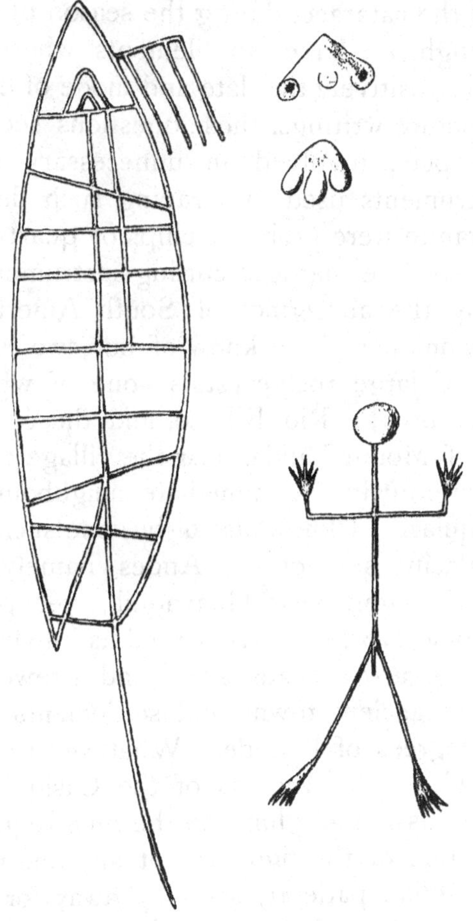

FIG. 22.—GROUP OF PICTURES AT JAUARITÉ CAXOEIRA, RIO UAUPÉS.

easy to explain these by natural causes as it is that of the panellas or pots, which are cylindrical holes frequently met with on the rocks of the falls of the Rio Negro and Uaupés; these have been worn—

from any accidental hollow at first—and then continually deepened by the pebbles and sand whirled round and round in them by the surging and eddying waves of the cataracts during the season of flood.[1]

Although we have no elements wherefrom to determine positively the date and mode of execution of the picture-writings, those questions seem to me to have been involved in unnecessary mystery. The instruments used in scraping such deep lines in the granite were probably chips of quartz crystal, which were the hardest cutting-instruments possessed by the aborigines of South America. In the Amazonian plain I know of but two extensive deposits of large rock-crystals—one of which is a good way up the Rio Branco, and the other is at the foot of Mount Duida, near the village of Esmeralda, therefore in the immediate neighbourhood of the Casiquiari. I know also of but one such deposit on the Pacific side of the Andes, namely, in the hills of Chongon near Guayaquil; yet pieces of quartz, some of which have served as knives, others as lance- or arrow-heads, are found strewed about the sites of ancient towns and settlements through several degrees of latitude. Whatever the instrument used by the Indians of the Casiquiari, it is difficult to assign any limit to the time required for the execution of the figures; but any one who has seen an Indian patiently scraping away for months at a bow or a lance before bringing it to the desired symmetry and perfection, or who knows that it has taken a lifetime to fashion and bore the white

[1] [The supposed tracks of animals are doubtless works of art like the other figures, probably due to a desire to imitate the well-formed impressions of feet that the hunter must continually meet with during his search for game.—Ed.]

stone which the Uaupés Indian wears suspended from his neck, will understand that *time* is no object to an Indian. I can fancy I see the young men and women sitting in the cool of the morning and evening, but especially in the moonlight nights, and amusing themselves by scratching on the rock any figure suggested by the caprice of the moment. A figure once sketched, any one, even a child, might aid in deepening the outlines. Indeed, the designs are often much in the style of—certainly not at all superior to—those which a child of five years old in a village school in England will draw for you on its slate; and the modern inhabitants of the Casiquiari, Guainia, etc., paint the walls of their houses with various coloured earths in far more artistic designs.

Having carefully examined a good deal of the so-called picture-writing, I am bound to come to the conclusion that it was executed by the ancestors of Indians who at this day inhabit the region where it is found; that their utensils, mode of life, etc., were similar to those still in use; and that their degree of civilisation was certainly not greater—probably less—than that of their existing descendants. The execution of the figures may have ranged through several centuries, a period which in the existence of a savage people is but a year in that of the highly-civilised nations of modern Europe. In vain shall we seek any chronological information from the Indian, who never knows his own age, rarely that of his youngest child, and who refers all that happened before his own birth to a vague antiquity, wherein there are no dates and rarely any epochs to mark the sequence of events.

[Among Spruce's miscellaneous notes, written during his voyage up the Rio Negro, the following passages serve to illustrate the questions above discussed:—]

I have never yet met with an Indian who knew his own age or how many years he had lived in his present house. My pilot on the Trombetas very gravely stated his age at a hundred years (he was evidently not more than fifty). I have asked an Indian the age of his daughter. "She may be twelve—she may be twenty—who knows? What matter do our ages make to us?"

. . . .

These picture-writings in Brazil and Spanish Guiana cannot be considered of remote antiquity, for (1) they sometimes show rude figures of lions and other objects belonging to the Old World; (2) some of them (and especially the Brazilian ones, *e.g.* at Monte Alegre, as stated by Mr. Wallace) have dates affixed, painted with the same colour and obviously of the same age as the pictures, which correspond very nearly with the dates of the establishment of the Portuguese towns of the Amazon, and not going back above a century or two.

CHAPTER XXVIII

A HIDDEN TREASURE OF THE INCAS

[THE following narrative forms one of the most curious pieces of genuine history in connection with the never-ceasing search for buried treasure in the territory of the Incas. We owe to the persevering exertions of Richard Spruce the discovery and the translation of one of the few remaining copies of the official order of the Spanish king to search for this treasure, with the accompanying detailed "Guide" to its locality. Still more are we indebted to his generally esteemed character and ingratiating manners for obtaining permission to copy the unique map of the district containing the treasure, and for undertaking the considerable labour of copying in the minutest detail so large and elaborate a map, without which both the "Guide" and the story of the search for the treasure would be unintelligible.

The essential portions of this map, containing the whole of the route described in the "Guide," as well as the routes of the various explorers (marked in red), have been reproduced here (see end of chapter). The portions farther east and south, which have no immediate relation to the quest for the treasure, having been omitted in order to make it more convenient for reference here. The scale of the map is, approximately, six miles to an inch.

In Dr. Theodore Wolff's *Geografia et Geologia de Ecuador* (1892), the region of Llanganati is still referred to as the most unknown part of the whole of Ecuador.]

A Hidden Treasure of the Incas, in the Mountains of Llanganati, Ecuador; an authentic Guide to its Locality; illustrated by a Map. The Map copied and the Guide translated by Richard Spruce

In the month of July 1857 I reached Baños, where I learnt that the snowy points I had observed from Puca-yacu, between Tunguragua and Cotopaxi, were the summits of a group of mountains called Llanganati, from which ran down to the Pastasa the densely-wooded ridges I saw to northward. I was further informed that these mountains abounded in all sorts of metals, and that it was universally believed the Incas had deposited an immense quantity of gold in an artificial lake on the flanks of one of the peaks at the time of the Spanish Conquest. They spoke also of one Valverde, a Spaniard, who from being poor had suddenly become very rich, which was attributed to his having married an Indian girl, whose father showed him where the treasure was hidden, and accompanied him on various occasions to bring away portions of it; and that Valverde returned to Spain, and, when on his death-bed, bequeathed the secret of his riches to the king. Many expeditions, public and private, had been made to follow the track indicated by Valverde, but no one had succeeded in reaching its terminus; and I spoke with two men at Baños who had accompanied such expeditions, and had nearly perished with cold and hunger on the paramos of Llanganati, where they had wandered for thirty days. The whole story seemed so improbable that I paid little attention to it, and I set to work to examine the vegetation of the adjacent volcano Tunguragua, at whose north-eastern foot the village of Baños is situated. In the month of September I visited Cotaló, a small village on a plateau at about two-thirds of the ascent of Guayrapàta, the hill in front of Tunguragua and above the confluence of the rivers Patate and Chambo. From Cotaló, on a clear night of full moon, I saw not only Tunguragua, El Altar, Condorasto, and the Cordillera of Cubilliú, stretching southwards towards the volcano Sangáy, but also to the eastward the snowy peak of Llanganati. This is one of the few points from which Llanganati can be seen; it appears again, in a favourable state of the atmosphere, a good way up the slopes of Tunguragua and Chimborazo.

At Baños I was told also of a Spanish botanist who a great many years ago lost his life by an accident near the neighbouring town of Patate, and that several boxes belonging to him, and containing dried plants and manuscripts, had been left at Baños, where their contents were finally destroyed by insects.

In the summers of the years 1858 and 1859 I visited Quito and various points in the Western Cordillera, and for many months the country was so insecure, on account of internal dissensions, that

I could not leave Ambato and Riobamba, where my goods were deposited, for more than a few days together. I obtained, however, indisputable evidence that the "Derrotero" or Guide to Llanganati of Valverde had been sent by the King of Spain to the Corregidors of Tacunga and Ambato, along with a Cedula Real (Royal Warrant) commanding those functionaries to use every diligence in seeking out the treasure of the Incas. That one expedition had been headed by the Corregidor of Tacunga in person, accompanied by a friar, Padre Longo, of considerable literary reputation. The Derrotero was found to correspond so exactly with the actual localities, that only a person intimately acquainted with them could have drawn it up; and that it could have been fabricated by any other person who had never been out of Spain was an impossibility. This expedition had nearly reached the end of the route, when one evening the Padre Longo disappeared mysteriously, and no traces of him could be discovered, so that whether he had fallen into a ravine near which they were encamped, or into one of the morasses which abound all over that region, is to this day unknown. After searching for the Padre in vain for some days, the expedition returned without having accomplished its object.

The Cedula Real and Derrotero were deposited in the archives of Tacunga, whence they disappeared about twenty years ago. So many people were admitted to copy them that at last some one, not content with a copy, carried off the originals. I have secured a copy of the Derrotero, bearing date August 14, 1827; but I can meet with no one who recollects the date of the original documents.

I ascertained also that the botanist above alluded to was a Don Atanasio Guzman, who resided some time in the town of Pillaro, whence he headed many expeditions in quest of the gold of Llanganati. He made also a map of the Llanganatis, which was supposed to be still in existence. Guzman and his companions, although they found no deposit of gold, came on the mouths of several silver and copper mines, which had been worked in the time of the Incas, and ascertained the existence of other metals and minerals. They began to work the mines at first with ardour, which soon, however, cooled down, partly in consequence of intestine quarrels, but chiefly because they became disgusted with that slow mode of acquiring wealth when there was molten gold supposed to be hidden close by; so the mines were at length all abandoned. This is said to have taken place early in the present century, but the exact date I can by no means ascertain. Guzman is reported to have met with Humboldt, and to have shown his drawings of plants and animals to that prince of travellers. He died about 1806 or 1808, in the valley of Leytu, about four leagues eastward of Ambato, at a small farmhouse called now Leytillo, but marked on his map San Antonio. He was a somnambulist, and having one night walked out of the house while asleep, he fell down a steep place and so perished. This is all I have been able to learn, and I fear no documents now exist which can throw any further light on the story of his life, though a botanical manuscript of his is believed to be still preserved in one of the archives of Quito. I made unceasing inquiries for the map, and at length ascertained that the actual possessor was a gentleman of Ambato, Señor Salvador Ortega, to whom I made application for it, and he had the kindness to have it brought immediately from Quito, where it was deposited, and placed in my hands; I am therefore indebted to that gentleman's kindness for the pleasure of being able to lay the accompanying copy of the map before the Geographical Society.

The original map is formed of eight small sheets of paper of rather unequal size (those of my copy exactly correspond to them), pasted on to a piece of coarse calico, the whole size being 3 feet 10½ inches by 2 feet 9 inches. It is very neatly painted with a fine pencil in Indian ink—the roads and roofs of houses red—but it has been so roughly used that it is now much dilapidated, and the names, though originally very distinctly written, are in many cases scarcely decipherable: in making them out I have availed myself of the aid of persons familiar with the localities and with the Quichua language. The attempt to combine a vertical with a horizontal projection of the natural features of the country has produced some distortion and dislocation, and though the actual outline of the mountains is intended to be represented, the heights are much exaggerated, and consequently the declivities too steep. Thus the apical angle of the cone of Cotopaxi (as I have determined it by actual measurement) is 121°, and the slope (inclination of its surface to the horizon) 29½°; while on Guzman's map the slope is 69¼°, so that the inclination is only three-sevenths of what he has represented it, and we may assume a corresponding correction needed in all the other mountains delineated.[1]

The whole map is exceedingly minute, and the localities mostly correctly named, but there are some errors of position, both absolute and relative, such that I suppose the map to have been constructed mainly from a simple view of the country, and that no angles and very few compass-bearings have been taken. The margins of the map correspond so nearly with the actual parallels and meridians, that they may be assumed to represent the cardinal points of the compass, as on an ordinary map, without sensible error.

The country represented extends from Cotopaxi on the north to the base of Tunguragua on the south, and from the plain of Callo (at the western foot of Cotopaxi) on the west to the river Puyu, in the forest of Canelos, on the east. It includes an area of something less than an equatorial degree, namely, that comprised between 0° 40′ and 1° 33′ S. lat., and between 0° 10′ W., and near 0° 50′ E. of the meridian of Quito. In this space are represented six active volcanoes (besides Cotopaxi), viz.—

1. El Volcan de los Mulatos, east a little south from Cotopaxi, and nearly on the meridian of the Rio de Ulva, which runs from Tunguragua into the Pastasa. The position of this volcano corresponds to the Quilindaña of most maps—a name which does not occur on Guzman's, nor is it known to any of the actual residents of the country. A group of mountains running to north-east, and terminating in

[1] The apical angle of Tunguragua—the steepest mountain I ever climbed—is 92½°, and the slope 43¾°.

the volcano, is specified as the Cordillera de los Mulatos: it is separated from Cotopaxi by the Valle Vicioso.

2. El Volcan de las Margasitas, south-east by east from Los Mulatos, and a little east of north from the mouth of the Rio Verde Grande. "Margasitas" (more properly Marquesitas) corresponds nearly to the term "pyrites," and is a general name for the sulphates of iron, copper, etc.

3. Zunchu-urcu, a smaller volcano than Margasitas, and at a short distance south-south-east of it. "Zunchu" is the Quichua term for mica or talc.

4. Siete-bocas, a large mountain, with seven mouths vomiting flame, south-west by south from Margasitas, west by south from Zunchu. Its southern slope is the Nevado del Atilis.

5. Gran Volcan del Topo, or Yurag-Llanganati, nearly east from Siete-bocas and south-west from Zunchu. A tall snowy peak at the head of the river Topo, and the same as I saw from Cotaló. It is the only one of the group which rises to perpetual snow, though there are many others rarely clear of snow; hence its second name Yurag (White) Llanganati.[1]

[This mountain is partly shown on the extreme right margin of the map here given.]

The last four volcanoes are all near each other, and form part of what Guzman calls the Cordillera de Yurag-urcu, or Llanganatis of the Topo.

North-east from the Volcan del Topo, and running from south-east to north-west, is the Cor-

[1] Villavicensio gives its height as 6520 varas (17,878 English feet) in his *Geografia del Ecuador*, from a measurement (as he says) of Guzman, but does not inform us where he obtained his information.

dillera de Yana-urcu, or the Llanganatis of the Curaray, consisting chiefly of a wooded mountain with many summits, called Rundu-uma-urcu or Sacha-Llanganati.

6. Jorobado or the Hunchback, south-south-west half west from Yurac-Llanganati, and between the river Topo and the head of the greater Rio Verde.

I have conversed with people who have visited the Llanganati district as far as forty years back, and all assure me they have never seen any active volcano there; yet this by no means proves that Guzman invented the mouths vomiting flame which appear on his map. The Abbé Velasco, writing in 1770,[1] says of Tunguragua, "It is doubtful whether this mountain be a volcano or not," and yet three years afterwards it burst forth in one of the most violent eruptions ever known. I gather from the perusal of old documents that it continued to emit smoke and flame occasionally until the year 1780. Many people have assured me that smoke is still seen sometimes to issue from the crater. I was doubtful about the fact, until, having passed the night of November 10, 1857, at the height of about 8000 feet on the northern slope of the mountain, I distinctly saw at daybreak (from 5½ to 6½ A.M.) smoke issuing from the eastern edge of the truncated apex.[2] In ascending on the same side, along the course of the great stream of lava that overwhelmed the farm of Juivi and blocked up the

[1] *Historia de Quito.*

[2] The same morning (Nov. 11), at 4 A.M., I observed a great many shooting-stars in succession, all becoming visible at the same point (about 40° from the zenith), proceeding along the arc of a great circle drawn through Orion's Belt and Sirius, and disappearing behind the cone of Tunguragua.

Pastasa, below the mouth of the Patate, for eight months, we came successively on six small fumaroli, from which a stream of thin smoke is constantly issuing. People who live on the opposite side of the valley assert that they sometimes see flame hovering over these holes by night. The inhabitants of the existing farm of Juivi complain to me that they have been several times alarmed of late (especially during the months of October and November 1859) by the mountain "bramando" (roaring) at night. The volcano is plainly, therefore, only dormant, not extinct, and both Tunguragua and the Llanganatis may any day resume their activity.

[Here follows a rather elaborate description of the various rivers and their tributaries as shown on the map, which, being of little interest to the general reader, are omitted. Of the map generally, Spruce makes the following observation :—]

As the great mineral districts of Llanganati, occupying the northern half of the map, was repeatedly travelled over by Guzman himself, it is fuller of minute detail than the rest; and I am assured by those who have visited the actual localities that not one of them is misplaced on the map; but the southern portion is much dislocated; and, as I have traversed the whole of it, I will proceed to make some remarks and corrections on this part of the map.

[As these corrections are accessible to all specially interested, and will no doubt be made use of in compiling future maps of Ecuador, I omit these also, and pass on to a description of the map itself, and to the remarkable document which it illustrates.]

The parts of the map covered with forest are represented by scattered trees, among which the following forms are easily recognisable:—

No. 1 is the Wax palm (Palma de Ramos of the Quitonians; *Ceroxylon andicola*, H. et B.), which I have seen on Tungüragua up to 10,000 feet. Nos. 2 and 3 are Tree-ferns (Helechos)—the former a Cyathea, whose trunk (sometimes 40 feet high) is much used for uprights in houses; the latter an Alsophila with a prickly trunk, very frequent in the forest of Canelos about the Rio Verde. No. 4 is the Aliso (*Betula acuminata*, Kunth), one of the most abundant trees in the Quitonian Andes; it descends on the beaches of the Pastasa to near 4000 feet, and ascends on the paramos of Tunguragua to 12,000. But there is one tree (represented thus), occupying on the map a considerable range of altitude, which I cannot make out, unless it be a Podocarpus, of which I saw a single tree on Mount Abitagua, though a species of the same genus is abundant at the upper limit of the forest in some parts of the Western Cordillera. A large spreading tree is figured here and there in the forest of Canelos which may be the Tocte—a true Walnut (Juglans), with an edible fruit rather larger than that of the European species. The remaining trees represented, especially those towards the upper limit of the forest, are mostly too much alike to admit of the supposition that any particular species was intended by them.

The abbreviations made use of in the map are: C° for Cerro (mountain), Cord^{a} for Cordillera (ridge), Mont^{a} for Montana (forest), A° for Arroyo (rivulet), L^{a} for Laguna, and C^{a} for Cocha (lake), Far^{n} for Farallón (peak or promontory), H^{a} for Hacienda (farm), and C^{l} for Corral (cattle or sheep-fold).

Mule-tracks (called by the innocent natives "roads") are represented by double red lines, and footpaths by single lines. I have copied them by dotted lines.

Having now passed in review the principal physical features of the district, let us return to the Derrotero of Valverde, of which the following is a translation. The introductory remark or title (not in very choice Castilian) is that of the copyist:

"The 'Derrotero' or Guide to the Hidden Treasure of the Incas. Translated by Richard Spruce."

TITLE

GUIDE OR ROUTE WHICH VALVERDE LEFT IN SPAIN, WHERE DEATH OVERTOOK HIM, HAVING GONE FROM THE MOUNTAINS OF LLANGANATI, WHICH HE ENTERED MANY TIMES, AND CARRIED OFF A GREAT QUANTITY OF GOLD; AND THE KING COMMANDED THE CORREGIDORS OF TACUNGA AND AMBATO TO SEARCH FOR THE TREASURE: WHICH ORDER AND GUIDE ARE PRESERVED IN ONE OF THE OFFICES OF TACUNGA

THE GUIDE

"Placed in the town of Pillaro, ask for the farm of Moya, and sleep (the first night) a good distance above it; and ask there for the mountain of Guapa, from whose top, if the day be fine, look to the east, so that thy back be towards the town of Ambato, and from thence thou shalt perceive the three Cerros Llanganati, in the form of a triangle, on whose declivity there is a lake, made by hand, into which the ancients threw the gold they had prepared for the ransom of the Inca when they heard of his death. From the same Cerro Guapa thou mayest see also the forest, and in it a clump of Sangurimas standing out of the said forest, and another clump which they call Flechas (arrows), and these clumps are the principal mark for the which thou shalt aim, leaving them a little on the left hand. Go forward from Guapa in the direction and with the signals indicated, and a good way ahead, having passed some cattle-farms, thou shalt come on a wide morass, over which thou must cross, and coming out on the other side thou shalt see on the left hand a short way off a jucál on a hill-side, through which thou must pass. Having got through the jucál, thou wilt see two small lakes called "Los Anteojos" (the spectacles), from having between them a point of land like to a nose.

"From this place thou mayest again descry the Cerros Llanganati, the same as thou sawest them from the top of Guapa, and I warn thee to leave the said lakes on the left, and that in front of the point or 'nose' there is a plain, which is the sleeping-place. There thou must leave thy horses, for they can go no farther. Following now on foot in the same direction, thou shalt come on a great black lake, the which leave on thy left hand, and beyond it seek to descend along the hill-side in such a way that thou mayest reach a ravine, down which comes a waterfall: and here thou shalt find a bridge of three poles, or if it do not still exist thou shalt put another in the most convenient place and pass over it. And having gone on a little way in the forest, seek out the hut which served to sleep in or the remains of it. Having passed the night there, go on thy way the following day through the forest in the same direction, till thou reach another deep dry ravine, across which thou must throw a bridge and pass over it slowly and cautiously, for the ravine is very deep; that is, if thou succeed not in finding the pass which exists. Go forward and look for the signs of another sleeping-place, which, I assure thee, thou canst not fail to see in the fragments of pottery and other marks, because the Indians are continually passing along there. Go on thy way, and thou shalt see a mountain which is all of margasitas (pyrites), the which leave on thy left hand, and I warn thee that thou must go round it in this fashion ⟲. On this side thou wilt find a pajonál (pasture) in a small plain, which having crossed thou wilt come on a cañon between two hills, which is the Way of the Inca. From thence as thou goest along thou shalt see the entrance of the socabón (tunnel), which is in the form of a church porch. Having come through the cañon and gone a good distance beyond, thou wilt perceive a cascade which descends from an offshoot of the Cerro Llanganati and runs into a quaking-bog on the right hand; and without passing the stream in the said bog there is much gold, so that putting in thy hand what thou shalt gather at the bottom is grains of gold. To ascend the mountain, leave the bog and go along to the right, and pass above the cascade, going round the offshoot of the mountain. And if by chance the mouth of the socabón be closed with certain herbs which they call 'Salvaje,' remove them, and thou wilt find the entrance. And on the left-hand side of the mountain thou mayest see the 'Guayra' (for thus the ancients called the furnace where they founded metals), which is nailed with golden nails.[1] And to reach the third mountain, if thou canst not pass in front of the socabón, it is the same thing to pass behind it, for the water of the lake falls into it.

"If thou lose thyself in the forest, seek the river, follow it on the right bank; lower down take to the beach, and thou wilt reach the cañon in such sort that, although thou seek to pass it, thou wilt not find where; climb, therefore, the mountain on the right hand, and in this manner thou canst by no means miss thy way."

[Having read this remarkable document, we shall better understand Spruce's account of the various attempts to discover the treasure, the chief routes followed being marked by red lines.]

With this document and the map before us, let us trace the attempts that have been made to reach the gold thrown away by the subjects of Atahuallpa as useless when it could no longer be applied to the purpose of ransoming him from the Spaniards.

Pillaro is a somewhat smaller town than Ambato, and stands on higher ground, on the opposite side

[1] [Query—sprinkled with gold.—ED.]

of the river Patate, at only a few miles' distance, though the journey thither is much lengthened by having to pass the deep quebrada of the Patate, which occupies a full hour. The farm of Moya still exists; and the Cerro de Guapa is clearly visible to east-north-east from where I am writing. The three Llanganatis seen from the top of Guapa are supposed to be the peaks Margasitas, Zunchu, and el Volcan del Topo. The "Sangurimas" in the forest are described to me as trees with white foliage; but I cannot make out whether they be a species of Cecropia or of some allied genus. The "Flechas" are probably the gigantic arrow-cane, *Gynerium saccharoides* (Arvoré de frecha of the Brazilians), whose flower-stalk is the usual material for the Indian's arrows.

The morass (Cienega de Cubillín), the Jucál,[1] and the lakes called "Anteojos," with the nose of land between them, are all exactly where Valverde places them, as is also the great black lake (Yana-cocha) which we must leave on the left hand. Beyond the lake we reach the waterfall (Cascada y Golpe de Limpis Pongo), of which the noise is described to me as beyond all proportion to the smallness of the volume of water. Near the water-fall a cross is set up with the remark underneath, "Muerte del Padre Longo"—this being the point

[1] Júco is the name of a tall, solid-stemmed grass, usually about 20 feet high, of which I have never seen the flower, but I take it to be a species of Gynerium, differing from *G. saccharoides* in the leaves being uniformly disposed on all sides and throughout the length of the stem, whereas in *G. saccharoides* the stem is leafless below and the leaves are distichous and crowded together (almost equitant) near the apex of the stem. The Júco grows exclusively in the temperate and cool region, from 6000 feet upwards, and is the universal material for laths and rods in the construction of houses in the Quitonian Andes.

from which the expedition first spoken of regressed in consequence of the Padre's sudden disappearance. Beyond this point the climate begins to be warm; and there are parrots in the forest. The deep dry quebrada (Quebrada honda), which can be passed only at one point—difficult to find, unless by throwing a bridge over it—is exactly where it should be; but beyond the mountain of Margasitas, which is shortly afterwards reached, no one has been able to proceed with certainty. The Derrotero directs it to be left on the left hand; but the explanatory hieroglyph puzzles everybody, as it seems to leave the mountain on the right. Accordingly, nearly all who have attempted to follow the Derrotero have gone to the left of Margasitas, and have failed to find any of the remaining marks signalised by Valverde. The concluding direction to those who lose their way in the forest has also been followed; and truly, after going along the right bank of the Curaray for some distance, a stream running between perpendicular cliffs (Cañada honda y Rivera de los Llanganatis) is reached, which no one has been able to cross; but though from this point the mountain to the right has been climbed, no better success has attended the adventurers.

"Socabón" is the name given in the Andes to any tunnel, natural or artificial, and also to the mouth of a mine. Perhaps the latter is meant by Valverde, though he does not direct us to enter it. The "Salvaje" which might have grown over and concealed the entrance of the Socabón is *Tillandsia usneoides*, which frequently covers trees and rocks with a beard 30 or 40 feet long.

Comparing the map with the Derrotero, I should

conclude the cañon, "which is the Way of the Inca," to be the upper part of the Rivera de los Llanganatis. This cañon can hardly be artificial, like the hollow way I have seen running down through the hills and woods on the western side of the Cordillera, from the great road of Azuáy, nearly to the river Yaguachi. "Guayra," said by Valverde to be the ancient name for a smelting-furnace, is nowadays applied only to the wind. The concluding clause of this sentence, "que son tachoneados de oro," is considered by all competent persons to be a mistake for "que es tachoneado de oro."

If Margasitas be considered the first mountain of the three to which Valverde refers, then the Tembladál or Bog, out of which Valverde extracted his wealth, the Socabón and the Guayra are in the second mountain, and the lake wherein the ancients threw their gold in the third.

Difference of opinion among the gold-searchers as to the route to be pursued from Margasitas would appear also to have produced quarrels, for we find a steep hill east of that mountain, and separated from it by Mosquito Narrows (Chushpi Pongo), called by Guzman "El Peñon de las Discordias."

If we retrace our steps from Margasitas till we reach the western margin of Yana-cocha, we find another track branching off to northward, crossing the river Zapalá at a point marked Salto de Cobos, and then following the northern shore of the lake. Then follow two steep ascents, called respectively "La Escalera" and "La Subida de Ripalda," and the track ends suddenly at the river coming from the Inca's Fountain (La Pila del Inca), with the remark, "Sublevacion de los Indios — Salto de Guzman," giving us to understand that the exploring party had barely crossed the river when the Indians rose against them, and that Guzman himself re-passed the river at a bound. These were probably Indians taken from the towns to carry loads and work the mines; they can hardly have been of the nation of the Curarayes, who inhabited the river somewhat lower down.

A little north and east of the Anteojos there is another route running a little farther northward and passing through the great morass of Illubamba, at the base of Los Mulatos, where we find marked El Atolladero (the Bog) de Guzman, probably because he had slipped up to the neck in it. Beyond this the track continues north-east, and after passing the same stream as in the former route, but nearer to its source in the Inca's Fountain, there is a tambo called San Nicolas, and a cross erected near it marks the place where one of the miners met his death (Muerte de Romero). Another larger cross (La Cruz de Romero) is erected farther on at the top of a basaltic mountain called El Sotillo. At this point the track enters the Cordillera de las Margasitas, and on reaching a little to the east of the meridian of Zunchu-urcu, there is a tambo with a chapel, to which is appended the remark, "Destacamento de Ripalda y retirada per Orden Superior." Beyond the fact thus indicated, that one Ripalda had been stationed there in command of a detachment of troops, and had afterwards retired at the order of his superiors, I can give no information.

There are many mines about this station, especially those of Romero just to the north, those of Viteri to the east, and several mines of copper and silver which are not assigned to any particular owner. Not far to the east of the Destacamento is another tambo, with a cross, where I find written, "Discordia y Consonancia con Guzman," showing that at this place Guzman's fellow-miners quarrelled with him and were afterwards reconciled. East-north-east from this, and at the same distance from it as the Destacamento, is the last tambo on this route, called El Sumadal, on the banks of a lake, near the Rio de las Flechas. Beyond that river, and north of the Curaray, are the river and forests of Gancaya.

Another track, running more to the north than any of the foregoing, sets out from the village of San Miguel, and passes between Cotopaxi and Los Mulatos. Several tambos or huts for resting in are marked on the route, which ends abruptly near the Minas de Pinel (north-east from Los Mulatos), with the following remark by the author—"Conspiracion contra Conrado y su accelerado regreso," so that Conrado ran away to escape from a conspiracy formed against him, but who he was, or who were his treacherous companions, it would now perhaps be impossible to ascertain.

Along these tracks travelled those who searched for mines of silver and other metals, and also for the gold thrown away by the subjects of the Inca. That the last was their principal object is rendered obvious by the carefulness with which every lake has been sounded that was at all likely to contain the supposed deposit.[1]

[1] The soundings of the lakes are in Spanish varas, each near 33 English inches.

The mines of Llanganati, after having been neglected for half a century, are now being sought out again with the intention of working them; but there is no single person at the present day able to employ the labour and capital required for successfully working a silver mine, and mutual confidence is at so low an ebb in this country that companies never hold together long. Besides this, the gold of the Incas never ceases to haunt people's memories; and at this moment I am informed that a party of explorers who started from Tacunga imagine they have found the identical Green Lake of Llanganati, and are preparing to drain it dry. If we admit the truth of the tradition that the ancients smelted gold in Llanganati, it is equally certain that they extracted the precious metal in the immediate neighbourhood; and if the Socabón of Valverde cannot at this day be discovered, it is known to every one that gold exists at a short distance, and possibly in considerable quantity, if the Ecuadoreans would only take the trouble to search for it and not leave that task to the wild Indians, who are content if, by scooping up the gravel with their hands, they can get together enough gold to fill the quill which the white man has given them as the measure of the value of the axes and lance-heads he has supplied to them on trust.

The gold region of Canelos begins on the extreme east of the map of Guzman, in streams rising in the roots of Llanganati and flowing to the Pastasa and Curaray,[1] the principal of which are the Bombonasa and Villano. These rivers and their smaller tributaries have the upper part of their course in

[1] The name Curaray itself may be derived from "curi," gold.

deep ravines, furrowed in soft alluvial sandstone rock, wherein blocks and pebbles of quartz are interspersed, or interposed in distinct layers. Towards their source they are obstructed by large masses of quartz and other rocks; but as we descend the stones grow fewer, smaller, and more rounded, until towards the mouth of the Bombonasa, and thence throughout the Pastasa, not a single stone of the smallest size is to be found. The beaches of the Pastasa consist almost entirely of powdered pumice brought down from the volcano Sangáy by the river Palora. When I ascended the Bombonasa in the company of two Spaniards who had had some experience in mining, we washed for gold in the mouth of most of the rivulets that had a gravelly bottom, as also on some beaches of the river itself, and never failed to extract a few fragments of that metal. All these streams are liable to sudden and violent floods. I once saw the Bombonasa at Pucayacu, where it is not more than 40 yards wide, rise 18 feet in six hours. Every such flood brings down large masses of loose cliff, and when it subsides (which it generally does in a few hours) the Indians find a considerable quantity of gold deposited in the bed of the stream.

The gold of Canelos consists almost solely of small particles (called "chispas," sparks), but as the Indians never dig down to the base of the wet gravel, through which the larger fragments of gold necessarily percolate by their weight, it is not to be wondered at that they rarely encounter any such. Two attempts have been made, by parties of Frenchmen, to work the gold-washings of Canelos systematically. One of them failed in consequence of a quarrel which broke out among the miners themselves and resulted in the death of one of them. In the other, the river (the Lliquino) rose suddenly on them by night and carried off their canoes (in which a quantity of roughly-washed gold was heaped up), besides the Long Tom and all their other implements.

I close this memoir by an explanation of the Quichua terms which occur most frequently on the map.

Spanish authors use the vowels *u* and *o* almost indiscriminately in writing Quichua names, although the latter sound does not exist in that language; and in some words which have become grafted on the Spanish, as spoken in Peru and Ecuador, the *o* has supplanted the *u* not only in the orthography but in the actual pronunciation, as, for instance, in Pongo and Cocha, although the Indians still say "Chimbu-rasu," and not "Chimborazo"—"Cutu-pacsi" or "Cutu-pagsi," and not "Cotopaxi." The sound of the English *w* is indicated in Spanish by *gu* or *hu*; that of the French *j* does not exist in Spanish, and is represented by *ll*, whose sound is somewhat similar; thus "Lligua" is pronounced "Jiwa." "Llanganati" is now pronounced with the Spanish sound of the *ll*, but whether this be the original mode is doubtful. An unaccented terminal *e* (as in Spanish "verde") is exceedingly rare in Indian languages, and has mostly been incorrectly used for a short *i*; thus, if we wish to represent the exact pronunciation, we should write "Casiquiari," "Ucayáli," and "Llanganati"—*not* Casiquiare, Ucayale, Llanganate.

"Llanganáti" may come from "llánga," to touch, because the group of mountains called by that name touches on the sources of the rivers all round; thus, on Guzman's map, we find "Llanganatis del Rio Verde"—"Llanganatis del Topo"—"Llanganatis del Curaray," for those sections of the group which respectively touch on the Rio Verde, the Topo, and the Curaray. The following are examples of the mode of using the verb "llanga." "Ama llangáichu!"—"Touch it not!" "Imapág llancángui?"—"Why do you touch it"; or "Pitag llåncaynírca?"—"Who told you to touch it?" And the answer might be "Llancanatág chári-cárca llancarcáni."—"[Thinking] it might be touched, I touched it."

It is to be noted that the frequent use of the letter *g*, in place of *c*, is a provincialism of the Quitonian Andes, where (for instance) they mostly say "Inga" instead of "Inca." But in

Maynas the *c* is used almost to the exclusion of the *g*; thus "yúrag," white, and "pítag," who, are pronounced respectively "yurac" and "pitac" in Maynas.

"Tungurágua" seems to come from "tungúri," the ankle-joint, which is a prominence certainly, though scarcely more like the right-angled cone of Tunguragua than the obtuse-angled cone of Cotopaxi is like a wen ("coto" or "cutu").

Of the termination "agua" (pron. "awa") I can give no explanation.

"Cungúri," in Quichua, is the knee; thus an Indian would say "Tungúri-mánta cungúli-cáma llustirishcáni urmáshpa," *i.e.* "In falling ('urmáshpa') I have scrubbed off the skin from the ankle to the knee."

Among rustics of mixed race, whose language partakes almost as much of Quichua as of Spanish, it is common to hear such expressions as "De tunguri á cunguri es una cola llaga."—"From the ankle to the knee is a continuous sore."

The following words occur repeatedly on the map:—

"Ashpa" (in Maynas "Allpa"), earth. "Urcu," mountain. "Rumi," stone. "Cócha (cucha)," lake.

"Yácu," river. "Ucsha," grass or grassy place ("Pajonál," Sp.). "Póngo (pungu)," door or narrow entrance.

"Cúchu," corner. "U'ma," head. "Paccha," cataract.

"Cúri," gold. "Cúlqui," silver. "Alquímia," copper. "Ushpa," ashes.

"Chíri," cold. "Yúnga," warm, from which the Spaniards have formed the diminutive "Yungúilla," warmish, applied to many sites where the sugar-cane begins to flourish.

"Yúrag," white. "Yána," black. "Púca," red. "Quílla," yellow.

"I'shcai," two; ex. "I'shcai-guáuqui," the Two Brothers, a cloven peak to the east of Los Mulatos. "Chunga," ten; ex. "Chunga-uma," a peak with ten points, a little to south of "Ishcai-guauqui." "Parca," double; thus a hill which seems made up of two hills united is called "Parca-urcu."

"Angas," a hawk. "Ambátu," a kind of toad.

"Sácha," forest. "Cáspi," tree. "Yúras," herb. "Quínua," the "Chenopodium Quinoa," cultivated for its edible seed. "Pujín," hawthorn (various species of Cratægus); thus "Montaña de Pujines," Hawthorn Forest; "Cerro Pujin el chico," Little Hawthorn-hill. "Cubiliín," a sort of Lupine, found only on the highest paramos. It gives its name to a long ridge of the Eastern Cordillera, mostly covered with snow, extending from Condorasto and El Altar towards Sangáy. "Totorra," a large bulrush from which mats are made; hence "Totorrál," a marsh full of bulrushes. "Sara," maize.

"Tópo" is the name given in Maynas to the Raft-wood trees,

species of Ochroma (of the N.O. Bombaceæ). They begin to be found as soon as we reach a hot climate, say from 3000 feet elevation downwards.

"Rundu," sleet; thus "Rundu-uma," Sleety Head. "Rásu" is snow, and occurs in "Chimbu-rasu," "Caraguai-rasu" (Carguairago), and many other names. The vulgar name for snow as it falls is "Papa-cara," *i.e.* potato peelings.

"Pucará" indicates the site of a hill-fort of the Incas, of which a great many are scattered through the Quitonian Andes.

Critical Note by the Editor

The preceding account of the various routes of the gold-seekers among the Llanganati Mountains leads to the conclusion that only the earliest—that led by the Corregidor of Tacunga and the friar Padre Longo—made any serious attempt to follow the explicit directions of the "Guide," since the others departed from it so early in the journey as the great black lake "Yana Cocha," going to the left instead of to the right of it. No doubt they were either deceived by Indian guides who assured them that they knew an easier way, or went in search of rich mines rather than of buried treasure. The first party, however, and those who afterwards followed it, kept to the route, as clearly described, to the sleeping-place beyond the deep ravine where Padre Longo was lost; but beyond this point they went wrong by crossing the river, and thus leaving the district of the three volcanoes, which twice at the beginning of the "Guide" are indicated as the locality of the treasure.

Although no route to these mountains is marked on the map, Spruce tells us that other parties did take the proper course, and found the "deep dry ravine" (marked on the map as "Quebrada honda"), and after it the mountain of Margasitas; but here they were all puzzled by the "Guide" directing them to leave the mountain on their left while the hieroglyph seems to leave it on the right, and following this latter instruction they have failed afterwards to find any of the other marks given by Valverde in his "Guide." Spruce himself suggests that the upper part of the Rivera de los Llanganatis (which is outside the portion of the map here given) is the "way of the Inca" referred to in the "Guide." But this is going quite beyond the area of the three mountains, so clearly stated as the objective of the "Guide."

It seems to me, however, that there is really no contradiction between the "Guide" and the map, and that the route so clearly pointed out in the former has not yet been thoroughly explored to its termination, as I will now endeavour to show. After crossing the deep dry ravine ("Quebrada honda" of the map), we are directed to "go forward and look for the signs of another sleeping-place." Then, the next day—"Go on thy way, and thou shalt see a mountain which is all of margasitas, the which leave on thy left hand." But looking at the map we shall see that the mountain will now be on the right hand, supposing we have gone on in the same direction as before, crossing the deep ravine. The next words, however, explain this apparent contradiction: they are—"and I warn thee that thou must go round it in this fashion," with the explanatory hieroglyph, which, if we take the circle to be the mountain and the right-hand termination of the curve the point already reached, merely implies that you are to turn back and ascend the mountain in a winding course till you reach the middle of the south side of it. So far you have been going through forest, but now you are told—" On this side thou wilt find a pajonál (pasture) in a small plain " (showing that you have reached a considerable height), " which having crossed thou wilt come on a cañon between two hills, which is the way of the Inca." This cañon is clearly the upper part of the " Chushpi pongo," while the " Encañado de Sacha pamba " is almost certainly the beginning of the " way of the Inca." The explorers will now have reached the area bounded by the three volcanoes of the " Guide " —the Margasitas will be behind them, Zunchu-urcu on his right, and the great volcano Topo in front, and it is from this point only that they will be in a position to look out for the remaining marks of the " Route "—the socabón or tunnel " in the form of a church porch," and evidently still far above them, the cascade and the quaking-bog, passing to the right of which is the way to " ascend the mountain," going " above the cascade " and " round the offshoot of the mountain " to reach the socabón. Then you will be able to find the Guayra (or furnace), and to reach the "third mountain," which must be the Topo, you are to pass the socabón " either in front or behind it, for the water of the lake falls into it." This evidently means the lake mentioned in the first sentence of the " Guide " as being the place where the gold prepared for the ransom of the Inca was hidden. The last sentence of the " Guide " refers to what must be done if you miss the turning shown by the hieroglyph, in which case you have to follow the river-bank till you come to the cañon (on the map marked " Chushpi pongo "), up the right-hand side of which you must climb the mountain, "and in this manner thou canst by no means miss thy way"; which the map clearly shows, since it leads up to the " Encañado," which is shown by the other and more easy route to be the " way of the Inca."

I submit, therefore, that the "Guide" is equally minute and definite in its descriptions throughout, that it agrees everywhere with Guzman's map, and that, as it is admitted to be accurate in every detail for more than three-fourths of the whole distance, there is every probability that the last portion is equally accurate. It will, of course, be objected that, if so, why did not Guzman himself, who made the map, also complete the exploration of the route and make the discovery? That, of course, we cannot tell; but many reasons may be suggested as highly probable. Any such exploration of a completely uninhabited region must be very costly, and is always liable to fail near the end from lack of food, or from the desertion of the Indian porters when there was doubt about the route. Guzman had evidently been diverted from the search by what seemed the superior promise of silver and gold mines, from which he may have hoped to obtain wealth enough to carry out the other expedition with success. This failing, he apparently returned home, and may have been endeavouring to obtain recruits and funds for a new effort when his accidental death occurred.

It is to be noted that beyond the point where the hieroglyph puzzled all the early explorers there is a complete absence of detail in Guzman's map, which contains nothing that might not have been derived from observations made from the heights north of the river, and from information given by wandering Indians.

It is also to be noted that only four sleeping-places are mentioned in the "Guide," so that the whole journey occupied five days. The last of the four sleeping-places is before reaching the spot where the path turns back round the Margasitas Mountain, so that the whole distance from this place to the "lake made by hand" must be less than twenty miles, a distance which would take us to the nearer slopes of the great Topo Mountain. In this part of the route the marks given in the "Guide" are so many and so well-defined that it cannot be difficult to follow them, especially as the path indicated seems to be mostly above the forest-region.

For the various reasons now adduced, I am convinced that the "Route" of Valverde is a genuine and thoroughly trustworthy document, and that by closely following the directions therein given, it may still be possible for an explorer of means and energy, with the assistance of the local authorities, to solve the interesting problem of the Treasure of the Incas. The total distance of the route, following all its sinuosities, cannot exceed ninety or a hundred miles at most, fully three-fourths of which must be quite easy to follow, while the remainder is very clearly described. Two weeks would therefore suffice for the whole expedition.

I have written this in the hope that some one who speaks Spanish fluently, has had some experience of the country, and is possessed of the necessary means, may be induced to undertake this very interesting and even romantic piece of adventurous travel. To such a person it need be but a few months' holiday.

GLOSSARY OF NATIVE NAMES[1]

ABACATE, AGUACATE. An oily fruit; cats fond of it; good for epilepsy.

ABILLA, JABILLA. A twiner with large seeds producing a bitter oil for lamps on the Huallaga river.

ACARICUARÁ. *Swartzia callistemon.* Curious perforated trunks; a dye from the bark.

AGUACATE. A tree (undetermined) of the fruit of which cats and many wild animals are very fond. It is very nutritious, and the seeds produce an oil very similar to that of olives.

AJARI. *Tephrosia toxicaria* (Leguminosæ).

ALCORNOQUES (cork trees). *Curatella Americana.*

ALDEA. A village.

ALGARROBO (Venez.)=JUTAHI (Braz.). Hymenæa sp. (Leg.). Fruit a remedy in asthma; seeds give a fine varnish; and incense.

ANAPÉ. The Jacaná, a long-toed water-fowl (*Parra jacana*).

ANAPÉ-YAPONA. *Victoria regia* (Nymphæaceæ). Jacaná's oven.

ANDIROBA OIL. From *Carapa Guianensis* (Meliaceæ).

ANGELIM. Andira sp. An excellent timber-tree.

ANIL. *Indigofera anil.* Produces blue colour used in painted cuyas.

APIRANGA. A fruit. *Mouriria Apiranga* (Melastomaceæ).

ARAPARI (tree). Fine wood for cabinet work, but small (*Nauclea guianensis*).

AREÇA. An acid berry. *Psidium ovatifolium* (Myrtaceæ).

ARIPECURÚ. A branch of the Trombetas river.

ARVORE DE CHAPETE. *Gustavia Brasiliensis.*

ASSAÍ. A drink from fruit of *Euterpe oleracea* (Palmaceæ).

BACÁBA. Œnocarpus sp. (Palmaceæ). Fruits yield a nutritious drink or food.

BACUARI-ASSU. *Platinia insignis* (Clusiaceæ). Edible fruit.

BAUNÁ. Root of a climber (Menispermaceæ), called also "maniocca açu" (great mandiocca), larger and more poisonous than mandiocca, but makes equally good farinha and cakes, and is much used on the Purús and Upper Amazon (see vol. i. p. 215).

BLACK PITCH. Clusiaceæ.

BOGA-BOGA (Peru), CAIWA (Maynas). Cucurbitaceæ. A gourd with seeds of an extraordinary rectangular shape.

BOMBONAJÉ. Carludovica sp. (Pandanaceæ). Leaves used for making Panama hats.

BRÊO BRANCO. White pitch. Icica sp.

BRUSCA (Venez.). *Cassia occidentalis.* Bitter root; good in fevers.

[1] This list comprises all the names I have met with in Spruce's Journals and MSS. They may be useful to other explorers or collectors.—ED.

CAAPÍ. *Banisteria caapi* (Malpighiaceæ). An intoxicant.
CAARURU. Podostemon sp. Used for food by the Indians; ashes give salt.
CAATINGA. Low forest—white forest.
CACHIMBO. A pipe.
CADÍ. Phytelephus sp. (Palmaceæ). Vegetable Ivory nut.
CAIMBÉ. *Curatella Americana* (Dilleniaceæ).
CAJU (= MEREY, Venez.). *Anacardium occidentale.* Cashew nut.
CAPOEIRAS. Second growth woods, on deserted farms, etc., in virgin forest.
CARAIPÉ. Licania sp. (Chrysobalaneæ). Pottery tree.
CARAJURÚ. *Bignonia chica.* A red dye.
CARAJURÚ PIRANGA. Bignonia sp. Produces red colour for cuyas.
CARANÁ. *Mauritia carana* (Palmaceæ).
CARANAÍ. *Mauritia aculeata* (Palmaceæ).
CARAPANÁS (L.G.). Mosquitoes.
CARIAQUITO. *Lantana Camara.* Leaves, root, and flowers medicinal.
CARIBÉ (Braz.). Cassava beer, on the Rio Negro.
CARIZA. A musical pipe.
CARTELHANA. *Yangua tinctoria* (Spruce). Gives a dye like that of indigo.
CASCARIA. Samydaceæ.
CASTANHA (Port.). *Bertholletia excelsa.* Brazil-nut tree.
CAURÉ. Perhaps *Kyllinga odorata*, from the roots of which a scented water is distilled by the Indians.
CAXIRÍ (L.G.). Mandiocca beer.
CEDAR. Icica sp. (Amyridaceæ). On the Amazon.
,, Phyllanthus sp. (Euphorbiaceæ). Quito.
CHICHA (Ven.). Cassava beer.
COCA. *Erythoxylon coca.*
COCUI. Agave sp. Root diuretic.
COCÚRA. Pourouma sp. (Artocarpeæ). Edible fruit.
COROZITO. Tree at Maypures.
CORUSI-CAÁ. *Calocophyllum coccineum* (Rubiaceæ). Sun-leaf. Very handsome flower-bracts.
COW-TREE. Mimusops sp. (Sapotaceæ). Produces wholesome milk.
,, Callophora sp. (Apocynaceæ). Produces wholesome milk.
,, Loureira sp. (Euphorbiaceæ). Yields milk.
CUIARÉ. *Elais melanococca* (Palm). Oil-producing.
CUMAÍ, CUMA-AÇU. Callophora (Apocynaceæ). Cow-trees.
CUMANDA-AÇU. *Campsiandra laurifolia* (Leg.). Beans grated used as an emetic.
CUMARÚ. *Dipteryx odorata* (Leguminosæ). Tonga bean, scent.
CUMARÚ-RANA. *Andira oblonga* (Leg.).
CUMATI. Myrcia sp. (hb. 1916) (Myrtaceæ). Bark gives a varnish used on cuyas.
CUNAMBI. *Icthyothera cunambi* (Compositæ). Roots used to stupefy fish.
CUNÚCO (Ven.). Mandiocca field in Venezuela.
CUNURÍ. Euphorbiaceæ. Seeds give an edible oil.
CUPANÁ (Ven.). *Paullinia cupana* (Sapindaceæ). An intoxicant.
CUPA-ÚBA. *Copaifera Martii* (Leg.). Yields balsam capivi.
CUPIM̃. Termites, white ants.
CUPU-ASSU. Theobroma sp. Pulp of fruit eatable.
CURAUÁ. *Bromelia Karatas* (Bromeliaceæ). Leaf fibres used in making hammocks.
CURUÁ. *Attalea spectabilis* (Palmaceæ).
CUSPARIA = CHUSPA. Galipea sp. Bark tonic and febrifuge.

CUYAS. Calabash basins.
CUYEIRA. Crescentia sp. Calabash tree.

EHEN (Ven.). A minute biting fly.
ESPIA (Braz.). A cable.

GAMALOTES. Panicum sp. Grasses in the Cinchona forests.
GAPÓ (L.G.). The flooded banks of rivers.
GENIPAPA. *Genipa Americana* (Cinchonaceæ). Fruit gives a black dye.
GUACO. Mikania sp. Supposed antidote to snake-bites.
GUAJARÁ. Lucuma sp. (Sapotaceæ). Cooked fruits eatable.
GUANABANO. *Anona muricata.* Said to be a powerful remedy in bilious fevers, dysentery, etc.
GUARANÁ. *Paullinia cupana*, stimulant from seeds of.

HOBO = JOVO. Same as Tapiriba (*q.v.*).

IGARAPÉ (L.G.). A small stream.
IMBAÚBA. Cecropia sp. Small white-leaved trees.
INAJÁ. *Maximiliana regia.* A lofty palm.
INGÁ (L.G.). Inga sp. (Mimoseæ). Small trees, produce varnish.
IPADÚ (L.G.). *Erythroxylon coca* (Erythroxylaceæ). Leaves stimulant.
IRAPAI. Carludovica sp. (Pandanaceæ). Peru.
ITA-ÚBA. Acrodiclidium sp. (Lauraceæ). Stone tree, hard wood, finely scented.
ITUÁ, ITUÁN. Gnetum sp. Fibre makes strong fishing-lines.
IÚ. *Astrocaryum acaule* (Palmaceæ).

JACITARA (L.G.). *Desmoncus macroacanthus.* A climbing palm.
JAPURA, YAPURA. *Erisma japura* (Vochysiaceæ).
JARÁ. Leopoldinia sp. Small graceful palms.
JARARACA-TUYA. Dracontium sp. (Araceæ). Stems snake-like.
JAUACÁNA. *Epeira falcata* (Cæsalpiniæ). Infusion of bark good for ague.
JAUARI (L.G.). *Astrocaryum jauari.* A tall prickly palm.
JUÇARA. Narrow strips or planks of shell of palms.
JUPATI. *Rhaphia tædigera.* A short-stemmed but noble palm with immense leaves.
JURUPARI (L.G.). Devil or demon of the Indians.
JUTAHI. Hymenæ sp. (Fabaceæ), Algaroba (Venez.), edible.

LAUREL AMARILLO. *Ocotea cymbarum* (Lauraceæ).
LECHEROTE. Asclepiadea? A twiner, with sweet, milky, wholesome juice, useful in coughs.

MACKRANDÚBA. Mimusops sp. (Sapindaceæ). The Pará cow-tree.
MARAJÁ. *Bactris maraja.* Small palm; fruit edible.
MARAYÁ. *Astrocaryum aculeatum* (Palmaceæ).
MARIMA. Trees producing eatable grubs.
MASUTO. Fermented yucas.
MATINHO. Second growth forest.
MATO VIRGEM (Port.). Virgin forest.
MAYACA, MAHICA. Mayacaceæ. Small bog plants.
MAYNAS. A province of N.E. Peru.

MIRA PIXUNA. *Swartzia grandiflora* (Cæsalpiniæ). Black wood.
MIRITI. Palms of the genus Mauritia.
MONKEY-PODS. Pithecolobium (Mimoseæ).
MOSQUITO (Span.). Sand-flies, etc.
MUCUIN. A small red tick.
MUCUJÁ. *Acrocomia lasiospatha.* Palm with eatable fruit.
MULATTO TREE. *Enkylista Spruceana* (Cinchonaceæ).
MULONGO. *Hancornia laxa* (Apocynaceæ). Cork wood.
MUMBACA. *Astrocaryum mumbaca.* Palm; fruit eatable.
MURIKITICA. A climber. Stem gives drinkable water.
MURIXI. *Byrsonima Poppigiana* (Malpighiaceæ). Bark for tanning.
MURUMURÚ. *Astrocaryum murumuru.* Palm; very spiny. Cattle eat the fruit.
MURURÉ. Floating plants.
MUTÚCA. Small biting flies.

NAMAÕ. *Carica Papaya* (Papayaceæ). The Papaw; fruit eatable.
NIOPO (Ven.). *Piptadenia Niopo.*

OANÁNI. Moronobœa sp. (Clusiaceæ). Black pitch.
OCUMO. Arum sp. Powder used in asthma; root contains half its bulk of fine starch.

PAACUA-RANA. Urania sp. An edible root.
PACÓVA. *Musa sapientiæ* (Musaceæ). Plantain fruit.
PACOVA-SOROROCA. *Alpinia Paco-seroca* (Jacq.). Gives a purple dye, not permanent.
PAJA MANIBA. *Cassia occidentalis.* Root bitter; good in fevers.
PAJUARÚ. Mandiocca beer, also called "caxirí."
PAO D' ARCO. Tecoma sp. Bows and cigar-holders made of this wood.
PAO DE LACRE. *Vismia guianensis* (Hypericaceæ). Yields sealing-wax.
PAO MULATTO. *Eukylista Spruceana* (Cinchonaceæ).
PAPAW. *Carica Papaya* (Papayaceæ). A fruit.
PARANA-MIRI. Side channels of the Amazon, small rivers.
PARATURÍ. Lauraceæ. Hard wood, on Upper Orinoco.
PARICA (L.G.). *Piptadenia Niopo* (Mimoseæ). Seeds make snuff.
PATAWÁ. *Œnocarpus Batawa* (Palm). Spines of leaf-stems used to make arrows for blowing-canes.
PAXIÚBA. *Iriartea exorhiza* (Palmaceæ).
PAXIÚBA-I. *Iriartea setigera* (Palm). Stem used for blowing-canes.
PIASSABA. *Leopoldinia piassaba* (Palmaceæ).
PIHIGUA. Eatable grub.
PINDÓBA. *Attalea compta* (Palmaceæ).
PIQUIÁ. Caryocar sp. (Rhizobolaceæ). Fruit with kernels like almonds.
PIRANHA-SIPO. A climber yielding drinkable water.
PIRARUCÚ. *Sudis gigas.* A large fish. When salted, a chief food on the Amazon.
PIRI-MEMBECKA. *Paspalum pyramidale* (Graminaceæ).
PITOMBA. *Sapindus cerasinus* (Sapindaceæ). Edible fruit.
PIUM̃ (L.G.). Small biting flies.
PUPUNHA. *Guilielma speciosa* (Palmaceæ). Peach palm.
PURU-PURU (L.G.). A leprous skin disease.
PUSKU-POROTO. A shrub with edible fruit (Papilionaceæ) cultivated in Tarapoto district.

RAIZ DE MATO. Aristolochia sp. A powerful tonic.
RETÁMA. *Thevetia neriifolia* (Apocynaceæ). Fruit eatable; seeds used for rattles.

SAMAÚMA. Eriodendron sp. (Sterculiaceæ). The Silk-cotton tree.
SAPUCAIA. Lecythis sp. Good ship timber.

TABATINGA. White earth, used in painting cuyas.
TABOCAL. A bamboo thicket.
TACUARI. *Mabea fistulosa* (Euphorbiaceæ). Stems make pipe-tubes.
TAMACOARÍ. Caraipa sp.? Produces a fine balsam; specific for itch.
TAMSHE. A liana used in the Andes.
TAPIIRA GUAYABA. Bellucia sp. (Melastomaceæ). A fruit.
TAPIRIBA. *Mauria juglandifolia* (Anacardiaceæ). A fruit, bark medicinal.
TAPUYAS. Indians semi-civilised.
TAUARÍ. Bark cloth. Tecoma sp. (Bignoniaceæ).
TERRA FIRME. Dry land, above floods.
TIMBO. *Paullinia pinnata* (Sapindaceæ). Roots used to stupefy fish.
TIMBO-TITICA. Heteropsis sp. Shields of Uaupés Indians made of this wood.
TRAGO (Barré). Native spirit, in the Rio Negro.
TUÇHAUA (L.G.). The chief of an Indian tribe.
TUCUM. *Astrocaryum vulgare* (Palmaceæ).
TUCUMÁ. *Astrocaryum tucuma* (Palmaceæ).
TUCUNDÉRA (L.G.). The large severely stinging ant.
TUPÍ. Indians who speak Lingoa Geral.
TURURI. Thick bark cloth.

UACÚ. Leguminosæ. Produces a bitter oil from seeds.
UARAMA. Marantaceæ. An edible root.
UARCA. Marantaceæ. An edible root.
UARÚMA. Maranta sp. Leaves used in making mats, baskets, etc.
UAUASSÚ. *Attalea speciosa.* Palmaceæ.
UBA, UBADA. Large dug-out canoes.
UBIM. Geonoma sp. Small forest palms.
UBUSSU. *Manicaria saccifera* (Palmaceæ).
UCU-ÚBA. *Myristica fatua* (Myristicaceæ). Fruit very oily.
UIRA (L.G.). *Gynerium saccharoides* (Granimeæ). Wild cane, much used in native houses in the Andes, and for arrows, etc.
UIRARI-RANA. *Strychnos Brasiliensis* (Loganiaceæ). A fruit, edible.
UMARÍ. Poraqueiba sp. Kernel eaten after steeping in water.
UMIRI. Humirium sp. Edible fruit.
UNI-BINI. Bignonia? Roots cure for ophthalmia.
URUBU. The Turkey-buzzard: a black vulture.
URUBU MARACAJÁ. *Passiflora fœtida*, fruit of.
URUCÚ. *Bixa orellana* (Flacourtiaceæ). Anatto, a dye.
URUCURÍ. *Attalea excelsa* (Palmaceæ). The fruit is burnt to smoke india-rubber.
URUPÉ. An edible agaric at Pará.

VIJAU. *Maranta Vijau.* Leaves used for making lids of baskets water-proof (on Pastasa river).

XERINGUE. Siphonia sp. (Euphorbiaceæ). India-rubber trees.
XIRIUBA. A tree at Tarapoto (*Uchpa chillca*), the ashes of which make the best lye for soap.

YACITÁRA. Desmoncus. Climbing palms.
YANGUA. *Yangua tinctoria* (Bignoniaceæ). Leaves produce a blue dye; bark a remedy for syphilis.
YENIPAPA. *Genipa macrophylla* (Cinchonaceæ). A fruit.
YUMURA CEEMI. Clusiaceæ. Sweet tree.
YUTAHI. Hymenæa sp., Peltogyne sp. (Fabaceæ). Seeds edible.

ZAMBO. A negro and Indian half-breed.
ZANAHOVIA. An edible root, like parsnips; near our carrot (*Daucus carota*).
ZANCUDOS. Mosquitoes.

NOTE.—The following terms also occur in Spruce's Journals or Notes, but I have been able to find no explanation of them:—

CAMAZAS (in Venezuela).
ISHPINGO.
JEBARIE.
RONDIN (see vol. ii. p. 114).
WINGO.

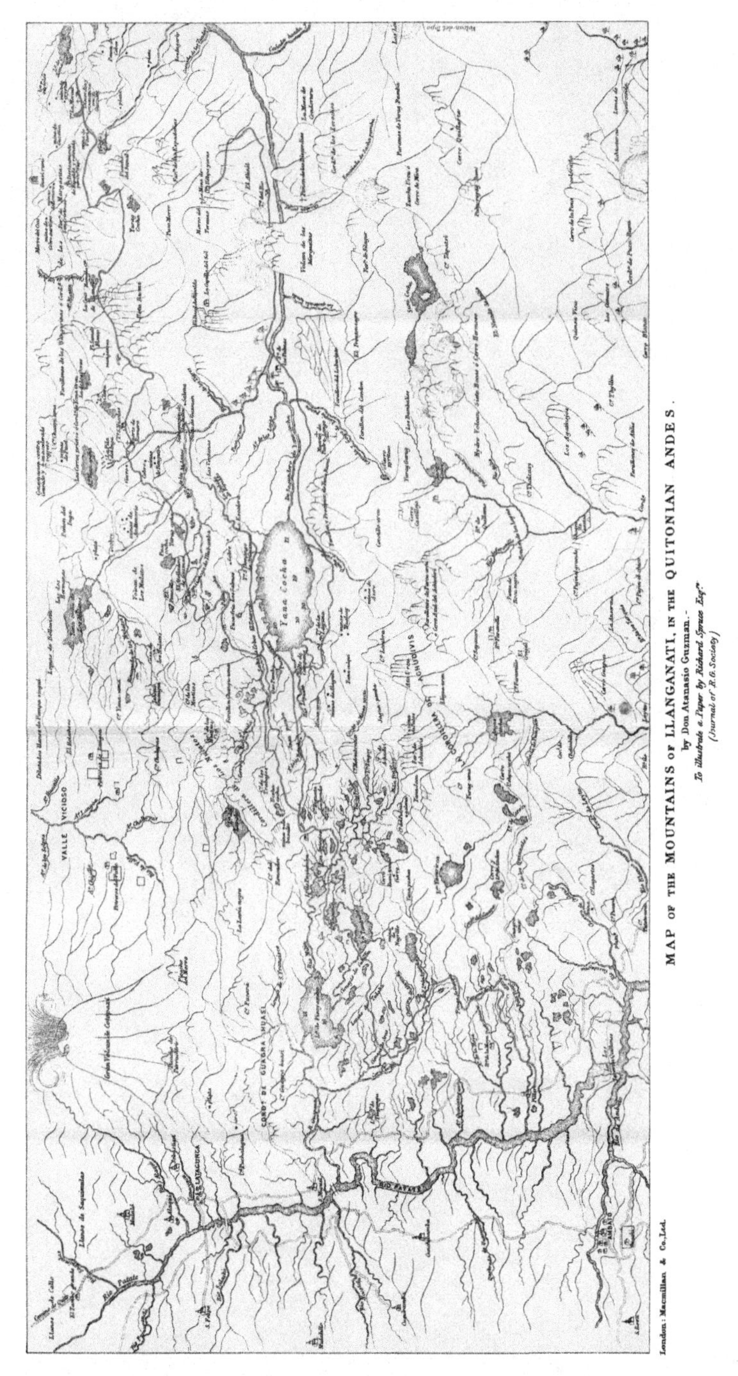

The material originally positioned here is too large for reproduction in this reissue. A PDF can be downloaded from the web address given on page iv of this book, by clicking on 'Resources Available'.

INDEX

THE END

Printed by R. & R. CLARK, LIMITED, *Edinburgh*.

Printed in the United States
By Bookmasters